# THE MACDONALD ENCYCLOPEDIA OF ROCKS AND MINERALS

D1321529

MACDONALD & CO
LONDON & SYDNEY

This edition first published in Great
Britain in 1983 by Macdonald & Co
(Publishers) Ltd
London & Sydney

Maxwell House
74 Worship Street, London EC2A 2EN

ISBN 0 356 09147 3

Printed in Italy by Officine Grafiche
Arnoldo Mondadori Editore, Verona

Text for the Italian edition
by Annibale Mottana,
Rodolfo Crespi and Giuseppe
Liborio. English translation
by Catherine Atthill, Hugh
Young and Simon Pleasance.

# CONTENTS

**PART ONE—MINERALS**

**PART TWO—ROCKS**

*The explanation of the visual symbols used in the body of the book will be found at the beginning of each section.*

# INTRODUCTION

# TO MINERALS

Minerals are normally defined as solid crystalline substances, formed by natural and usually inorganic processes. They are the essential materials which make up the Earth and those extraterrestrial bodies of which man has samples (meteorites and the Moon). A mineral is characterized by homogeneous physical properties; that is, any two parts of a piece of a particular mineral will have the same set of properties. However, these properties may vary with direction in the crystal; for example, the mineral may be harder on one side than on another or may grow faster in one direction than in another, but this characteristic will be the same for all pieces of that mineral. Each mineral has a distinctive chemical composition which may vary within certain limits but is always clearly defined. Above all, minerals have distinct crystal structures: the atoms which make up a mineral are arranged in a regular and repeating three-dimensional network. No matter where we sample the crystal, the arrangement is the same. This is basic to the crystalline state. The general arrangement of the atoms may be the same in different minerals though the constituent atoms are not of the same type; this phenomenon is known as *isotypism*. If some of the atoms in two isotypous minerals are so similar in size and chemical properties that they can take each other's place without greatly deforming the crystal structure, the minerals may form what is known as a *solid solution*. Solid solutions are families of minerals with chemical compositions which vary continuously between two or more extreme compositions. Substances may also have the same atoms combined in the same proportion (i.e., be chemically identical) but have different crystal structures; that is, the atoms may be arranged in a different fashion. This phenomenon is known as *polymorphism*. The particular polymorph of a substance depends on the temperature and pressure where the mineral forms. This makes polymorphic minerals valuable as indicators of the geological conditions which existed where the minerals were located.

The definition of minerals given above does however admit certain exceptions or, more precisely, extensions. For example, although mercury normally occurs as a liquid, it is considered a mineral by many, whereas obsidian, which is solid in appearance, is not considered a mineral, even though they resemble each other in not having the regular arrangement of atoms peculiar to the crystalline state.

Not all minerals form from inorganic processes: for example, many types of limestone——rocks formed essentially from calcite——originate from the accumulation of skeletons of animal and vegetable marine organisms. But this calcite is usually considered to be a mineral because (1) it is usually identical with calcite formed by inorganic processes and (2) deformation and recrystallization usually occur in formation of limestone, both inorganic processes. All substances synthesized in the laboratory cannot be considered minerals, because regardless of whether they form by an organic or inorganic process, they are man made. However, when the substance which was synthesized in the laboratory is found in nature and there is proof that it was not the product or by-product of man, plant or animal, it is classified as a mineral and given a name. Interestingly, in order to specify the physical properties accurately, it may be preferable to use a synthetic product rather than a natural one. Thus some natural substances which a chemist would consider "organic" (like certain crystalline heavy hydrocarbons, oxalates and paraffins) but which have become consolidated, modified and *recrystallized* by natural geological processes will be called minerals. Natural substances which do not fit any of the mentioned criteria are sometimes called *mineraloids.* Some examples are amber, a solidified resin from coniferous trees; opal, an amorphous variety of hydrated silica; and limonite, an amorphous hydrated iron oxide.

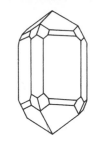

*Zircon crystal resulting from the combination of several simple crystal forms.*

## CRYSTALLOGRAPHY AND CRYSTAL CHEMISTRY

The external expression of the arrangement of the atoms which form a mineral is known as a *crystal.* A crystal is a polyhedral form, a geometric solid, with a specific set of faces, edges and corners which is consistent with the geometric packing of the atoms within the crystal. Since the number of unique arrangements of atoms possible within a crystal is not infinite, and since the variety of regular three-dimensional shapes which can be stacked into a larger volume is limited, the possible types of crystals in the mineral world are fairly limited. If crystals appear complicated to the casual observer, this is because in many cases the few basic forms combine in various ways into different and irregular forms, making it hard to recognize the primary ones.

### Morphology of crystals

Given that crystals are geometric bodies, they can be studied from a purely descriptive geometric point of view; in other words, without taking into account how the atoms inside them are oriented. This approach represents the first stage of the development of mineralogy as a science (1500–1912).

**9**

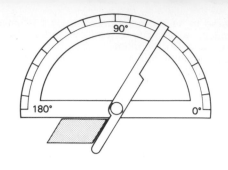

Contact goniometer (above) and diagram of the principle of a reflecting goniometer. In the first type, it is obvious that the hinged axis at the center (alidade) indicates on the graduated circle the value of the dihedral angle between the two faces of the crystal placed between the base and the alidade. In the second, the angular rotation necessary for the reflection of the light source (S) visible at the eyepiece (O), to be repeated on the two faces AB and BC, has the value $\alpha'$, the angle between the normals to the faces themselves, and supplementary to the dihedral angle $\alpha$.

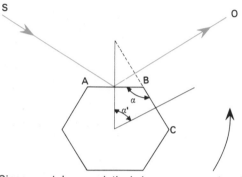

Since crystals are relatively large compared to the "invisible" atoms from which they are made, the first observations about their morphology could be made with the naked eye, perhaps with the aid of a *contact goniometer* (a simple device for measuring the angles between crystal faces); later, lenses and microscopes were used for smaller crystals. It was soon realized that dimension and size are of little importance in crystals, because their form remains constant no matter how small they are. In fact, the smaller a crystal is, the clearer and brighter are its faces, the more acute are its edges and vertices and the better are the observations which can be made of it.

The great leap forward in the field of *morphological crystallography* (the science which studies the forms of crystals) occurred when W. H. Wollaston in 1809 applied to a microscope a device for measuring angles, thus producing the *reflecting goniometer*. But even before this happened, the two basic rules of crystallography had already been set down. The first, outlined in 1667 by Niels Stensen, physician at the court of the Medici in Florence, says: *in all the crystals of the same substance at the same temperature and pressure, the dihedral angles between equivalent faces are equal.* Thus, in crystals, the relative size of the individual faces is not important: but the angles be-

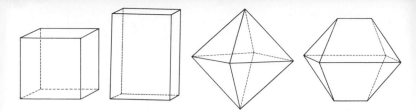

tween them are. So, from a crystallographic point of view, a quartz crystal 1 meter (3 feet) long will have the same angular measurements as a quartz crystal that is only 1 millimeter (.04 inches) long. However, it will be quite different from, say, an orthoclase crystal, whatever the latter's size.

The second law of crystallography, issued by the French abbot R. J. d'Haüy (1782), says that *if we take as the axes of reference of a crystal three straight lines parallel to three edges which are convergent but not coplanar, the relations between the intercepts made on these same axes by any two faces of the crystal will be to one another like three whole and usually small rational numbers.* In other words if we call the intersections on the axes of the first face *a, b* and *c,* and those of a second face *a', b'* and *c',* we find that

$$\frac{a}{a'} : \frac{b}{b'} : \frac{c}{c'} := h : k : l.$$

The ratio $h : k : l$ will be three small numbers. Ideally, in a practical study the largest face that inter-

*The constant nature of the dihedral angle. In regular and disproportionate crystals the equivalent dihedral angles are always constant: here the parallel translations of a disproportionate form are shown relative to the corresponding regular form.*

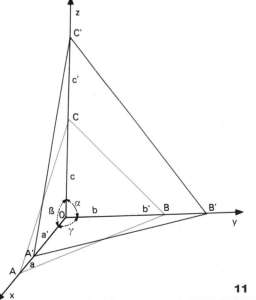

*The axial frame of reference x, y, z, with angles α, β, γ. The face ABC, which has intercepts a, b, c on the axes of reference, is the fundamental face (blue) and, in accordance with the formula derived from Haüy's law, the parameters a', b', c' of the face A'B'C' (red) are related to a, b, c respectively by a simple ratio.*

11

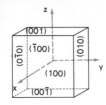

*Symbols of the faces yielding the form |100| of the isometric system: the cube.*

sects all three lines or axes is chosen as the *fundamental* or *unit face,* as it should have values of *a, b* and *c* that always yield simple fractions. Or, if all fractions are multiplied by a small integer *n,* they become the integers *h, k* and *l.* The integers *h, k* and *l* are the *indices* of the face and when enclosed by parentheses, (*hkl*), represent the face itself. For the fundamental face the indices will necessarily be 1, 1, 1, or (111). Faces which intersect one of the axes negatively will have the corresponding negative index written as a bar placed above the axis: for example, ($h\bar{k}l$); faces parallel to one or two axes (i.e., with an infinite intersection) will have the corresponding index equal to zero: e.g., (*hk*0) or (0*kl*).

In the case of two infinite intersections, the third is inconsequential and is set at unity, so the indices of the coordinated planes for *xy,* where *a* = ∞, *b* = ∞, *c* = *c'*, are (001), for *xz* (010), and for *yz* (100).

A direction in the crystal is defined as perpendicular to a plane, and the indices are written between brackets [100]. In most crystals, several faces are related to each other by some type of symmetry in the crystal structure. These equivalent faces can be related to a single face and be derived from it. The single face is called the *simple form* and is denoted by the index of the face enclosed by braces |*hkl*|. The index of each face is derived from the index of the form by exchanging positions and switching the sign. For example: The cubic form, written |100|, is given by the overall set of the six faces: (100) (010) (001) ($\bar{1}$00) (0$\bar{1}$0) (00$\bar{1}$).

## Elements of symmetry

Everyone uses the term "symmetry," but what kinds of symmetry are there and how does symmetry "work" in crystals?

The elements of symmetry are geometric operations which determine the repetition of geometrically and physically equivalent properties. The relation can be either by rotation around a direction (*axis, A*) or by reflection across a surface (*plane, P*) or by inversion through a point (*center, C*).

A center of symmetry is the midpoint common to the lines which connect equivalent points in a crystal: it thus coincides with the center of gravity of a perfect crystal. In a crystal which has a center there will be pairs of parallel faces, parallel edges or opposite vertices.

The *plane of symmetry* divides the crystal into two symmetrical parts; the operation is called reflection. Since the parts are mirror images of each other, the plane is called a *mirror plane.* It may be absent in some crystals, or there may be a number of equivalent or nonequivalent mirror planes. All the planes divide the crystal into two symmetric

*The center of symmetry (C) divides the equivalent joined elements of the polyhedron in half.*

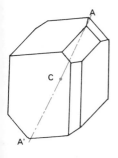

**12**

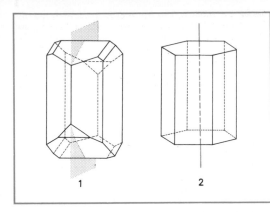

Elements of symmetry: (1) *The plane of symmetry divides the crystal into two mirror images.* (2) *Around the 6-fold axis of symmetry* (blue) *the six faces of the hexagonal prism are reproduced six times, at equal angular intervals.*

halves, but in equivalent planes the pairs of two halves also have the same shape.

The *axis of symmetry* is a direction or imaginary line; when the crystal rotates through a 360° angle, we see a certain number of equivalent positions. The number of equivalent positions equals the *serial number* ($n$) of the axis (hence an $n$-fold axis), and by dividing the 360° angle by $n$ we obtain the angle measurement of each rotation necessary to obtain an equivalent position.

There are only a limited number of possible rotation axes because the rotation has to be consistent with being able to build a crystal from blocks having the $n$-fold symmetry. This building operation is called *translation*.

The axes which are consistent with translation are of the order 1 (all bodies have these, inasmuch as all bodies can be rotated completely on their own axis) ($A_1$, 360°/1 = 360°); 2 ($A_2$, binary/2-fold, 360°/2 = 180°); 3 ($A_3$, ternary/3-fold, 120°); 4 ($A_4$, quaternary/4-fold, 90°) and 6 ($A_6$, senary/6-fold, 60°).

*Axes of symmetry: Left to right, 2-fold, 3-fold, 4-fold and 6-fold.*

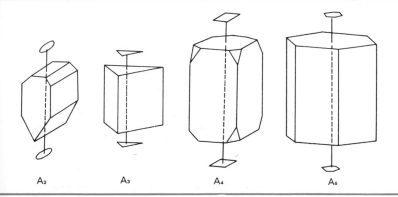

| $A_2$ | $A_3$ | $A_4$ | $A_6$ |

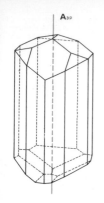

$A_{3p}$

*Polar axis: In a tourmaline crystal the 3-fold axis is polar because it joins together elements which arenot equivalent from a geometrical point of view; the two ends of the crystal are not equivalent.*

Some crystals do not have axes of symmetry (excluding $A_1$, of course). Others have one (single axis) and still others have more than one of the same order or of a different order. The first may be equivalent or nonequivalent. Some axes encounter equivalent points of the crystal at their two extremities, and others encounter nonequivalent points. The latter are known as *polar axes* and they have noteworthy physical features. They are represented by the symbol $A_p$.

Finally, crystals may have a compound element of symmetry, known as the *axis of inversion.* In this case points are related to each other by a rotation and then an inversion through a point on the axis. The symbol for the axis is $A_{\bar{n}}$, e.g., $A_{\bar{4}}$.

In a single crystal, several different elements of symmetry may exist. Once the presence of two elements has been determined, it is always possible to predict or exclude the presence of a third. It is also possible to predict the angular relationship of the various elements. Thus for every crystal there is a characteristic formula of symmetry which expresses its degree of symmetry.

### Degree of symmetry

The *degree of symmetry* of a crystal is given by the set of symmetry elements common to all its properties. It is necessary to distinguish the apparent symmetry (or geometrical symmetry), which can be deduced from an examination of the crystal's outward geometrical forms, from the real (or true) symmetry. This can only be revealed by a series of chemical and physical experiments and it may not be the same as the geometrical symmetry. The true symmetry of a crystal is the *minimum* symmetry present and common to all its properties.

In order to determine the crystal's true symmetry, one must examine *all* the features of the faces including the presence of growth striations (to be distinguished from striations on cleavage surfaces), different brightness or luster and natural corrosion figures. Various sorts of chemical tests can be made, and experiments can be carried out on the crystal ascertaining its piezoelectricity and pyroelectricity (see below under "Electrical Properties") to confirm the presence of polar axes and

*Apparent symmetry: Examples of striations on an octahodral sphalerite crystal and on a cubic pyrite crystal.*

various other properties. For example, in a sphalerite octahedron we find four lustrous faces and four striated or streaked faces: this crystal is formed by the interpenetration of two tetrahedra, and the true symmetry is that of the hextetrahedral class and not of the hexoctahedral class, within the cubic system. Likewise, the striation on the cube faces of a pyrite crystal makes it possible to assign it to a lower symmetry class than that of the geometric cube, i.e., that of the diploidal class within the cubic system.

If there are no striations on the faces, one can proceed to examine the natural or experimentally produced corrosion figures: for example, their form and arrangement on a hexagonal prismatic calcite crystal show that the mineral is not hexagonal but rhombohedral of the scalenohedral class. Once we have defined the total number of symmetry elements consistent with all the properties of the mineral, we can establish the formula of symmetry valid for all the crystals of that mineral. Many minerals may have the same formula of symmetry. In other words, although their chemical composition differs, they have the same symmetric distribution of physical properties which, though differing in their absolute values, show similar orientation. Such minerals all belong to the same *class of symmetry,* which is defined as a fixed and distinctive grouping of symmetry elements. In all, there are thirty-two known classes of symmetry, including one which is totally without symmetry, and each includes a greater or lesser number (down to zero) of minerals and substances. These classes are labeled either on the basis of the most distinctive crystalline form (that which presents the greatest number of faces) or on the basis of its most typical mineral. Thus $CA_4$ $2A_2^!$ $2A_2^{!!}$ $P2P'$ $2P''$ describes the ditetragonal dipyramidal class, or the cassiterite class.

Examination of the thirty-two classes yields six crystal systems. Each is characterized by a reference coordinate base defined by three axes in three-dimensional space; the relative lengths of the axes and the angles between them are unique for each system. There is one system that has only one degree of variation in axial length and hence is *isometric.* Two systems have two degrees of axial variation (hexagonal and tetragonal) and thus are *dimetric,* while the rest have all three axes of different lengths and are *trimetric.* The six systems and thirty-two classes with their relative equations of symmetry and typical minerals are listed synoptically in the following tables. These tables include the possible *simple forms* for each case. (We should remember that by simple form we mean the combination of all the faces which, starting from a single face, can be obtained by the operation of

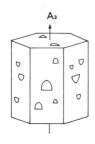

Apparent symmetry: An example of corrosion figures on a hexagonal prismatic calcite crystal.

# THE ISOMETRIC SYSTEM

crystallographic constants $\begin{cases} \alpha = \beta = \gamma = 90° \\ a = b = c \end{cases}$

| CLASS AND DEGREE OF SYMMETRY | SIMPLE CRYSTALLINE FORMS | MINER |
|---|---|---|
| **HEXOCTAHEDRAL** (holohedral) $4/m\overline{3}2/m$ $C \cdot 3A_4 \cdot 4A_3' \cdot 6A_2'' \cdot 3P \cdot 6P''$ | * or trigonal trisocta-hedron † or tetragonal trisocta-hedron ‡ or rhombic-dodecahedron | *copper silver gold platinu galena* **halite** **fluorite** *uranini spinels garnets* |

hexoctahedron    trisoctahedron *    trapezohedron †

octahedron    tetrahexahedron    dodecahedron ‡    cube

| **GYROIDAL** $432$ $3A_4 \cdot 4A_3' \cdot 6A_2''$ | gyroid trisoctahedron trapezohedron octahedron tetrahexahedron dodecahedron cube | |

gyroid

| **HEXTETRAHEDRAL** $\overline{4}3m$ $3A_2 \cdot 4A_{3\,p}' \cdot 6P$ | hextetrahedron deltoid dodecahedron tristetrahedron tetrahedron tetrahexahedron dodecahedron cube | **sphale** *tetrahe sodalite* |

hextetrahedron    deltoid dodecahedron

tristetrahedron    tetrahedron

| **DIPLOIDAL** $2/m\overline{3}$ $C \cdot 3A_2 \cdot 4A_3' \cdot 3P$ | diploid pentagonal dodecahedron or pyritohedron trisoctahedron octahedron dodecahedron cube | **pyrite** *hauerite skutteru* |

diploid    pentagonal dodecahedron or pyritohedron

| **TETARTOIDAL** $23$ $3A_2 \cdot 4A_{3\,p}'$ | tetartoid pyritohedron deltoid dodecahedron tristetrahedron tetrahedron dodecahedron | *ullmann* |

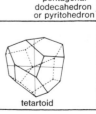

tetartoid

# HEXAGONAL SYSTEM, HEXAGONAL DIVISION

crystallographic constants $\begin{cases} \alpha = 120°; \beta = \gamma = 90° \\ a = b \neq c \end{cases}$

| CLASS AND DEGREE OF SYMMETRY | SIMPLE CRYSTALLINE FORMS | MINERALS |
|---|---|---|
| **DIHEXAGONAL DIPYRAMIDAL**<br><br>**6/m 2/m 2/m**<br>(holohedral)<br>C·A₆·3A₂'·3A₂''·P·3P'·3P'' | dihexagonal dipyramid<br>hexagonal dipyramid (first and second order)<br>dihexagonal prism<br>hexagonal prism (first and second order)<br>basal pinacoid<br><br> <br>hexagonal dipyramid    dihexagonal dipyramid<br><br> <br>dihexagonal prism + basal pinacoid    hexagonal prism + basal pinacoid | *graphite*<br>*covellite*<br>*molybdenite*<br>*pyrrhotite*<br>***beryl*** |
| **HEXAGONAL DIPYRAMIDAL**<br><br>**6/m**<br>C·A₆·P | hexagonal dipyramid<br>hexagonal prism<br>basal pinacoid<br><br><br>hexagonal dipyramid | ***apatite***<br>*pyromorphite*<br>*mimetite*<br>*vanadinite* |
| **DIHEXAGONAL PYRAMIDAL**<br><br>**6mm**<br>A₆ₚ·3P'·3P''· | dihexagonal pyramid<br>hexagonal pyramid<br>dihexagonal prism<br>hexagonal prism<br>pedion<br><br>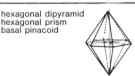<br>dihexagonal pyramid + pedion | ***wurtzite***<br>*greenockite*<br>*zincite* |
| **HEXAGONAL TRAPEZOHEDRAL**<br><br>**622**<br>A₆·3A₂ₚ'·3A₂ₚ'' | trapezohedron (right and left)<br>hexagonal dipyramid<br>dihexagonal prism<br>hexagonal prism<br>basal pinacoid<br><br><br>trapezohedron | *quartz* |
| **HEXAGONAL PYRAMIDAL**<br><br>**6**<br>A₆ₚ | hexagonal pyramid<br>hexagonal prism<br>pedion<br><br><br>hexagonal pyramid + pedion | *nepheline* |

# HEXAGONAL SYSTEM, TRIGONAL OR RHOMBOHEDRAL DIVISION

crystallographic constants $\begin{cases} \beta = \gamma = 90°; \alpha = 90° \\ a = b \neq c \end{cases}$

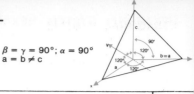

| CLASS AND DEGREE OF SYMMETRY | SIMPLE CRYSTALLINE FORMS | MINERAL |
|---|---|---|
| **HEXAGONAL SCALENOHEDRAL**<br><br>$\bar{3}2/m$<br>$C \cdot A_3 \cdot 3A_2' \cdot 3P$ | scalenohedron<br>rhombohedron<br>hexagonal dipyramid<br>dihexagonal prism<br>hexagonal prism<br>basal pinacoid<br><br>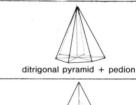<br>scalenohedron    hexagonal prism + basal pinacoid<br>rhombohedron   hexagonal dipyramid   dihexagonal prism + basal pinacoid | *bismuth*<br>*corundum*<br>*hematite*<br>*brucite*<br>**calcite**<br>*magnesite*<br>*siderite*<br>*smithsonite* |
| **DITRIGONAL PYRAMIDAL**<br><br>$3m$<br>$A_{3p} \cdot 3P$ | ditrigonal pyramid<br>trigonal pyramid<br>hexagonal pyramid<br>ditrigonal prism<br>trigonal prism<br>hexagonal prism<br>pedion<br><br><br>ditrigonal pyramid + pedion | *proustite*<br>*pyrargyrite*<br>**tourmaline**<br>*millerite* |
| **TRIGONAL TRAPEZOHEDRAL**<br><br>$32$<br>$A_3 \cdot 3A_2p$ | trapezohedron<br>rhombohedron<br>trigonal dipyramid<br>ditrigonal prism<br>hexagonal prism<br>trigonal prism<br>basal pinacoid<br><br>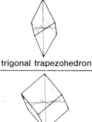<br>trigonal trapezohedron | **quartz**<br>*cinnabar* |
| **RHOMBOHEDRAL**<br><br>$\bar{3}$<br>$C \cdot A_3$ | rhombohedron<br>hexagonal prism<br>basal pinacoid<br><br>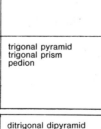<br>rhombohedron | *ilmenite*<br>**dolomite**<br>*willemite*<br>*dioptase*<br>*phenakite* |
| **TRIGONAL PYRAMIDAL**<br><br>$3$<br>$A_{3p}$ | trigonal pyramid<br>trigonal prism<br>pedion<br><br>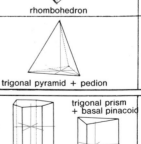<br>trigonal pyramid + pedion | *gratonite* |
| **DITRIGONAL DIPYRAMIDAL**<br><br>$\bar{6}m2$<br>$A_3 \cdot 3A_{2p}' \cdot P \cdot 3P'$ | ditrigonal dipyramid<br>trigonal dipyramid<br>hexagonal dipyramid<br>ditrigonal prism<br>hexagonal prism<br>trigonal prism<br>basal pinacoid<br><br><br>trigonal prism + basal pinacoid<br>ditrigonal dipyramid   ditrigonal prism + basal pinacoid | *benitoite* |
| **TRIGONAL DIPYRAMIDAL**<br><br>$\bar{6}$<br>$A_{3p} \cdot P$ | trigonal dipyramid<br>trigonal prism<br>basal pinacoid<br><br><br>trigonal dipyramid | |

# TETRAGONAL SYSTEM

crystallographic constants $\begin{cases} \alpha = \beta = \gamma = 90° \\ a = b \neq c \end{cases}$

| CLASS AND DEGREE OF SYMMETRY | SIMPLE CRYSTALLINE FORMS | MINERALS |
|---|---|---|
| **DITETRAGONAL DIPYRAMIDAL** (holohedral) <br> **4/m 2/m 2/m** <br> $A_4 \cdot 2A_2' \cdot 2A_2'' \cdot P \cdot 2P' \cdot 2P''$ | ditetragonal dipyramid <br> tetragonal dipyramid <br> ditetragonal prism <br> tetragonal prism <br> basal pinacoid <br><br> ditetragonal dipyramid <br> tetragonal dipyramid — ditetragonal prism + basal pinacoid — tetragonal prism + basal pinacoid | *rutile* <br> **cassiterite** <br> *anatase* <br> *zircon* <br> *vesuvianite* <br> *apophyllite* |
| **TETRAGONAL DIPYRAMIDAL** <br> **4/m** <br> $C \cdot A_4 \cdot P$ | tetragonal dipyramid <br> tetragonal prism <br> basal pinacoid <br><br> tetragonal dipyramid | **scheelite** <br> *scapolite* |
| **DITETRAGONAL PYRAMIDAL** <br> **4mm** <br> $A_{4p} \cdot 2P \cdot 2P'$ | ditetragonal pyramid <br> tetragonal pyramid <br> ditetragonal prism <br> tetragonal prism <br> pedion <br><br> ditetragonal pyramid + pedion | *diaboleite* |
| **TETRAGONAL TRAPEZOHEDRAL** <br> **422** <br> $A_4 \cdot 2A_{2p}' \cdot 2A_{2p}''$ | trapezohedron <br> tetragonal dipyramid <br> tetragonal prism <br> basal pinacoid <br><br> tetragonal trapezohedron | *cristobalite* |
| **TETRAGONAL SCALENOHEDRAL** <br> **$\bar{4}2m$** <br> $A_2 \cdot 2A_{2p} \cdot 2P$ | scalenohedron <br> tetragonal dipyramid <br> ditetragonal prism <br> disphenoid <br> tetragonal prism <br> basal pinacoid <br><br> tetragonal scalenohedron — tetragonal disphenoid | *chalcopyrite* <br> *stannite* |
| **TETRAGONAL PYRAMIDAL** <br> **4** <br> $A_{4p}$ | tetragonal pyramid <br> tetragonal prism <br> pedion <br><br> tetragonal pyramid + pedion | **wulfenite** |
| **TETRAGONAL DISPHENOIDAL** <br> **$\bar{4}$** <br> $\cancel{P}A_{4p}$ | disphenoid <br> tetragonal prism <br> basal pinacoid <br><br> tetragonal disphenoid | *cahnite* |

# ORTHORHOMBIC SYSTEM

crystallographic constants $\begin{cases} \alpha = \beta = \gamma = 90° \\ a \neq b \neq c \end{cases}$

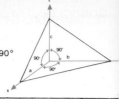

| CLASS AND DEGREE OF SYMMETRY | SIMPLE CRYSTALLINE FORMS | MINERA |
|---|---|---|
| **RHOMBIC DIPYRAMIDAL** (holohedral) $2/m\ 2/m\ 2/m$ $C \cdot A_2 \cdot A_2^{\prime} \cdot A_2^{\prime\prime} \cdot P \cdot P^{\prime} \cdot P^{\prime\prime}$ | dipyramid prism pinacoid<br><br>rhombic dipyramid          rhombic prisms + pinacoids | *sulfur* stibnite aragonite cerussite barite celestite anglesite anhydrite olivine topaz |
| **RHOMBIC PYRAMIDAL** $mm2$ $A_{2p} \cdot P \cdot P^{\prime}$ | pyramid prism dome pinacoid pedion<br><br>rhombic pyramid + pedion          domes | *hemimor* prehnite |
| **RHOMBIC DISPHENOIDAL** $222$ $A_{2p} \cdot A_{2p}^{\prime} \cdot A_{2p}^{\prime\prime}$ | dome disphenoid prism pinacoid<br><br>rhombic disphenoid | *epsomite* olivenite |

# MONOCLINIC SYSTEM

crystallographic constants $\begin{cases} \alpha = \gamma = 90°; \beta \neq 90° \\ a \neq b \neq c \end{cases}$

| CLASS AND DEGREE OF SYMMETRY | SIMPLE CRYSTALLINE FORMS | MINERALS |
|---|---|---|
| **PRISMATIC** (holohedral) $2/m$ $C \cdot A_2 \cdot P$ | prism pinacoid <br><br> monoclinic prisms + pinacoids | *realgar* *orpiment* *azurite* *malachite* *gypsum* *epidote* *pyroxenes* *amphiboles* *micas* ***orthoclase*** |
| **DOMATIC** $m$ $P$ | dome pinacoid pedion <br><br> dome | ***hilgardite*** |
| **SPHENOIDAL** $2$ $A_2 P$ | sphenoid pinacoid pedion <br><br> sphenoid | ***mesolite*** *natrolite* |

# TRICLINIC SYSTEM

crystallographic constants $\begin{cases} \alpha \neq \beta \neq \gamma \neq 90° \\ a \neq b \neq c \end{cases}$

| CLASS AND DEGREE OF SYMMETRY | SIMPLE CRYSTALLINE FORMS | MINERALS |
|---|---|---|
| **PINACOIDAL** (holohedral) $\bar{1}$ $C$ | pinacoid <br><br> pinacoid | *kyanite* *rhodonite* ***albite*** *anorthite* *pyroxmangite* |
| **PEDIAL** $1$ No element of symmetry | pedion <br><br> pedion | ***parahilgardite*** |

the symmetry elements in each class.) As the degree of symmetry increases (a greater number of symmetry elements), the number of faces for any form also increases. The more general the form (nonzero, nonequivalent *h*, *k*, and *l*), the more faces exist. Thus in the triclinic system only forms with one face (*pedion*) or two faces (*pinacoid*) are possible; in the monoclinic system forms with four faces are also possible (*prism*), in the orthorhombic system forms may exist with eight faces (*bipyramid*) and so on up to the cubic system. In the class with the greatest degree of symmetry we find a form with fully forty-eight equivalent faces (*hexoctahedron*).

Another and relatively simple method of labeling the crystal classes is with *Hermann-Mauguin symbols*, which are used in preference to a list of symmetry elements; these are given for each symmetry class in the accompanying tables. This convention uses a number to express a rotation axis (6 for a 6-fold axis), a number with a line over it for a rotary inversion axis ($\bar{6}$) and *m* for a mirror plane. If a slash appears between a number and *m* (6/*m*) it means the mirror plane is perpendicular to the rotation axis. The first element of symmetry usually refers to the *z* axis [001] of the reference system (except the *y* axis in the monoclinic system). The second element is equivalent to the *x* axis [100] for rotation axes or (100) for mirror planes except in the isometric system, where it is perpendicular to the octahedron face (111) or the [111] direction. The third element lies symmetrically between the *x* and *y* axes for dimetric classes, being either the [110] axis or the (110) plane, or parallel to *b* for the orthorhombic classes.

### Structure of minerals

We have already mentioned that minerals are formed by homogeneously distributed periodically

*A linear lattice arrangement: A row of particles (atoms) set apart by a period of identity* a.

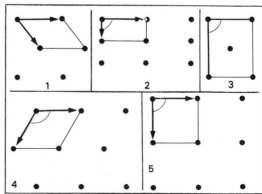

*Simple plane lattices: (1) simple oblique lattice; (2) simple rectangular lattice; (3) simple centered rectangular lattice; (4) simple hexagonal lattice; (5) simple square lattice.*

22

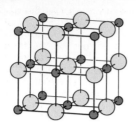 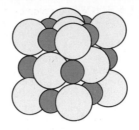

spaced groupings of atoms, the visible expression of which is the crystal. The rules for the arrangement of atoms into a crystalline solid are based on a combination of mathematical and physical constraints and were outlined in a theoretical form by M. A. Bravais in 1850, before it was possible to demonstrate directly the existence of atoms or the atomic constitution of minerals (this was not proved until Max von Laue x-rayed crystals in 1912).

A crystal structure is formed by regular groupings of atoms that are homogeneous, having the same characteristics throughout the crystal, and periodic inasmuch as equivalent atoms are located at fixed and repeated intervals within the crystal. Regular repetition in three dimensions defines a lattice.

To understand the concept of a lattice, first imagine an infinitely long line with equally spaced points or nodes on it, called a *linear lattice*. Now move all points the same distance off the line, keeping track of their first position, and do this an infinite number of times. This is a *plane lattice*, defined by the row spacings (*translations*) and the angle between the directions of translation. There are five possible types of simple plane lattices (see diagram).

Three linear lattices or rows not lying on the same plane produce the *space lattice,* determined by the three periods of translation peculiar to each row $a, b, c,$ and by the angles between the rows. In the space lattice it is possible to define the minimal volume defined by eight lattice points, and by means of the periodic translations one can reconstruct the entire lattice: this minimum volume is called the *unit cell.*

There are fourteen types of unit cells, from which it is possible by using translation to reconstruct the entire spatial distribution of the points of a crystal lattice: these are known as the *Bravais lattices.* Seven of these lattices are *simple* or *primitive* and are formed by points arranged at the vertices. Seven others, called *centered lattices,* are derived by means of the translation and interpenetration of two or more simple lattices. These have nodes not only at the vertices but also inside or on the

# THE 14 BRAVAIS LATTICES

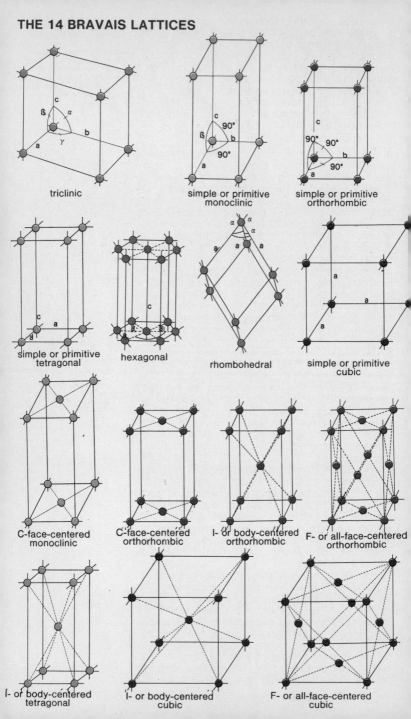

triclinic

simple or primitive monoclinic

simple or primitive orthorhombic

simple or primitive tetragonal

hexagonal

rhombohedral

simple or primitive cubic

C-face-centered monoclinic

C-face-centered orthorhombic

I- or body-centered orthorhombic

F- or all-face-centered orthorhombic

I- or body-centered tetragonal

I- or body-centered cubic

F- or all-face-centered cubic

faces. The symmetry of Bravais lattices is the same as that of only seven of the crystal classes, i.e., those with a maximum possible symmetry in each system (*holohedral classes*). In order to explain the symmetry found in the classes with less symmetry (*merihedral classes*), some of the contents of a unit cell must have a lower symmetry than that of the particular Bravais lattice. An interesting result is that other elements of symmetry are manifested, valid only at the atomic level and produced by the combination of the operations of pure symmetry, with the process of translation. Two of these hybrid symmetry elements operate within crystals: the *axis of symmetry with translation* (or screw axis), in which an angular rotation is associated with a translation half that of the lattice translation in the direction of the axis itself; and the *plane of symmetry with translation* (or glide plane), in which a reflection across a plane of symmetry is associated with a 50 percent translation parallel to the plane.

By combining the independent elements of space symmetry (24) with the elementary lattices (14) we obtain 230 distributions of points in space, which represent all the possible distributions that atoms can assume in minerals. These are called the 230 *space groups* and are often represented with symbols similar to Hermann-Mauguin symbols.

The crystal structure described above (developed in its crystal-chemical implications by P. H. von Groth) was determined experimentally by Max von Laue. With the assistance of W. Friedrich and P. Knipping, in 1912 he obtained a diffraction spectrum by x-raying a sphalerite crystal. The periodicity of the atoms in the crystal causes diffraction of x-rays; an amorphous material would merely scatter the radiation. Within just a few years, thanks mainly to the work of W. L. Bragg, the structures of several substances were determined, and at this time about 15,000 structure determinations have been made. Thus it was possible to demonstrate that the constituents of minerals obey the rules of a crystal lattice.

The forces that hold atoms and hence crystal structures together (called *bonding forces*) are electrostatic. These bonds may be ionic, covalent, metallic, residual (van der Waal's) and hydrogen in type, and in any one mineral there can be bonds of a single type (*isodesmic compounds*) or of two or more different types (*anisodesmic compounds*). Metallic bonds are present only in a few native elements and in their alloys. In native metals, the nuclei and inner electrons form a lattice held together by a cloud of moving valence or outer electrons. The resulting structure is very compact, with markedly dense packing of the atoms, and

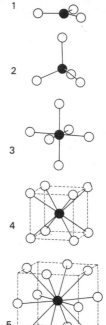

*Coordinating polyhedra:*
*(1) planar trigonal*
*( coordination number 3);*
*(2) tetrahedral (c.n. = 4);*
*(3) octahedral (c.n. = 6);*
*(4) cubic (c.n. = 8); (5)*
*dodecahedral (c.n. = 12).*

From top to bottom: *crystal structures characterized by a metallic bond (copper), covalent bond (diamond) and ionic bond (fluorite).*

the density is usually high. The mobility of the electrons produces high thermal and electrical conductivity and opacity; the indefinite extension of the bond and its ability to dissolve and immediately reform give rise to the properties of ductility, malleability and sectility, and also to filamentary, curly, dendritic, arborescent, laminar or stellar crystalline forms.

Crystals with ionic bonding consist of alternating cations and anions (or anionic groups held together by covalent bonds, like $CO_3^{2-}$, $SO_4^{2-}$, etc.). Ions are atoms that have lost or acquired one or more electrons in their outer orbits; cations have lost electron(s) and are considered to be positively (+) charged, while anions have gained electron(s) and are negatively (−) charged. The force which generates the bond is the electrostatic attraction between adjacent opposite charges, and each ion tends to surround itself with the greatest possible number of ions with an opposite charge compatible with its dimensions and the dimensions of the nearby ions. This form of reciprocal matching, in which neighboring ionic spheres must be in mutual contact, yields *coordinating polyhedra*. The form of a polyhedron depends exclusively on the relation between the ionic radius of the cation (or anion) which occupies the center of the polyhedron and the anions (or cations) which represent its vertices. The most common coordinating polyhedra are those with four vertices (tetrahedron), six vertices (octahedron or trigonal prism) and twelve vertices (cubo-octahedron and straight hexagonal prism), in that order, as the coordinated cation grows relative to the coordinating anions. When the cation is extremely small, we find a plane coordination with four vertices (square) or with three vertices (triangular), or even a linear coordination.

In order to interpret the behavior of minerals with an ionic bond, it is necessary to know the dimensions of the ionic radii in their various states of ionization. As a general rule, anions have a much larger ionic radius than cations, so an ionic structure can be imagined as a regular arrangement of large, negatively charged spheres with small, positively charged spheres in its interstices. Ionic crystals (of which halite or rock salt [NaCl] is an example) generally have low thermal and electrical conductivity because the electrons are held in the orbits of the single ions and are not free to move. They are transparent and often variable in color, hard and have a high melting point.

Crystals with primarily covalent bonding have atoms which share one or more outer electrons with the neighboring atoms in a stable configuration. These shared peripheral electrons remain in their characteristic orbits, which is why the bonds are oriented in the same direction as that of the orbit in question. The coordinating polyhedra which result are strongly bound together but do not fill the space perfectly because tangency between the spheres is not generally maintained. The general density of the crystal is thus lowered. The maximum coordination number is limited to four (tetrahedron). Covalent bonding results in minerals with a high degree of hardness, a high melting point, extremely low thermal conductivity, a high index of refraction and medium or low density. In many elements the covalent bond may take on two or more configurations, thus minerals with the same chemical composition may exhibit very different physical properties (e.g., diamond and graphite).

Van der Waals bonds are never the only type of bonding (this is only possible for the rare gases in the solid state) but arise as the result of covalent bonds in which the arrangement of the various atoms is asymmetrical and there is a concentration of positive charges on one side and negative charges on the other. In this way dipoles are formed, several of which may join together because of their weak electrostatic forces. The crystals containing these residual bonds generally have a low melting point, are not very hard and make excellent thermal and electric insulators.

The hydrogen bond usually develops when hydrogen is bonded to oxygen. The single hydrogen electron is taken into the oxygen orbits, leaving an exposed hydrogen nucleus which attracts negatively charged ions. Hydrogen bonding is essentially restricted to hydrous minerals (bearing $H_2O$ or $OH^-$) and is usually insignificant. Ice is one of the few exceptions.

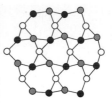

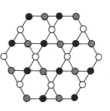

*Schematic representation of the two structures of quartz projected down the C- or Z-axis which shows how, in the passage from the upper form to the lower, the silicon atoms are more symmetrically spaced. For the sake of simplicity only the silicon ions are shown: black, gray and white circles represent ions with different structural levels.*

## Polymorphism

Polymorphism is the phenomenon whereby a substance (an element or a compound) may exist with more than one distinct crystal structure depending upon temperature and/or pressure. Minerals which have two crystal structures are called *dimorphic,* those with three *trimorphic* and those with more than three *polymorphic.* The structures themselves are called *polymorphs* or *polymorphic modifications.* Each of these is stable in a given range of temperature or pressure; in fact, each one represents the atomic configuration that uses the minimum energy possible for the particular conditions. Finding polymorphs in rock enables us to make deductions about the temperature and pressure conditions under which the rock was formed.

The transition from one polymorph to another may be quick or slow, reversible or irreversible, and is facilitated by temperature or pressure. Often cer-

*Example of pseudomorphism: limonite after pyrite (Brazil).*

tain transformations are possible in the laboratory only in the presence of appropriate catalysts. For this reason it is presumed that similar chemical conditions must occur in nature to create certain polymorphs. The chemical environment may encourage the appearance of polymorphs even outside their physical field of stability (their range of pressure and temperature), i.e., in *metastable conditions.* This is a dangerous source of error in the geological use of polymorphic systems, because it may suggest the wrong condition for the formation of a particular rock. For example, many deposits of travertine and many stalactites are not formed of calcite (the modification of $CaCO_3$ stable at atmospheric conditions) but of aragonite (the modification stable at room temperature only at pressures of about 3000 atmospheres). It would be a mistake to think pressure of this sort is present in mineral-forming environments such as springs or caves. The explanation is that the chemical environment contains such large concentrations of $Sr^{2+}$, $Mg^{2+}$, etc., in solution that it favors the precipitation of the variant in which $Ca^{2+}$ has a 9-fold coordination instead of a 6-fold. But all it needs is a slight heating or a slight compression (*diagenesis*) to convert all the aragonite into calcite.

However, it must be realized that our ability to find a high-temperature or high-pressure mineral at atmospheric conditions normally presumes that it is in a metastable state. This is true of most minerals exposed at the Earth's surface, notably for diamond, which is only stable 40 km (25 miles) within the Earth or deeper.

By changing the pressure and temperature, a mineral may undergo a polymorphic transformation without a modification of the external form of the crystal. For example, in lava it is common to find trigonal $\alpha$ quartz, which is stable at room temperature in its prismatic-dipyramidal hexagonal form, $\beta$ quartz, which is only stable above 573° C (1063° F), or even the lamellar hexagonal form, tridymite, which is stable above 870° C (1598° F). This phenomenon is called *paramorphism,* and it is useful in deciphering the evolution of a rock. Linked to this is the phenomenon of *pseudomorphism,* in which a mineral is substituted for not by one of its polymorphs, but by a totally different mineral with a different though perhaps similar chemical composition (e.g., pyrite by limonite, or azurite by malachite). (Sometimes pseudomorphism occurs with a complete change of composition, which is not an instance of polymorphism but is due to chemical exchanges.)

Polymorphism is very widespread: as well as the cases noted, we should mention the three polymorphs of $Al_2SiO_5$, andalusite, kyanite and sillimanite, indicators respectively of low pressure, high pressure and high temperature; or the well-known case of carbon (C), which at high pressure crystallizes with tetrahedral covalent bonds and forms transparent crystals which are very hard and excellent electrical insulators (diamond), while at low pressure it crystallizes with triangular planar covalent bonds and yields molecular crystals which are perfectly opaque, very soft and excellent electrical conductors (graphite).

## Isomorphism

Just as several different crystalline structures may be chemically identical, so one structure may be shared by different compounds. In other words, two minerals of different chemical composition can crystallize into the same structure. This phenomenon is known as *isotypism* when the structural equivalence of the typical mineral and the mineral in question is complete, with respect both to positions of the atoms and to structure (e.g., NaCl-PbS); and as *homeotypism* when the positions of the atoms are the same but the chemical formula changes (C = diamond versus ZnS = sphalerite). Two substances are called *isostructural* when they have structures which are identical or just slightly different due to small displacements of their atomic sites. When the relative dimensions and the characteristics (electronegativity, ionization potential, etc.) of the atoms which occupy the same structural positions are very similar, a phenomenon known as *solid solution* may develop, in which there are a variety of possible compositions limited by two or more compositions called *end-members*. In the past, solid solution was wrongly called *isomorphism* because in solid solutions external morphology is commonly uniform. An example will make this phenomenon easier to understand, since solid solutions are very important to mineralogy, enabling us to identify series and groups of minerals, and facilitating their classification. Magnesite, $MgCO_3$, and siderite, $FeCO_3$, are two isotypous minerals, and $Mg^{2+}$ and $Fe^{2+}$ have similar ionic radii and the same ionization potential so that one completely replaces the other. This relationship produces a solid solution, written as $(Mg,Fe)CO_3$, in which (in the parentheses separated by a comma) we see the elements which can be substituted in the structure. The varieties of the mineral series which result take different names depending on the ratio of iron to magnesium (breunnerite, mesitite, pistomesite), but the stoichiometric ratio of divalent cations for $CO_3^{2-}$ remains fixed (1:1). The series may become more complicated when other cations with similar dimensions and bonding characteristics (zinc, manganese, cobalt) enter into the solid solution. We then have ternary mixtures, quaternary mixtures, etc., whose stability and extent is directly related to the affinity the new cations have with magnesium and iron. Solid solubility is a very complex phenomenon, present in all minerals at least to some extent, because there is practically no element for which there is not a similar one that can replace it, even if in trace amounts.

Substitutions are also possible between elements in different oxidation states (e.g., $Na^+$ instead of $Ca^{2+}$), but in order to maintain the mineral's electrostatic neutrality it is necessary to find a substitute with a higher valence in another site of the structure (e.g., $Si^{4+}$ instead of $Al^{3+}$). This is the case for a very common mineral series, the plagioclase feldspars, in which we find the coupled substitution $Na^+Si^{4+} \leftrightarrows Ca^{2+}Al^{3+}$, and whose general formula is

$$(Na, Ca) Al(Al, Si)Si_2O_8.$$

The phenomenon of solid solution is closely linked with the pressure and even more often the temperature of formation. We frequently find substances which exist as solid solutions at high temperatures, whereas at a lower temperature they develop *exsolution,* i.e., the unmixing of two or more types of crystals from a single one with the two phases intergrown with a precise crystallographic orientation. For example, in the alkali feldspars, sanidine $(KAlSi_3O_8)$ and albite $(NaAlSi_3O_8)$ are soluble in all proportions at 1000° C (1832° F), but at a temperature of about

600° C (1112° F) they begin to unmix, producing textures called perthitic intergrowths. The host crystal is $KAlSi_3O_8$ and the "guest" crystal is $NaAlSi_3O_8$ in the form of thin lamellae oriented in certain crystallographic planes of the host (near ($\bar{6}01$)). The composition of the two mixed parts depends on the temperature; the two phases approach end-member composition with decreasing temperature until a minimum value is reached, where the structures can accept the remaining solid solution. The phenomena of solid solution and exsolution thus offer another possibility of reconstructing the pressure and temperature history and, in this case, the conditions of evolution of a particular mineral- or rock-forming environment.

In solid solutions the physical properties usually vary continuously with composition. This variation is often used as a means of identification, especially in petrography, where optical properties are measured on a petrographic (polarizing) microscope. But it can also be a source of possible error if we forget that there are few strictly binary (two end-member) solid solutions in nature. The value of a physical property may be influenced not only by one element in solid solution, but by two or three, some of which may have nothing to do with the petrographic problem in question.

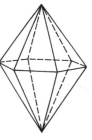

*Open and closed forms: (1) Open form: hexagonal pyramid. (2) Closed form: hexagonal dipyramid.*

## IDENTIFICATION AND STUDY OF MINERALS

Every mineral is defined by its chemical composition and by its manner of crystallization. In order to recognize a mineral, however, it is not always necessary to resort to a thorough analysis: in most cases it suffices to examine the external form (which reflects the crystal symmetry) and several physical properties (which depend simultaneously on the chemical composition and on the crystal structure) to make a quick identification.

### Crystal form

The term *form* has a very precise crystallographic meaning: it is the set of the physically equivalent faces of a crystal, whose presence is controlled by the symmetry of the crystal class. Some forms are *closed*, i.e., they completely enclose an area of space, and some are *open*: these cannot exist on their own but are always associated with others. The form of crystals, as seen by the naked eye, is usually not given by a single crystallographic form but by a *combination* of various forms, each developed to a greater or lesser degree. A characteristic combination of forms is called the *trait* of the crystal, while the term *habit* is used to describe appearance determined by the predominant form

Skeletal crystals: Dendrites
of manganese oxides
which start from a fissure
in a calcareous rock
(Solnhofen, Germany).

in relation to the others. The possible faces of a crystal are governed by the internal arrangement of the atoms, and thus a given trait reflects relations of growth-speed among the individual forms. The habit, on the other hand, represents an adaptation of a crystal to the growth environment and depends not only on the pressure and the temperature acting at the moment of crystallization, but also on the chemical environment, on the presence of oriented stress, the presence of impurities and also on the simultaneous growth of other crystals. So crystals of the same substance can grow with different habits in different places and yet grow at different rates for a single place.

The external form of a crystal is determined by its rate of growth. Interestingly, the most rapidly growing faces are those which show the least; the slower growing ones, on the contrary, grow larger and tend to make the others disappear. Growth may occur by successive *strata,* i.e., by the addition of structural layers over the entire surface, or in *spiral* form, with the addition of rows of particles starting from a linear discontinuity on a particular face; or it may be *skeletal,* with the addition of lattice blocks in just one direction, which is preferable for a rapid rate of growth. In this latter case we find the formation of *dendrites*—needlelike (acicular), ramified (hornlike) and reticulated crystals. Their growth is particularly common from the vertices of preexisting crystals. In particular, the habit of crystals is greatly influenced by the presence of impurities (even in very small amounts) in the midst of the growth medium. These impurities can

Opposite: *color zonations
in a section of a tourmaline
crystal (Malagasy
Republic).*

**32**

be deposited on growth surfaces (crystal faces) and prevent their development, although they may not be enclosed in the crystal. In other cases they may become enclosed in the crystal but only on specific planes or in certain directions. Fairly regular arrangements are thus formed of inclusions which are typical of certain minerals, such as the "hourglass" in augite and the "iron cross" in andalusite (chiastolite). Certain chemical impurities can be accepted in the crystal structure in the form of a solid solution and in some cases cause not only a change in color but also a change in the manner of growth. The case of Elba and Malagasy tourmaline is typical: here, while a color changes from pink ($Mn^{2+}$) to green ($Fe^{2+}$) while maintaining trigonal forms, the change to brown ($Fe^{2+}$, $Mg^{2+}$) involves a change to pseudohexagonal forms. A variation in form may also be due to the temperature of crystallization: for example, fluorite has an octahedral or cubic form when it grows at high temperatures, whereas it takes on gradually more and more complicated forms as the temperature drops. This may also be connected with pressure; in attempts to synthesize diamonds it has been observed that the crystals obtained at more than 60,000 atmospheres are octahedral, and those obtained at lower pressures are cubic.

*Hopper faces in a quartz crystal (St. Gotthard, Switzerland).*

34

Furthermore, the external form of minerals depends on the imperfections that crystals have on their surfaces. Some are due to irregular growth and appear, when growth is very fast, as small *reliefs* or *stairs* which often produce "hopper" crystals or striated crystals (the cube-face striations of pyrite). Similarly, depressions or hollows may form, like the triangular pits ("trigons") on the octahedral faces of the diamond, or the triangular hollows on the rhombohedral faces of quartz, and the square tiers on the cubic faces of halite or rock salt. Other imperfections are due to dissolution brought about by fluids circulating after the crystal has been formed. In some cases these *corrosion figures* are produced artificially to facilitate the identification of the true symmetry, because these figures are only the same on crystallographically equivalent faces.

*Parallel association of crystals (quartz).*

## Crystal aggregates and twinning

It is rare to find isolated crystals in nature. Minerals usually occur in groups or aggregates and in regular or irregular associations, the commonest of which are rocks. Irregular aggregates are typical mainly of minerals with an acicular (needlelike) or lamellar (platelike) habit. Among the former, typical examples are the *bladed* aggregates of

*Epitaxy: An example of a regular association of rutile crystals on hematite (Grigioni, Switzerland).*

*Examples of twins:* (1, 2) *Interpenetrating twins of fluorite and orthoclase;* (3, 4) *contact twins of gypsum* (swallowtail) *and cassiterite* (intergrowth); (5) *polysynthetic contact twin of plagioclase.*

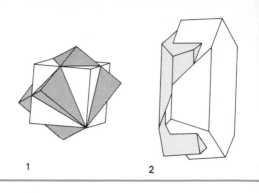

1          2

stibnite, the *acicular* aggregates of epidote and the *fibrous* aggregates of asbestos minerals like chrysolite and crocidolite. Among the latter we find the *foliated* (leaflike) aggregates of the micas (called *micaceous* in this case), the *tabular* aggregates of gypsum and barite, and the *radiating-spherulitic* aggregates of wavellite. Even more regular are such associations as the *rosette* associations of lamellar hematite, the *selliform* (saddle-shaped) associations of dolomite, the *spiral* (gwindle) associations of quartz, the associations of aragonite in *coral* and the *dendritic* associations of psilomelane.

In intrusive igneous rocks and more rarely in extrusive ones, we find associations of crystals of one or more mineral species coating a matrix which is often crystal terminations itself. These are called *druses* or, when association occurs in a closed cavity, *geodes*. Then there are regular associations in which a mineral, or several minerals, intergrow with a precise crystallographic orientation. Among the *parallel associations* of individuals of the same species are those of quartz (including the "scepter"-like association), barite, calcite and copper; among those of individuals of different species are the *epitaxes,* or precisely oriented intergrowths of rutile with hematite, kyanite with straurolite and zircon with xenotime. However, the most significant are the *perthites,* intergrowths of albite and orthoclase or microline due to exsolution.

Twinned crystals are a special case of association between two or more individuals of the same species in accordance with a rule (the law of twinning) which puts two crystallographic elements together. There can be twins along a plane or an axis, but the orientations cannot coincide with elements of symmetry of the two twinned parts, because in such a case we would have a parallel

36

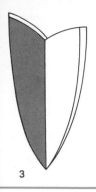
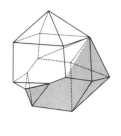
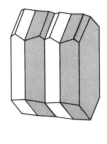

3                    4                    5

association. Twinning can occur by simple *contact* on a flat surface, in which case it is easy to see the reentrant angles (these are impossible in single crystals), or by *interpenetration* when the two crystals are associated internally or through one another. Twinning may also be *multiple* or *polysynthetic*. In some instances it may even simulate a symmetry which is greater than that of the crystal (mimicry or pseudosymmetry).

Some twinning is characteristic of certain minerals: for example, "swallowtail" crystals of gypsum, small "v"'s in intergrowing crystals of cassiterite and the "kneelike" (geniculated) crystals of rutile. Conversely the "Greek cross" and "St. Andrew's cross" crystals of staurolite are twinned by interpenetration, as are the "cross" crystals of arsenopyrite and the "iron cross" crystals of pyrite. Multiple twinning includes the "sixlings" of aragonite simulating a hexagonal prism with reentrant angles (a clear case of pseudosymmetry), the "star" or "snowflake" crystals

*Mimicry: Pseudo-hexagonal aragonite cyclic twin or "sixling" (Aragon, Spain).*

**37**

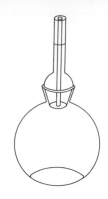

The pycnometer, an instrument for determining the specific gravity of a solid. The operation requires three weighings: $M$ = weight of a certain quantity of the solid in question; $P$ = weight of the pycnometer containing distilled and degassed water up to the reference line; $P'$ = weight of the pycnometer containing the solid and water up to the reference line. The equation for finding the specific gravity is

$$s.g. = \frac{M}{P + M - P'},$$

in which the denominator represents the weight of the volume of water displaced by the solid.

of cerussite, chrysoberyl and arsenopyrite. Calcite, the feldspars and quartz all have distinctive and complicated twins.

## Physical properties

As well as form, which may be completely absent in mineral fragments, other physical properties constitute a very useful element for the recognition and identification of minerals. Some of these can be immediately determined, others require simple measurement and others still need complex and expensive instruments which are only available in specialized laboratories.

The physical properties of solid bodies can be divided into nondirectional and directional groups. The first include the *scalar properties,* which can be expressed quantitatively, and certain other properties that are hard to define numerically and may depend on a subjective evaluation based on the human senses. The second are known as *vectorial properties,* where it is necessary not only to express a numerical value (modulus) but also to specify the direction in which the measurement has been made. In crystals with lower than isometric symmetry there is always at least one vectorial property which changes in a discontinuous way, i.e., with an abruptly variable modulus, with slight deviations in the direction of measurement.

## Scalar properties

The *density* of a body is its mass per unit of volume ($g/cm^3$), numerically equal to the *specific weight* or *gravity,* which indicates how many times the body weighs more than an equal volume of water. Density is directly related to atomic weight and closeness of packing of the atoms in the lattice, and is therefore high in compounds with a high coordination number (metals) and low in those with a lower coordination number (compounds with van der Waals and covalent bonds). Usually one does not measure the density but the specific gravity (s.g.), based on the well-known Archimedes principle and using very simple instruments such as the pycnometer or specific-gravity bottle, the hydrostatic (Jolly) balance or liquids with a predetermined specific gravity. It is useful to give an approximate value by simply weighing the sample. In this book the various minerals are classified as follows.

|  | s.g. |  |
|---|---|---|
| very light | < 2 | (carnallite, borax) |
| light | 2–3 | (quartz, calcite) |
| heavy | 3–5 | (barite, diamond) |
| very heavy | 5–10 | (cinnabar, scheelite) |
| extremely heavy | > 10 | (uraninite, gold) |

## Vectorial properties

The most important vectorial property is rate of growth, which is directly related to the geometry and the type of bonding in the crystal structure, and hence varies in a discontinuous anisotropic way. Linked with this property are the cohesive properties, likewise anisotropic and discontinuous. However, those determined by the transmission of various forms of energy (light, heat, electricity, etc.) through the body are clearly directional but are continuously variable.

## Cohesive properties

When broken into fragments, a crystal may produce irregularly shaped pieces (*fractures*) or pieces with essentially flat surfaces parallel to crystallographic planes (faces), or *cleavage*. The first is typical of substances with bonds that are all of roughly equal strength (isodesmic) and distributed fairly equally in all directions. Fracture may occur on irregular surfaces, curved surfaces (conchoids), fibrous or splintery surfaces or hackly surfaces. Cleavage occurs in substances in which there are planes across which the bonds are relatively weak and there is minimal cohesion. The outcome is forms of cleavage which, in the crystal classes with a high degree of symmetry, may even be closed: octahedra for fluorite and diamond, cubes for rock salt, dodecahedra for sphalerite, etc. More often than not the cleavages are distinct and straightforward—excellent on a basal plane with one or more less well developed planes at an angle with the first, so that they simulate complex forms (prisms, pseudo-cubes etc.). The character-

*Example of cleavage: rhombohedral cleavage of calcite (Iceland).*

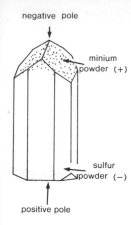

negative pole

minium
powder (+)

sulfur
powder (−)

positive pole

*Pyroelectricity. A tourmaline crystal, heated and then left to cool slowly, is charged with electricity with opposite signs at the two extremities. The polarity is made clear with electrified sulfur (yellow) and minium or red lead (red) powder.*

istic angle between the planes of cleavage may enable us to identify a specimen, as in the case of pyroxenes and amphiboles. In the first case the angle between the planes is about 90° while in the second case the angle is approximately 120°

*Tenacity* or *toughness* is another property which depends on the structural cohesion and defines the way in which a mineral is mechanically deformed or distorted. Some minerals are *fragile,* easily breaking into pieces either by cleaving (kyanite) or by fracturing (sulfur); others are *malleable* and flatten out in sheets without breaking (gold); others are *ductile* and can be drawn in very thin threads (gold); or *sectile* and can be cut with a blade to produce shavings (cerargyrite and gypsum). Other minerals are *flexible* (talc, molybdenite); others *elastic* so that if bent and released, they reassume their initial shape (mica).

*Hardness* is the property of resisting abrasion or scratching. This too depends on the structural strength but in different ways than toughness: for example, diamond, the hardest substance known, is fragile and if struck cleaves easily. Hardness is measured by referring to the *Mohs scale,* a hierarchy of minerals each of which can scratch the previous one and can be scratched by the one following it. But the classification used here is more empirical and practical. Minerals which can be scored with a fingernail are defined as *very soft* (hardness of 1–2 on the Mohs scale), those which can be scored with a piece of copper wire or a copper coin are called *soft* (hardness of 2–3), those which can be easily scored with a pocketknife *semihard* (3.5–4.5), those not easily scored with a pocketknife *hard* (5–6.5) and those which cannot be scored with a steel file, *very hard* (7–10). In determining the degree of hardness it is important to test a fresh, clean surface, if possible a cleavage plane, on which one should always trace parallel lines. The degree of hardness changes markedly with direction, especially in certain minerals in which the bond strengths show a definite orientation. A typical case of this is kyanite, which is semihard (5) parallel to the elongation [001] but is very hard (7) perpendicular to it.

## Electrical properties

The conductivity of minerals varies with direction and, in particular, with the type of bonding in that direction. The metallic elements, certain sulfides and a few oxides are good conductors. Most oxides are mediocre conductors; and almost all minerals with ionic or covalent bonds are very poor conductors (dielectrics or insulators). Among the nonconducting minerals are some which are capable of developing an electrical potential if subjected to directional mechanical stresses such as compression, traction or torsion (piezoelectricity) or thermal stresses (pyroelectricity). These minerals have polar axes and hence do not have a center of symmetry. Quartz, tourmaline and hemimorphite all exhibit this property, among other minerals. There are many industrial applications of this phenomenon, and some are of great importance (manometers, transducers and frequency couplers used in radio, and crystal oscillators, for example for "quartz" watches).

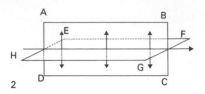

*Polarized light: In a normal ray of light (1) the vibrations occur in every possible direction. In a polarized ray (2) the light vibrates in a single plane which contains the direction of movement (ABCD). In conventional terms we say that it is polarized in the plane (EFGH) normal to the plane of vibration.*

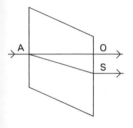

Above: *double refraction. Penetrating the calcite rhombohedron (A), the light ray splits into two rays: one ordinary (O), which does not deviate, and one extraordinary (S), which does. When the crystal is rotated the ordinary ray does not shift, while the extraordinary ray rotates with the crystal.*

## Magnetic properties

It is easiest to examine the magnetic properties of minerals when they are placed in an external magnetic field, usually generated by a permanent magnet. Some turn out to be strongly attracted (ferromagnetic), some just slightly attracted (paramagnetic) and others repelled (diamagnetic). Magnetite and pyrrhotite are the only common minerals of the first type, both iron bearing. Furthermore, magnetite is not only ferromagnetic but in some cases is *magneto-polar,* because it maintains an induced magnetism for a long period of time and thus becomes a magnet itself. Other metallic minerals containing iron and also certain iron-bearing silicates (biotite, hornblende, black tourmaline, etc.) can be magnetized to a lesser degree. All minerals are very slightly diamagnetic but this quality is too weak to be detected with a magnet.

## Radioactive properties

Some minerals (autuntite, carnotite, monazite, torbernite, uraninite, zircon, etc.) are radioactive, spontaneously emitting particles capable of excit-

Right: *double refraction in calcite.*

ing a Geiger counter or of turning photographic film black. The dangers of radioactivity require special handling (lead shielding, decontamination, etc.) when large quantities of such materials are being worked with or stored.

## Optical properties

Regarding its interaction with light energy, matter can be divided into two major categories: singly refractive (*isotropic*) and doubly refractive (*birefringent*). In the former light travels at the same speed in every direction within the substance, although it slows down and may change its direction slightly (refraction) compared with its velocity and direction through the air. Gaseous bodies, most liquids, amorphous substances, and (among the crystalline solids) those substances belonging to the isometric system are all isotropic. Their index of refraction, which is the ratio of the speed of light in a vacuum to that in the substance in question ($A = c/c_m$), is constant in every direction. On the other hand, a ray of light which enters a birefringent substance (dimetric or trimetric crystals) splits into two linearly polarized rays of light, which vibrate in planes which are at right angles to one another and travel at different speeds (*birefringence* or *double refraction*). For this reason a birefringent body has two different indices of refraction for specific directions of light movement. But there is one direction in the dimetric crystals, and two in the trimetric crystals, in which the phenomenon of double refraction does not occur. These directions of single refraction are called *optic axes*. Dimetric crystals are known as uniaxial birefringent crystals; trimetric crystals are called biaxial birefringent crystals. The phenomenon of double refraction is particularly evident in calcite: if you look through a cleavage rhombohedron of this mineral at a point drawn on a sheet of paper, the image of the point is doubled. If you rotate the rhombohedron, one of the images stays still while the other turns around the first, tracing a circle. The first is caused by the *ordinary ray,* which obeys the laws of refraction, the second by the *extraordinary ray,* which behaves abnormally with respect to these laws. The two rays travel at different speeds and thus require different indices of refraction.

*Luster* measures the degree to which light is reflected by the surfaces of a crystal and thus is linked not only with the refractive index but also with the perfection of the faces and with chromatic absorption. First we must distinguish between *metallic* luster, typical of opaque minerals in which there is no light transmission, and *nonmetallic* luster, typical of transparent and translucent minerals. There is also an intermediate degree called *submetallic* luster found in minerals which are normally opaque in mass but transparent if sliced into thin scales (e.g., cinnabar, cuprite). Minerals with a nonmetallic luster show various degrees of luster.

The brightest luster is called *adamantine*, diamondlike (also typical of cassiterite and zircon); a variety of this luster, typical of bodies with a high refractive index and a color tending to yellow, is *resinous* luster (sulfur, sphalerite, etc.). Most minerals have a *vitreous* or glassy luster, as on quartz crystal faces. Varieties of this luster arecalled *greasy*, typical of nepheline or quartz on fracture surfaces; *Mother-of-pearl* or pearly (the micas and talc), which is slightly iridescent because part of the light is dispersed through the microlayers of the mineral; and *silky*, typical of fibrous minerals like gypsum and asbestos. Minerals that reflect only a little are those with a *waxy* luster (chalcedony) and, in particular, those with an *earthy* luster. In this latter case the mineral does not reflect at all, either because it has very rough surfaces or because it is a porous aggregate of minute crystals (bauxite).

*Color* is another very obvious optical property in minerals, but it is less diagnostic. Though some minerals always have the same color (*idiochromatic*), many others have various colors depending on the impurities or imperfections they contain (*allochromatic*).

The color of opaque minerals is due to the reflection of a particular band or bands of light in the visible spectrum and is almost always idiochromatic: pyrite is pale yellow, chalcopyrite is brass yellow, copper is red, pyrrhotite is bronze colored, and so on. The color of transparent minerals is due to the fact that the mineral selectively absorbs only a part of the light it receives and transmits another part. It is thus colored by the color complementary to that of the wavelength absorbed. Some such colors are idiochromatic, like azurite and lapis lazuli (deep blue), malachite and dioptase (green), sulfur and orpiment (yellow), cuprite and cinnabar (red). Others are allochromatic: this applies, as a rule, to minerals which are colorless in the pure state, but take on a coloration of variable intensity either because (1) they absorb chemical impurities during their formation (*chromophorous ions,* mainly transitional metals), (2) because of physical impurities (the presence of minute inclusions or air bubbles which affect the light) or (3) because of specific lattice defects (*color centers*). In birefringent minerals the color changes with direction in the crystal, yielding *pleochroism*. There are two extreme ranges in the dimetric minerals (*dichroic*)

and three in the trimetric minerals (*trichroic*).
When making color observations it is always necessary to use a fresh patina-free fracture and indirect lighting. The powder of an allochromatic mineral is usually whitish or light gray, whereas the powder of an idiochromatic mineral retains the color of the body in mass, although it is usually paler. This observation is made using the *streak* test, that is, by rubbing the mineral against a piece of unglazed white porcelain.

*Asterism in enstatite.*

Some minerals have specific optical properties, often related to the presence of impurities. *Chatoyancy* is due to the presence of minute cavities, or inclusions or fibers of another mineral oriented in a single direction, which cause light to concentrate in an oscillating and iridescent, thin, light band. This can be seen in chrysoberyl (cat's eye), in quartz with asbestos inclusions (tiger's eye), in adularia (moonstone), gypsum, etc. Related to chatoyancy, *asterism* is due to the presence of crystallographically controlled inclusions in the form of small criss-cross needles (often of rutile) and appears as a four- or six-pointed shining star. This phenomenon produces star rubies and sapphires, but also occurs in rose quartz, in phlogopite and in certain pyroxenes. *Labradorescence*, typical of labradorite, is due to diffraction from closely spaced planar intergrowths. When viewed from a specific angle the mineral produces a distinctive bright blue sheen.
*Luminescence* is an emission of light with a specific wavelength (color) as the result of a stress which may be mechanical such as rubbing or friction (*triboluminescence*, for example, of fluorite, sphalerite and lepidolite) or physical such as heating (*thermoluminescence* as in celestite, barite, apatite and spodumene) or chemical (*chemiluminescence*). Most commonly seen is *photoluminescence*, where the material emits visible light when illuminated with higher-energy light (usually ultraviolet). Photoluminescence is divided further into *fluorescence*, if it stops when the stimulating rays stop, and *phosphorescence*, if it continues for a certain time after. There are some minerals which produce photoluminescence only by means of short ultraviolet wavelengths, others by longer wavelengths and still others by both, in which case the color may remain the same or change. Some minerals are always luminescent, others only if they contain lattice defects or specific chemical impurities known as *activators*: willemite as a rule is not fluorescent, but willemite from Franklin, New Jersey, which contains manganese, is markedly so in the green-yellow range; scheelite is always fluorescent in the blue-white range except when it contains molybdenum, in which case it is yellow.

*A sample of scheelite, coming from Traversella (Turin, Italy), in natural light (above) and exposed to ultraviolet rays (below). Under UV light, photoluminescence (fluorescence) appears.*

Other minerals which fluoresce as the result of activators are scapolite, calcite, aragonite, zircon and fluorite.

## Sensate properties

Only some minerals have these properties because they depend on the presence of fairly weak ionic bonds. They have to do with *taste*, like salty for halite, bitter or sour for sylvite, drying for alum etc., typical of certain water-soluble salts, and *smell*, typical of certain minerals which vaporize easily or have volatile contents (the earthy aroma of kaolin and the garlic aroma of heated arsenic compounds).

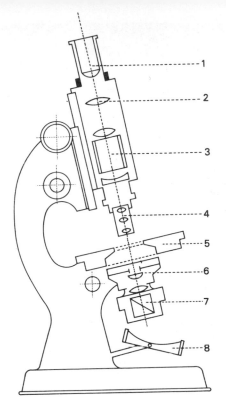

Left: *diagram of a polarizing microscope.*
(1) *eyepiece;* (2) *Bertrand lens;* (3) *analyzer;*
(4) *objective lens;*
(5) *graduated rotating stage;* (6) *condenser;*
(7) *polarizer;* (8) *mirror.*

## Using the microscope

The *mineralogical* or *petrographic microscope* makes it possible to study the optical properties of minerals. It is equipped with polarizers capable of producing polarized light (*Nicol* prisms from calcite or artificial *polaroids*). One of these is mounted beneath a revolving plate and is called the *polarizer* because it transmits polarized light to the prepared slide or specimen; the other, situated between the lens and the eyepiece, is called the *analyzer*, because it is used to analyze the light which has passed through the specimen. The analyzer is movable and can be inserted into the path of the rays or removed.

Using the polarizer alone, it is possible to determine the refractive index of isometric crystals and, with appropriately oriented sections, the refractive indices of dimetric crystals and of trimetric crystals. With colored birefringent crystals it is also possible to make observations of pleochroism, that is, the different colors these crystals show in different directions.

When the analyzer is inserted across the path of

Above: *diagram of a polarizing Nicol prism.* ABCD *is a rhombohedron of transparent calcite (Iceland spar) cut along the lesser diagonal and glued with Canada balsam.* Red *shows the natural incident light;* blue *shows the extraordinary ray, slightly deviated, which passes through the second half of the Nicol and emerges polarized in the plane of the drawing;* yellow *shows the ordinary ray which, being more deviated, is totally reflected by the layer of Canada balsam and then absorbed by the surface* BD, *blackened with lampblack.*

*Glaucophane crystals with characteristic pleochroism in a thin section of Piedmont eclogite as viewed through a polarizing microscope: greenish-yellow (left), bluish-violet (center), Prussian blue (right).*

the light rays, placed perpendicularly to the polarizer (*crossed Nicols*), and there is no slide on the microscope plate, the field of observation appears dark or *extinguished* (extinct). No light can pass through because the analyzer suppresses all the light issuing from the polarizer. Similarly, if one places an amorphous substance or an isotropic crystal between the two polarizers the field is also dark. By placing a birefringent material between them, however, light usually passes through, its color depending on the difference of the refractive indices and on the thickness of the body itself (*interference color*). This is due to a complex interplay of light waves within crystals and to interference beyond the analyzer. By turning the plate of the microscope through a 360° angle, the birefringent material in general takes on four positions of extinction which are perpendicular to one another and four positions of maximum luminosity, likewise perpendicular to one another and at an angle of 45° from the extinction.

Observations in convergent light are made with the mineral placed between the two crossed polarizers and lit by a beam of light converging at the exact center of the crystal by means of a lens called the *condenser*, mounted below the specimen. It is also necessary to insert a second lens (*Bertrand lens*) between the analyzer and the eyepiece. Under these conditions characteristic *interference figures* can be observed for uniaxial crystals, consisting of a series of concentric rings, alternately lit up and extinguished, known as *isochromatic* rings, intersected by a dark cross with 90° arms (called *isogyres*). For biaxial crystals, the typical interference figures are more complicated. Last, it is possible to observe the phenomena of dispersion and rotatory polarization (optically active crystals).

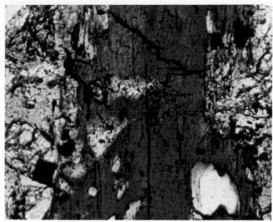

## X-ray analysis

X-ray diffraction techniques are the surest method of identifying a mineral. X-rays make it possible, within certain limits, to determine the crystal class and the unit-cell parameters, and where relevant the position of the various atoms inside the crystal can be determined. All this can be done without destroying the specimen, and often from a very small one. As with observations through the microscope, this method examines the interactions of electromagnetic radiation, in this case not light but rays with a much shorter wavelength (roughly of the same size as the lattice dimensions), and with deep penetrating ability. In particular, it is possible to determine the distances between the atomic planes of the lattice and the type and distribution of the atoms on these planes, two typical and unique properties for each mineral species. Although more complex and refined methods exist, the one still most commonly used in the identification of minerals is the relatively simple *powder* method. This consists of placing a finely ground sample of the specimen in a narrow beam of monochromatic x-rays and measuring the angular dispersion and the intensity of the diffracted x-rays, either with photographic film or by an electronic device which measures and records them with pen traces on graph paper (these are the camera and diffractometer methods). The position of the lines of reflection depends on the geometry of the unit cell, in other words on its linear and angular dimensions. The intensity depends on the type of atoms present in the various positions of the structure and approximately on the number of reflections allowed from the symmetry of the cell itself. By consulting tables which list more than

*Interference figures of trimetric crystals cut at right angles to the optic axis (1) and of crystals cut at right angles to the acute bisecting line of the optic axes (2).*

1

2

30,000 reference substances and their diffraction characteristics, it is usually possible to match one of these to the mineral in question.

**Chemical analysis**
Quantitative analysis of a mineral is a lengthy, complicated and difficult procedure. Also, the results can be misleading because it is hard to separate the pure mineral in sufficient quantities. For small samples the best methods use instrumental analysis, for example an electron microprobe (a device which can measure extremely small quantities of inorganic material), but this is a sophisticated device found only in large mineralogical or geological research laboratories. There are, however, qualitative chemical tests which require only a very few grains of the mineral but are nevertheless accurate and inexpensive. They provide useful data about the composition of a mineral and narrow the range of identification possibilities considerably.

*Flame tests* consist in heating the mineral in a gas flame (about 1000°C; 1832°F), which may sometimes be reinforced with oxygen or pressurized air blown from a blowpipe (about 1500°C; 2732°F). The flame is made up of an inner, reducing, flame, ranging from bluish or gray-blue to dark blue, and by an outer, oxidizing, flame which is yellow to almost colorless and much hotter. First we measure the *fusibility* of the mineral, a property which cannot be easily defined in a physical sense because it is bound up not only with the melting point but also with the capacity of the mineral to disperse or transmit heat in various directions. A small piece of the mineral, held with a pair of tweezers or laid on a piece of charcoal, is heated and compared with the *Kobell scale*, which is empirical and based on the following minerals:

1. Stibnite       Melts in wax flame (approx. 525° C; 977° F)
2. Chalcopyrite   Melts in gas flame (approx. 800° C; 1472° F)
3. Almandine     Melts in gas flame but only in thin slivers (approx. 1050° C; 1922° F)
4. Actinolite       Does not melt in blowpipe; becomes rounded at edges (approx. 1200° C; 2192° F)
5. Orthoclase     In the blowpipe the sliver barely becomes rounded (approx. 1300° C; 2372° F)
6. "Bronzite"     In the blowpipe only the finest sliver becomes rounded (approx. 1400° C; 2552° F)
7. Quartz         Not fusible

During heating various phenomena can be observed: some minerals swell and emit bubbles (zeolite group, epidote group, scapolite group), some shed flakes (vermiculite, pyrophyllite), some decrepitate and break up (carbonates, barites), some melt giving the flame a specific color (yellow from sodium, crimson from strontium, violet from potassium, etc.), some produce irritant or toxic vapors (sulfides, arsenic minerals), some leave a magnetic black globule (certain iron minerals) or red (copper minerals) or gray (lead minerals). Fusion may be helped by special fluxes like borax. In such cases a glassy bead is obtained; its color again depends on the element present and varies accordingly if obtained with an oxidizing flame or with a reducing flame. It may also change again when the bead is cold.

*Wet tests* are qualitative analytical chemical tests, carried out by dissolving the mineral in various acids for examination. The most common consist in checking to see whether the mineral is water soluble (halides, nitrates and certain sulfates), then whether it is soluble in hydrochloric acid, diluted and cold (calcite) or hot (magnesite), or concentrated and cold (dolomite) or hot (siderite), and finally in nitric acid and in aqua regia. It is also helpful to keep an eye on the development of gas bubbles (effervescence), and on the color or clarity of the solution.

*Instrumental methods* are typical chemico-physical methods, generally used only in research laboratories. Briefly, there is the *spectroscope,* used for qualitative analysis of the elements contained in a mineral on the basis of the presence of absorption lines in the light which has passed through the specimen and then been dispersed by a prism. The *infrared* (IR) *spectrograph* measures the vibrational frequencies of atomic groups (especially of water or the hydroxyls) present in a mineral. *Differential thermal analysis* (DTA) measures the loss of water and other gases, and changes of physical state linked to temperature variations.

## PRINCIPLES OF CLASSIFICATION

In current mineralogical literature there are about 3000 different mineral species, and the list is being added to at the rate of about 60 each year, with a small number being discredited after reexamination.

Such a large number of species, plus a much larger number of other names (some 14,000 synonyms, varieties or old-fashioned and commercial names) makes it necessary to create a classifica-

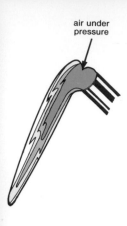

air under
pressure

*Cross section of an
oxidizing flame of a
blowpipe with its typical
colorations depending on
temperature.*

tion system, i.e., an organized form of cataloging based on fixed principles. Classification systems vary, and in the last two centuries different systems have been proposed grouping minerals according to a fundamental principle, which may be *chemical, physical, genetic, structural* or *crystallochemical*. This book follows the crystallochemical criterion, whereby the basic systematic unit, the mineral species, is defined as *all those individual minerals characterized by an identical structural motif and by a chemical composition which is variable within defined limits, generally quite restricted, and which are in apparent thermodynamic equilibrium at conditions effectively realized on Earth or in extraterrestrial bodies accessible to man.* A species may have *varieties*, characterized by specific properties (chemical or physical) which are very homogeneous but do not appear throughout the species. The varieties may be divided up on the basis of color, form, the presence of a particular substitution (diadochy), of a structural difference (polytypism), of inclusions and so forth.

The basis of most modern systematics in mineralogy is the work of James B. Dana and Edward S. Dana, who, starting in 1837 with the *System of Mineralogy* by the former, laid down a chemical and crystal-morphological classification. As more minerals were discovered and as structural data became available, the *System* was updated through the 19th century. The crystallochemical classification we follow here is an elaboration started by Hugo Strunz, a German mineralogist, in 1938. Since then it has undergone a constant process of updating as structural identifications were gradually refined and chemical compositions specified. We have used *Mineralogische Tabellen* (5th ed; Leipzig, 1970), which divides the mineral species (then about 2000 classified ones) into nine classes:

| | | |
|---|---|---|
| I. | Native elements (with alloys, carbides, nitrides, phosphides) | approx. 50 |
| II. | Sulfides (with selenides, tellurides, arsenides, antimonides and bismuthides) | approx. 300 |
| III. | Halides | approx. 100 |
| IV. | Oxides and hydroxides | approx. 250 |
| V. | Nitrates, carbonates and borates | approx. 200 |
| VI. | Sulfates (with chromates, molybdates and wolframates) | approx. 200 |
| VII. | Phosphates, arsenates and vanadates | approx. 350 |

However, some modifications from Strunz have been made. The convention adopted in this book is that names and chemical formulas which are approved by the International Mineralogical Association (IMA) are taken from the *1975 Glossary of Mineral Species*, by Michael Fleischer. References to synonyms have been deleted, as the IMA discourages their use. Names referring to substances that are not proper mineral species are given in quotation marks. Examples are a mineraloid, "amber" or "opal"; an amorphous mixture of minerals, "limonite"; a habit, as "cleavlandite" of albite; or a field term like "wolframite" for the huebnerite-ferberite series. If a mineral is part of an isostructural mineral group, the group is listed in parentheses after the mineral class: Almandine, Silicates (nesosilicates, garnet group).

A specific problem is posed by the *silicates*, which are fairly widespread and differentiated minerals accounting for 80 percent of the minerals at the Earth's surface. They are classified not so much on chemical as on structural bases, and more specifically on the way in which the basic unit $[SiO_4]^{4-}$ is linked with the neighboring ions. The bonding occurs solely at the apices of the tetrahedral group $[SiO_4]^{4-}$, and the differentiation of the groups depends on the way in which the tetrahedra are joined together. We thus have the following subclasses:

1) *Nesosilicates*, with isolated tetrahedral groups $[SiO_4]^{4-}$, interconnected by a cation which is not silicon; as a rule they have a compact habit and a high degree of hardness, a high specific gravity and a high refractive index, e.g., forsterite.

2) *Sorosilicates*, with two tetrahedra joined by a vertex forming a group $[Si_2O_7]^{6-}$; these groups are coordinated with cations that are not silicon by the three free vertices on the one tetrahedron and three opposite vertices on the other, e.g., epidote.

3) *Cyclosilicates*, with tetrahedra which are joined together into rings with three, four or six elements. They are characterized by prismatic, trigonal, tetragonal and hexagonal habits, the most common being those with rings of six tetrahedra, e.g., beryl.

4) *Inosilicates*, with tetrahedra joined together in indefinite chains. The most common cases are single open chains, typical of the pyroxene group, or double or closed chains, typical of the amphibole group. Needlelike (acicular) or fibrous habits are predominant, and as a rule there is also a characteristic, easy cleavage parallel to the elongation.

5) *Phyllosilicates*, the tetrahedra of which are

# CHEMICAL ELEMENTS

| Symbol | Name | Atomic Number | Symbol | Name | Atomic Number |
|--------|------|---------------|--------|------|---------------|
| Ac | Actinium | 89 | Mn | Manganese | 25 |
| Ag | Silver | 47 | Mo | Molybdenum | 42 |
| Al | Aluminum | 13 | N | Nitrogen | 7 |
| Am | Americium | 95 | Na | Sodium | 11 |
| Ar | Argon | 18 | Nb | Niobium | 41 |
| As | Arsenic | 33 | Nd | Neodymium | 60 |
| At | Astatine | 85 | Ne | Neon | 10 |
| Au | Gold | 79 | Ni | Nickel | 28 |
| B | Boron | 5 | No | Nobelium | 102 |
| Ba | Barium | 56 | Np | Neptunium | 93 |
| Be | Beryllium | 4 | O | Oxygen | 8 |
| Bi | Bismuth | 83 | Os | Osmium | 76 |
| Bk | Berkelium | 97 | P | Phosphorus | 15 |
| Br | Bromine | 35 | Pa | Protactinium | 91 |
| C | Carbon | 6 | Pb | Lead | 82 |
| Ca | Calcium | 20 | Pd | Palladium | 46 |
| Cd | Cadmium | 48 | Pm | Promethium | 61 |
| Ce | Cerium | 58 | Po | Polonium | 84 |
| Cf | Californium | 98 | Pr | Praseodymium | 59 |
| Cl | Chlorine | 17 | Pt | Platinum | 78 |
| Cm | Curium | 96 | Pu | Plutonium | 94 |
| Co | Cobalt | 27 | Ra | Radium | 88 |
| Cr | Chromium | 24 | Rb | Rubidium | 37 |
| Cs | Cesium | 55 | Re | Rhenium | 75 |
| Cu | Copper | 29 | Rh | Rhodium | 45 |
| Dy | Dysprosium | 66 | Rn | Radon | 86 |
| Er | Erbium | 68 | Ru | Ruthenium | 44 |
| Es | Einsteinium | 99 | S | Sulfur | 16 |
| Eu | Europium | 63 | Sb | Antimony | 51 |
| F | Fluorine | 9 | Sc | Scandium | 21 |
| Fe | Iron | 26 | Se | Selenium | 34 |
| Fm | Fermium | 100 | Si | Silicon | 14 |
| Fr | Francium | 87 | Sm | Samarium | 62 |
| Ga | Gallium | 31 | Sn | Tin | 50 |
| Gd | Gadolinium | 64 | Sr | Strontium | 38 |
| Ge | Germanium | 32 | Ta | Tantalum | 73 |
| H | Hydrogen | 1 | Tb | Terbium | 65 |
| He | Helium | 2 | Tc | Technetium | 43 |
| Hf | Hafnium | 72 | Te | Tellurium | 52 |
| Hg | Mercury | 80 | Th | Thorium | 90 |
| Ho | Holmium | 67 | Ti | Titanium | 22 |
| I | Iodine | 53 | Tl | Thallium | 81 |
| In | Indium | 49 | Tu | Thulium | 69 |
| Ir | Iridium | 77 | U | Uranium | 92 |
| K | Potassium | 19 | V | Vanadium | 23 |
| Kr | Krypton | 36 | W | Tungsten | 74 |
| La | Lanthanum | 57 | Xe | Xenon | 54 |
| Li | Lithium | 3 | Y | Yttrium | 39 |
| Lu | Lutetium | 71 | Yb | Ytterbium | 70 |
| Lw | Lawrencium | 103 | Zn | Zinc | 30 |
| Md | Mendelevium | 101 | Zr | Zirconium | 40 |
| Mg | Magnesium | 12 | | | |

joined together by three vertices producing indefinite, hexagonal and in rare cases tetragonal flat sheets. Between the sheets are bonds made only by the oxygens at the fourth vertex, all oriented in the same direction; the result is a lamellar (leaved) habit with excellent cleavage parallel to the base; low specific gravity and hardness, e.g., muscovite.

6) *Tectosilicates*, the tetrahedra of which are joined together by all four vertices, producing indefinite "cages" or frameworks with complex structures. The presence of aluminum in the place of silicon at the center of certain tetrahedra makes it possible for stably bonded cations to be present in the framework structures, as in the feldspars. Mostly of low density and with a compact habit, except in certain cases where the cages are oriented in such a way as to give elongated forms, or forms which are developed in two directions.

In the silicates the rigid structural motif caused by the great strength of the Si-O bond and an easy substitutability of the other cations produce complex solid solutions with two or more extreme limits, which are called *groups* (olivine, pyroxene, amphibole, feldspar, etc.). There are no set rules for mineral nomenclature. Names for minerals are derived from their chemical composition (uraninite), with other names drawn from the first place in which a mineral was discovered (benitoite, San Benito County, California), or dedicated to a scholar (dolomite named for Déodat de Dolomieu, an 18th-century French geologist), or to a patron (zoisite for Baron Sigismund Zois von Edelstein of Slovenia), or indicating a physical particularity (axinite for its axelike crystals or orthoclase for its right-angle cleavage).

## GENESIS OF MINERALS

### Formational environments

Minerals are the product of a complex series of chemical, phase reactions (crystallization) through which, starting from disordered atoms, pieces of matter are derived which are tightly organized in a periodical or homogeneous manner. At the moment of formation, and in the course of growth, the mineral is attempting to achieve a state of equilibrium with its environment, which is why every change in the conditions (pressure, temperature, acidity of solutions, etc.) either interrupts growth (in certain cases even triggering dissolution) or manifests itself by changes in the inner organization of the mineral (structural defects, presence of inclusions, chemical zonations or divisions, etc.). Some minerals are formed in very short periods of time (gypsum crystals half a meter in length have actually been *seen* growing over a matter of hours

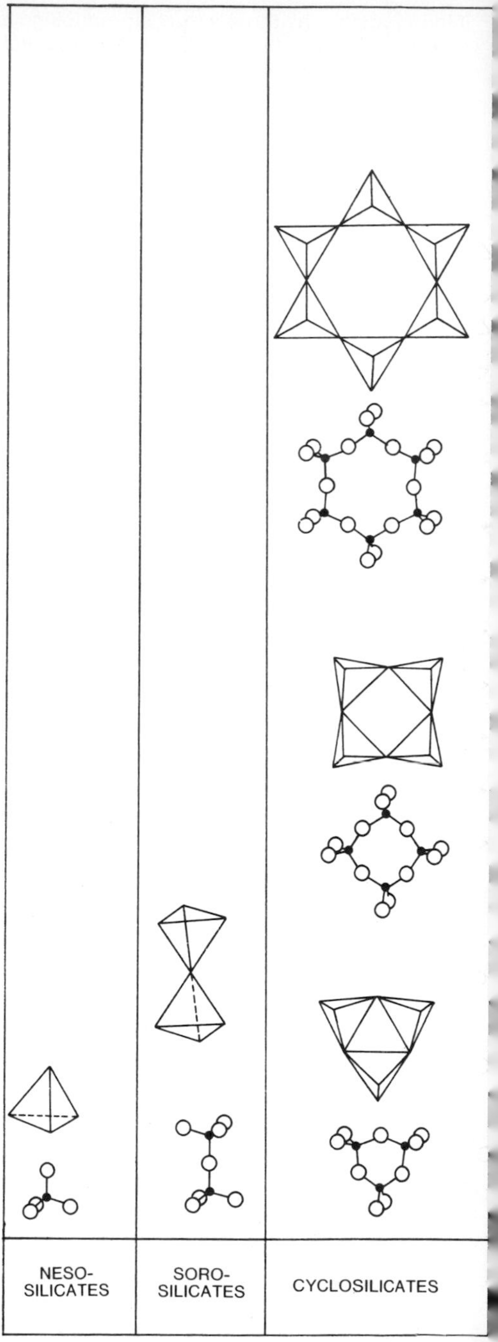

The structure of silicates. The centers of the atoms (black = silicon or aluminum; white = oxygen) and the tetrahedral coordinating polyhedra are shown here. In the nesosilicates the tetrahedron is isolated. In the sorosilicates two of these tetrahedra are linked by a vertex; in the cyclosilicates three, four or six of these form a ring, because they are all linked by vertices; in the inosilicates there are chains with tetrahedra linked by two opposite vertices (a single chain is characteristic of the pyroxene family and a band of two facing chains is typical of the amphibole family); in the phyllosilicates tetrahedra sharing three vertices form exact or nearly hexagonal plane lattices; in the tectosilicates the tetrahedra share all four of the vertices.

NESO-SILICATES

SORO-SILICATES

CYCLOSILICATES

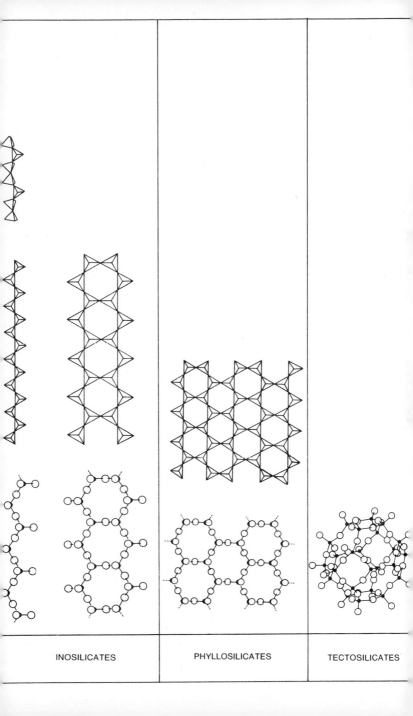

INOSILICATES     PHYLLOSILICATES     TECTOSILICATES

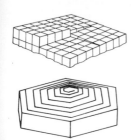

*Crystal growth by development of a screw dislocation and formation of a growth spiral.*

in desert lagoons), others have grown at geological rates. For example, it has been calculated that certain quartz crystals from alpine cavities took 200,000–300,000 years to reach their present-day dimensions.

The process of crystallization is divided into two phases: nucleation and growth. *Nucleation* is the formation of a nucleus or kernel of a mineral from which it can enlarge by the process of *growth*. The nucleus forms by the assemblage of randomly scattered constituent materials of a mineral into at least several joined unit cells, sometimes on the surface of an impurity. *Growth* and *enlargement* will only proceed if the conditions are favorable; otherwise the nucleus may break up. *Growth* proceeds by the accretion of successive layers of matter, or by the successive addition of rows of atoms, starting from an irregularity of the nuclear crystal surface. This second type of growth is frequently spiral and is visible with an electron microscope and sometimes even with the naked eye. Experience has taught us that it takes only very slight environmental variations to produce a considerable effect on the growth pattern. In saturated solutions we see that a gradual cooling produces only a few nuclei with uniform growth into large crystals, while rapid cooling, conversely, produces many nuclei with many small crystals at the completion of growth. An equivalent effect is produced by oversaturation of solutions: the higher the oversaturation, the easier the nucleation and growth of many small crystals. The final form of crystals also depends a great deal on this. In a slowly cooling solution the few, large crystals that develop easily assume their typical pattern (*euhedral* crystals), whereas in the second case the numerous crystals tend to hamper one another during growth and are often distorted. A fast-cooling solution may give rise to interpenetrating aggregates along curved surfaces (*anhedral* crystals).

The environments for the nucleation and growth of crystals are the same as for rocks and will thus be dealt with in that section. It will suffice to mention certain mineral characteristics which are found in the various environmental petrologic associations, although we must bear in mind that only a few minerals have a field of stability so restricted that they can only crystallize in one environment.

**Igneous environments.** In rocks which have formed slowly during the principal stage of crystallization of a body of molten rock (magma), silicates are commonly found (olivine, pyroxene, amphibole, mica, felspars, quartz and zircon), as are phosphates (apatites), sulfides (pyrite, pyrrhotite), oxides (magnetite, chromite) and a few native elements (platinum, diamond). Pegmatites contain many rare minerals, sometimes in huge crystals, the growth of which is helped by the fluidity of the gas-filled environment: silicates (topaz, beryl, tourmalines, micas, spodumene, etc.), phosphates (apatites, monazite), oxides (cassiterite, columbite, corundum, uraninite, thorianite), halides (cryolite) and sulfides (molybdenite, arsenopyrite). In hydrothermal veins which have formed in fissures as a result of precipitation from solutions, one can find: silicates (feldspars, quartz, epidote), sulfates (barite), carbonates (calcite, ankerite, rhodochrosite), oxides (hematite, ilmenite, uraninite, rutile), halides (fluorite) and a large number of sulfides (galena, sphalerite, chalcopyrite, stibnite, pyrite, etc.). In fumarolic deposits that have formed as a result of the sublimation of volcanic fumes, we find: silicates (zeolites), sulfates (alunite), oxides (hematite), sulfides (pyrite, cinnabar, realgar, bismuthinite, stibnite) and also certain native elements (sulfur, arsenic and mercury).

**Sedimentary environments.** The minerals present in sedimentary rocks are largely the remains of preexisting rocks, only rarely derived from the chemical transformation of these. Sometimes the sedimentary minerals precipitate directly from solution. Among the *secondary minerals,* the particularly important ones are associated with the weathering of mineral deposits, and they fall into two types: one superficial and oxidized, the other deeper and reduced. Here we find: silicates (chrysocolla, hemimorphite), sulfates (anglesite), phosphates and vanadates (carnotite, · vanadinite, pyromorphite), carbonates (malachite, azurite, cerussite, smithsonite), a few oxides (cuprite) and certain native elements (gold, silver, copper).

Among the products of the evaporation of water, especially of sea water, we have: sulfates (gypsum), carbonates (calcite, dolomite), borates (borax), nitrates (trona) and above all halides (rock salt or halite, sylvite). Other widespread and important minerals are produced by the activity of animal and vegetable organisms (sulfur, aragonite, apatites, oolitic limonite) and left as an insoluble residue (clays, bauxite, limonite, chalcedony).

**Metamorphic environments.** Here the minerals in question are formed by solid-state transformations, at temperatures and pressures often different than the original mineral-forming ones. Few are really typical. Among these are silicates (serpentines, chlorites, garnets, micas, kyanite, andalusite, sillimanite, amphiboles, staurolite, vesuvianite, scapolite), carbonates (magnesite), oxides (rutile, corundum, ilmenite, spinels), hydroxides (brucite), sulfides (pyrite and pyrrhotite) and the native element graphite.

# COLLECTIONS

It is not enough to know in what environment minerals are formed in order to be sure of finding some during an outing. You need to know where to look——the places in which it is possible (even though it is rarely easy) to find some hidden and handsome group of crystals. In mountainous regions the potential mineralogical sites are those where the rock is fresh and exposed, and possibly partly broken up: landslides, both where they start and end, valley floors, the bottom of mountain streams and glacial moraines. Other possible mineralogical sites are manmade cuts in rocks: mines, caves and the exposures obtained by building roads, parking lots and all sorts of structures. The first zones that a collector in the field should visit are mine dumps and quarry rubble heaps, where there will always be the odd fine specimen that has been left behind or overlooked by previous collectors, or where you may be able to obtain specimens from the workmen. The equipment required to collect is very inexpensive. As well as normal hiking equipment, which varies depending on altitude and climate, you will need a good crack hammer (0.8–1.0 kg; 2–2.5 lbs) with a broad head and/or a smaller one with a pointed end, commonly called a rock hammer, a series of chisels ranging from very small to medium, a stiff brush, a 10X hand lens, newspaper for packing the specimens and a knapsack to carry everything in. It is always a good idea to carry safety goggles or glasses to protect your eyes from flying rock and steel splinters when using the hammers and chisels. Useful items, though not absolutely necessary, include: a lever or crowbar for prying open fissures and cracks without causing too many splinters, a small sledge hammer, shovel, hard hat (if necessary), labels and cotton for wrapping up very fragile specimens. The collecting outing should be preceded by a period of preparation and documentation, including consultation of the topographical and geological maps of the area in question and of descriptions, both scientific and amateur, of the minerals already found there.

## Cataloging and filing

Once back from an outing the material collected must be sorted, carefully examined, identified and at last filed. Identification is undoubtedly the hardest, especially for beginners. It can be carried out with the help of a good hand lens or a binocular microscope, making fullest use of the external features of the specimen. In other words, you must carefully examine the form, type of cleavage, luster and color and possibly work out the hardness and the streak. If these tests do not furnish an identification, you can move on to more subtle ones, such as measuring the optical characteristics using grains of the specimen or preparations of powder immersed in a drop of mineral oil on a slide and looked at through the polarizing microscope, flame tests, and so on.

The second operation is the preparation of the specimens for filing. You must free the minerals of crusts and patinas and sometimes also of those adjacent crystals which might prevent you making the most of the selected crystal's appearance. Cleaning is done with brushes, large and small, of different degrees of hardness; the crust is removed with small steel picks (like dentist tools) or with carefully selected acid treatments. The final stage of cleaning is to wash the specimen with a weak detergent solution. Some inexperienced collectors think they should completely isolate their crystals: this is a serious mistake because the presence of a natural base (*matrix*) is not only essential from a scientific point of view but also enhances the aesthetic appearance of the mineral.

Specimens should always be labeled, showing not only the name of the mineral but also the year it was collected, the place of origin and, in the case of exchanged or purchased specimens, the name of the supplier. A mineral loses much of its value if its place of origin is not known. It is thus a good idea to keep an up-to-date record book containing all the data relating to every specimen in the collection, including observations made or tests carried out to identify the species. A good idea for labeling is to number your specimens and the description cards so that, if the card and specimen become separated, you will be able to reidentify the specimen. This is easily achieved by first painting on a small white patch (typewriter correction fluid is very good for this) and then writing on the number with pen and indelible ink. As far as filing is concerned, you can follow the crystallo-chemical criterion, which is that used in the systematic part of this book. Other criteria, which are less commonly used, are *geographical and regional, crystallographic, geological, economic* and *alphabetical.*

Single-species collections are also very common, concentrating on a single species or group, with all its various forms and places of origin. *Micromount* collections are also becoming popular: a series of very small specimens (0.5–1.0 cm; 0.2–0.4 in), often beautifully crystallized, designed to be looked at with binocular microscopes.

# MINERALS

## KEY TO SYMBOLS

CHARACTERISTIC CRYSTAL FORM OF MINERAL

SPECIFIC GRAVITY OF MINERAL
(and possible range of variation)

RARITY OF MINERAL

very rare

rare

common

very common

*Note on the captions: "ca. x 1.5" means "magnification about 1.5 times actual size."*

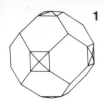

## 1 COPPER

NATIVE ELEMENTS
Cu (Copper)

**System**   Isometric.

**Appearance**   Tetrahexahedral or octahedral crystals, usually twinned, rare. Generally occurs in compact masses, sometimes of considerable size, or in dendritic and filiform masses. Characteristic copper-red color on fresh surfaces, more often with a greenish film of malachite or a blackish or iridescent film. Sometimes occurs as a pseudomorph after calcite, aragonite or cuprite.

**Physical properties**   Fairly soft (2.5–3), very heavy, ductile, malleable, no cleavage, hackly fracture. Opaque with metallic luster. Very thin sheets are translucent, letting through weak, greenish light. Excellent conductor of heat and electricity. Dissolves easily in nitric acid, staining the solution pale-blue when excess ammonia is added. Fuses at 1082° C (1980° F).

**Environment**   A typical mineral formed by chemical processes in reducing conditions in the oxidation zone of sulfide deposits. Also occurs in cavities of basalts and conglomerates, sometimes in considerable quantities. Often found in old mines subject to periodic flooding by water containing copper sulfates, appearing as crusts on iron objects or replacing fibers of wooden supports.

**Occurrence**   The finest crystals of native copper, measuring up to 3 cm (1.18 in), come from the Keweenaw Peninsula (Lake Superior, USA) where masses weighing up to 400 metric tons and natural alloys of copper and silver known as ''halfbreeds'' have also been found. There are other deposits with fine crystals of native copper in Germany, and Bisbee, Arizona (USA) and in the manganese skarns of Långban (Sweden) and Franklin, New Jersey (USA). Dendrites and masses are very common in many deposits (USSR, Zambia, Chile). In Europe, small deposits are found near Pisa and Florence (Italy).

**Uses**   Native copper rarely occurs in large enough quantities to be worth exploiting commercially. The metal has been important in human history, second only to iron. Nowadays its chief use is in electrical engineering (electric cables and wires) and for alloys (brass, bronze and a new alloy with 3 percent beryllium which is particularly vibration resistant).

▲ Native copper (ca. ×1). New Mexico.

▶ Native copper (ca. ×1.5). Lake Superior, Michigan.

▼ Dendritic native copper (ca. ×1). Lake Superior, Michigan.

## 2 SILVER
NATIVE ELEMENTS
Ag (Silver)

**System** Isometric.

**Appearance** Rare, cube-shaped or octahedral crystals, always small, usually displaying stepped faces. Compact masses, dendrites and wire-like forms of a silvery, gray-white color. Arborescent aggregates with small individual branches at right angles or star-shaped aggregates are common.

**Physical properties** Fairly soft (2.5–3), very heavy, ductile and malleable. Opaque with bright metallic luster, though almost always dulled by a blackish film caused by surface chemical alteration. Fuses at a low temperature (960° C; 1760° F). Soluble in nitric acid. Tarnishes if exposed to fumes of hydrogen sulfide. The best known conductor of heat and electricity.

**Environment** Formed by reduction of sulfides in the lower part of lead, zinc and silver deposits. Sometimes also a primary mineral, either in low-temperature hydrothermal veins associated with calcite or in high-temperature veins associated with nickel or cobalt sulfides and uraninite. Frequently associated with copper.

**Occurrence** The finest dendritic and wirelike crystals come from Köngsberg (Norway). Other famous localities are Freiberg (DDR) and San Luis Potosì (Mexico). Large amounts of silver, though not fine crystals, are found at Chañarcillo (Chile), Cobalt, Ontario (Canada), Broken Hill (Australia) and Redbeds, Colorado (USA). The largest blocks are from Aspen, Colorado (USA), where one weighing 380 kg (844 lb) was mined. However, the highest level of production has been from the Guanajuato mine (Mexico), about 500 billion kilos (460,000,000 tons) from the year 1500 to the present day. Found in southern Europe on the island of Sardinia.

**Uses** An excellent ore of the metal silver, but rare. Silver is used in photography, chemistry, jewelry and in electronics because of its very high conductivity. In the USA and some other countries it is still used as currency, generally in some form of alloy.

▲ Native silver on calcite (ca. ×1). Köngsberg, Norway.

▲ Native silver and copper (ca. ×1.5). Mexico.

## 3 GOLD
### NATIVE ELEMENTS
### Au (Gold)

**System** Isometric.

**Appearance** Very rare, small, octahedral, cubic and dodecahedral crystals. Normally occurs in very small, shapeless grains, sheets and flakes. Dendrites rare. In placers (alluvial or glacial deposits) nuggets are common. Yellow color, varying in brightness depending on the impurities present.

**Physical properties** Fairly soft (2.5–3), very heavy, ductile and malleable. Opaque with bright metallic luster. Very thin sheets let through feeble, greenish light. Medium fusion point (1061° C; 1942° F). An excellent conductor of heat and electricity. Its insolubility in acids (except aqua regia) and its specific gravity distinguish it from yellow sulfides and from the small altered plates of biotite often found in sands, where it is associated with pyrite, arsenopyrite and pyrrhotite and with tellurides and selenides of gold. Also occurs in various volcanic rocks and tuffs, associated with chalcedony and manganese minerals. Large concentrations known as bonanzas are formed by the erosion and redeposition of gold-bearing lavas.

**Environment** Occurs primarily in high-temperature hydrothermal quartz veins in extrusive rocks. Frequently found as a natural alloy with silver (electrum) and less often with palladium (porpezite) and rhodium (rhodite). However, most gold is obtained from concentrations of sedimentary origin (placers), both recent (river sand) and fossil deposits (conglomerate matrix), where it is accompanied by other heavy minerals. Gold flakes are also found in the cementation zones of sulfide, selenide and telluride deposits, formed at high temperature under hydrothermal conditions.

**Occurrence** The main gold-bearing districts are the Witwatersrand (South Africa), the Mother Lode (California, USA), the Yukon (Alaska, USA), Porcupine (Northwest Territory, Canada) and the USSR. Formerly mined in a small district near Monte Rosa (Italy). The main source of the commercial metal, used mainly as a monetary standard, in jewelry, in dentistry and for scientific and electronic instruments.

▲ Native gold crystals (ca. ×2). Vorospatak, Hungary.

▶ Native gold nugget with quartz (ca. ×2). Ural Mountains, USSR.

▼ Native gold nugget (ca. ×2). Alaska.

## 4 MERCURY

NATIVE ELEMENTS
Hg (Mercury)

**System** Hexagonal (at −39°C; −38°F).

**Appearance** The only mineral liquid at normal temperatures. Takes the form of tiny drops or impregnations of a silvery-white color, usually associated with cinnabar.

**Physical properties** A very soft (indeterminate hardness), very heavy liquid metal. Volatilizes at a low temperature (350°C; 662°F), solidifies at −39°C (−38°F). Bright metallic luster. Excellent conductor of electricity and heat.

**Environment** In the reduction zones of cinnabar deposits, either in bituminous calcareous rocks or more commonly in volcanic rocks and thermal fields. Occasionally pockets containing considerable amounts of mercury collect, so that miners are literally showered as they work.

**Occurrence** Occasionally found at Idria (Yugoslavia), Almadén (Spain) and several American localities including Terlingua, Texas. Sometimes found filling geodes (San José, California, USA). Found in quartzite lenses near Serravezza (Lucca, Italy) in deposits of commercial importance.

**Uses** In amalgams for the extraction of gold and silver. In the manufacture of explosives, in electrical engineering for large rectifiers and in physics equipment. In the pharmaceutical industry for medical purposes (calomel).

13.60

## 5 NICKEL-IRON

NATIVE ELEMENTS
Ni,Fe (Nickel iron)

**System** Isometric.

**Appearance** Normally occurs in gray or blackish masses, sheets or tiny grains contained in other minerals. Never in distinct crystals.

**Physical properties** Medium hard (4–5), hardness increases with the percentage of nickel, very heavy. Opaque, with metallic luster. Malleable. Strongly magnetic. Infusible. Soluble in strong acids.

**Environment** In terrestrial rock formed by the reduction of iron oxides contained in basalts that have come into contact with carbonate rocks. Common in meteorites, where regular associations of alloys with a low nickel content (kamacite, dark) and a high nickel content (taenite, pale) occur parallel to the octahedral faces. The regular intergrowths of the two phases are shown by chemical etching on a polished section (Widmanstatten pattern; see *opposite*).

**Occurrence** Terrestrial native iron with a low nickel content (2–3 percent) is found at Ovifak on Disko Island (Greenland) in economic quantities. Iron meteorites are found all over the world, but the most famous is at Canyon Diablo (Arizona, USA).

**Uses** Excellent raw material for the iron and steel industries, but not used much because of its rarity.

7.0
7.8

▲ Native mercury globules (ca. ×2). Indria, Yugoslavia.

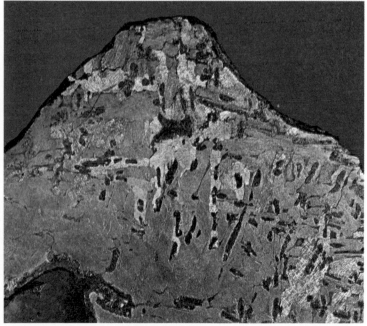

▲ Nickel-iron meteorite (ca. ×1). Canyon Diablo, Arizona.

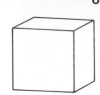

## 6 PLATINUM
NATIVE ELEMENTS
Pt (Platinum)

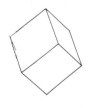

**System** Isometric.
**Appearance** Rarely in small, poorly-formed cubic crystals. Grains, plates or even more rarely nuggets. Silver-gray color.
**Physical properties** Medium hard (4–4.5), extremely heavy, ductile and malleable, no cleavage. Opaque with metallic luster not altered by contact with the atmosphere. Fuses at 1775°C (3227°F). Insoluble in acids, except aqua regia. Slightly magnetic when small impurities of iron are present. Good conductor of heat and electricity.
**Environment** Its primary origin is in the early segregation phase of mafic and ultramafic rocks (dunite and serpentinite). Mainly concentrated in river and marine placers, recent or fossil.
**Occurrence** Plentiful in rivers running down from the Urals (the famous Perm deposit in the USSR). Primary deposits at Sudbury (Canada), in the Bushveld Complex (South Africa), in the USA, Colombia and Peru.
**Uses** Jewelry, precision mechanical and electrical instruments and laboratory equipment. Also used as a catalyst.

## 7 ARSENIC
NATIVE ELEMENTS
As (Arsenic)

**System** Hexagonal.
**Appearance** Rare, pseudo-cubic crystals, normally in microcrystalline masses, often mamillated, with concentric bands. Black with whitish incrustations.
**Physical properties** Semi-hard (3.5), heavy, perfect cleavage, fragile. Opaque and dull. Streak gray or tin-white. Infusible, but volatilizes at a low temperature (450°C; 842°F), giving off white fumes with a strong smell of garlic.
**Environment** Hydrothermal veins; associated with arsenides and sulfides of silver, nickel and cobalt, probably through reduction. Sometimes reniform masses have been found in metamorphosed dolomitic rocks.
**Occurrence** Large masses at Gikos (Siberia). Crusts and small veins in the Harz (Germany), Erzgebirge (DDR), France, Italy, Romania and Czechoslovakia. In the USA at Washington Camp, Santa Cruz County, Arizona.
**Uses** The mineral is of little use, as the metal is usually extracted as a by-product of the treatment of arsenopyrite.

▲ Native platinum nugget (ca. ×3). Siberia, USSR.

▲ Native arsenic (ca. ×1). Sondalo, Italy.

## 8 ANTIMONY
NATIVE ELEMENTS
Sb (Antimony)

**System** Hexagonal.
**Appearance** Silver-white, encrusted, granular masses or radiate nodules. Rarely in platy aggregates.
**Physical properties** Soft (3–3.5), very heavy, perfect cleavage. Opaque with bright metallic luster. Fuses at a low temperature (630°C; 1166°F) and burns in air, giving off white fumes and coloring the flame blue-green. Insoluble.
**Environment** In hydrothermal lodes containing sulfo-arsenides and sulfo-antimonides, especially of silver, as a reduction product of these minerals.
**Occurrence** Small quantities in Sarawak (Borneo), Sala (Sweden), Andreasberg (Germany), Coimbra (Portugal), Val Cavargna (Italy), Sardinia, New Brunswick (Canada) and Kern County, California (USA).
**Uses** The metal is extracted from stibnite, so the mineral is of interest only to scientists and collectors.

---

## 9 BISMUTH
NATIVE ELEMENTS
Bi (Bismuth)

**System** Hexagonal.
**Appearance** Crystals rare and imperfect. Almost always occurs in lamellar masses and dendritic and skeletal aggregates of a rose-pink color tarnished with an iridescent film.
**Physical properties** Soft (2–2.5), very heavy, perfect cleavage. Sectile but not malleable. Opaque with bright metallic luster. Fuses at a very low temperature (270°C; 518°F). Dissolves easily in nitric acid, forming a white precipitate which appears after the solution is diluted. Very diamagnetic.
**Environment** In hydrothermal veins, associated with nickel, cobalt, silver, tin and uranium sulfides.
**Occurrence** Common in the bismuthinite deposits at Oruro and Tasna (Bolivia), the galena deposits of the Erzgebirge (DDR), Cornwall (England), Köngsberg (Norway) and the cobaltite deposits of Ontario (Canada). In Italy there have been occasional finds at Perda Majori (Sardinia) and Traversella (Turin). Rarely found near Monroe, Connecticut, and in Summit and Boulder counties, Colorado (USA).
**Uses** Main source of the metal used for alloys with a low fusion point, antifriction and in the pharmaceutical and cosmetics industries.

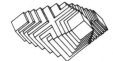

▲ Massive native antimony (ca. ×1.5). Borneo.

▲ Native bismuth crystals (ca. ×1.5). Queensland, Australia.

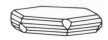

## 10 GRAPHITE
NATIVE ELEMENTS
C (Carbon)

2.23

**System**  Hexagonal.

**Appearance**  Foliated masses or small loose sheets, opaque black, sometimes in six-sided crystals with triangular markings on the basal plane.

**Physical properties**  Dimorphic with diamond. Very soft (1–2), greasy feel, very light, perfect cleavage. Opaque with submetallic, sometimes dull luster. Gray-black streak. Infusible and insoluble. Excellent conductor of electricity. There is an orthorhombic polytype which is less common in nature.

**Environment**  In high-grade metamorphic rocks as a final product of the carbonization of organic substances. Probably also a primary magmatic substance in some pegmatites and hydrothermal veins.

**Occurrence**  Large deposits in Sri Lanka, Malagasy Republic, the USSR, South Korea, Mexico, Czechoslovakia. Economic deposits at Val Chisone (Italy). Small hexagonal crystals in marble at Ogdensburg, New Jersey (USA) and in gneiss at Edison, New Jersey (USA).

**Uses**  In Industry, for refractories and diaphragms with a high fusion point, electrodes and dry lubricants, pencils and dyes.

## 11 DIAMOND
NATIVE ELEMENTS
C (Carbon)

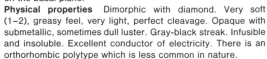

3.52

**System**  Isometric.

**Appearance**  Usually octahedral, rounded crystals. Colorless, yellow, brown, gray, green and black. Blackish microcrystalline masses (carbonado variety). Sometimes rounded with a fibrous, radiate structure (bort variety).

**Physical properties**  Dimorphic with graphite. The hardest mineral (10 on the Mohs scale). Heavy, fragile, with perfect cleavage. Transparent with adamantine luster and strong dispersion of light. Insoluble and infusible. Burns with difficulty, giving off carbon dioxide. Often fluorescent in ultraviolet light.

**Environment**  In ultramafic rocks, especially kimberlite breccias, and in detrital sedimentary deposits derived from them (river and marine placers).

**Occurrence**  Splendid crystals occur in the kimberlites of South Africa, Yakutia (USSR), Murfreesboro, Arkansas (USA), Brazil, Zaire, Sierra Leone and Ghana. Small, nongem-quality crystals are found in Brazil, Venezuela, Zaire and other countries.

**Uses**  Colorless and attractively colored varieties are famous as gemstones. Diamonds are used in industry as an expensive, high-quality abrasive for cutting and drilling equipment.

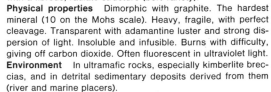
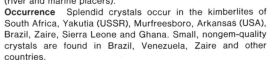

▲ Massive graphite (ca. ×1.5). Val Chisone, Italy.

▲ Octahedral diamond crystal in kimberlite (ca. ×2); *(inset)* octahedral diamond crystal (ca. ×10). South Africa.

## 12 SULFUR
NATIVE ELEMENTS
S (Sulfur)

**System**   Orthorhombic.

**Appearance**   Frequently in fine, dipyramidal crystals, granular aggregates, impregnations and incrustations. Color varies from the characteristic lemon-yellow, when the mineral is pure, to brown and black.

**Physical properties**   Sulfur is trimorphous ($\alpha$-sulfur, $\beta$-sulfur and $\gamma$-sulfur; rosickyite). $\alpha$-sulfur is stable at room temperature, while monoclinic $\beta$-sulfur is stable at temperatures between 95°C and 119°C (203°–246°F). Very soft to soft (1.5–2.5), very light, fragile, with poor cleavage. Transparent to translucent with resinous to greasy luster, depending on the amount of bitumen present. Fuses at a very low temperature (119°C; 246°F) and burns at 275°C (527°F), giving off toxic sulfur dioxide fumes. A poor conductor of heat—sometimes a warm hand is enough to cause cracks inside the crystals. Charged with electricity by friction (triboelectric).

**Environment**   A mineral associated with sedimentary deposits of the evaporite type and with oil-bearing deposits. Believed to be a product of the dissociation of sulfates (especially gypsum) caused by the action of certain bacteria. The intermediate stage of dissociation probably yields hydrogen sulfide which is then oxidized to elementary sulfur. Deposits of this type are very common in Texas and Louisiana (USA) as crusts on top of salt domes buried under impermeable layers of clay. Also very common in the gypsum and sulfur-bearing formation in the outer arc of the Apennines, particularly in Sicily and the Romagna in Italy. Also occurs as alteration product of some sulfides, such as galena, stibnite and pyrite. The monoclinic polymorph ($\beta$-sulfur) is formed by sublimation from hydrogen sulfide given off by volcanic exhalation (solfataras).

**Occurrence**   The finest orthorhombic sulfur crystals come from Sicily (Agrigento and Caltanissetta) and the Romagna (Italy). $\beta$-sulfur crystals come from the Campi Flegrei (Campania, Italy) and Japan.

**Uses**   Nowadays most commercial sulfur is a by-product of the oil industry (produced by desulfuration of hydrocarbons) and is the raw material used in the manufacture of sulfuric acid. It is also used in the vulcanization of rubber and the manufacture of explosives, fungicides and fertilizers. Sicily (Italy) used to be the world's main producer of sulfur, but since 1900 it has been overtaken by the USA. Salt domes are exploited by drilling holes and pumping down superheated water. Production from solfataras (Japan and Indonesia) is relatively unimportant.

▲ Sulfur crystals on matrix (ca. ×1). Sicily, Italy.

▶ Sulfur crystals with celestite (ca. ×1). Sicily, Italy.

▼ Sulfur crystal (ca. ×1). Belisio, Italy.

## 13 CHALCOCITE
SULFIDES
Cu₂S (Copper sulfide)

**System** Orthorhombic.
**Appearance** Rare, tabular, pseudo-hexagonal, striated crystals. Usually in granular aggregates of a dull gray color, altered on the surface to black and green.
**Physical properties** Soft (2.5–3), very heavy, almost malleable, with conchoidal fracture. Opaque with metallic luster. Streak shiny dark gray. Fusible. Dissolves easily in nitric acid. Colors flame green and gives off irritating sulfur dioxide fumes.
**Environment** In ore veins in hydrothermal sulfide deposits, or concentrated in the reduction zone of low-tenor copper deposits, where it is often associated with cuprite, malachite and azurite.
**Occurrence** Splendid pseudo-hexagonal crystals found in Cornwall (England), the Transvaal (South Africa) and at Bristol, Connecticut (USA). Sizable masses at Tsumeb (Namibia), Butte, Montana (USA), in Chile, Peru, Mexico and the USSR. Found at Montecatini in Val di Cecina (Pisa, Italy), Calabona (Sardinia) and Libiola (Liguria, Italy).
**Uses** An important copper mineral.

5.5
5.8

## 14 BORNITE
SULFIDES
Cu₅FeS₄ (Copper iron sulfide)

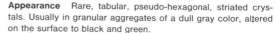

**System** Isometric.
**Appearance** Rarely in cubic, octahedral or dodecahedral crystals. Compact granular masses of a reddish-bronze color, which quickly tarnishes on the surface showing an iridescent purple and blue film (peacock ore).
**Physical properties** Soft (3), very heavy, fragile, with poor cleavage. Opaque with metallic luster. Gray-black streak. Soluble in strong acids, with separation of sulfur. Fuses fairly easily, giving a magnetic globule.
**Environment** In mafic rocks as a mineral of magmatic segregation, in high-temperature pegmatite and hydrothermal veins and in the oxidation zone of copper deposits, associated with malachite.
**Occurrence** Fine crystals are found at Butte, Montana (USA), Bristol, Connecticut (USA), in Cornwall (England) and at Tsumeb (Namibia). Large masses found in Mexico, Peru, Chile, Australia, Zambia and the United States.
**Uses** One of the most important industrial copper ores.

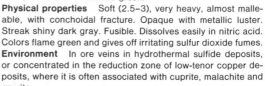

5.7

▲ Chalcocite crystals (ca. ×1.5). Cornwall, England.

▲ Massive bornite with malachite (ca. ×1.5). Montecatini, Italy.

# 15 ARGENTITE — ACANTHITE

SULFIDES

Ag₂S (Silver sulfide)

**System**   Isometric (argentite), monoclinic (acanthite).

**Appearance**   Groups of distorted, pseudo-cubic crystals, of a shiny, lead-gray color, usually blackened on the surface. Dendritic aggregates, masses and incrustations.

**Physical properties**   Cubic argentite is stable above 179°C (354°F). At normal temperatures it is replaced pseudomorphically by a monoclinic form (acanthite). Consequently, specimens labeled "argentite" in mineral collections are really acanthite. Soft (2–2.5), very heavy, malleable, sectile and plastic. Opaque with bright metallic luster, but fresh surfaces turn black quickly. If exposed to strong light it loses reflective power. Soluble in acids. Fuses easily to a globule of metallic silver, giving off irritating sulfur fumes. Cubic crystals are similar to galena, but with poor cleavage. Dendritic aggregates look like native silver, but when broken turn dull black instead of silvery-white.

**Environment**   In low-temperature hydrothermal veins, either associated with other silver minerals or disseminated in galena as minute drops or plates arranged parallel to the cubic (100) planes of the host (silver-bearing galena). Also forms in the cementation zone of lead and zinc deposits, associated with cerussite, chlorargyrite, native silver and other minerals.

**Occurrence**   Splendid crystals at Köngsberg (Norway), Freiberg (DDR), many mines in Mexico (Pachuca, Zacatecas and Guanajuato), Bolivia, Honduras and Czechoslovakia (Pibram, Kutnshora, Schemnitz and Kremnitz). Found in the Sarrabus mining district in Sardinia (Italy) and in some lead and zinc-bearing deposits in the eastern Italian Alps. Sizeable masses are found in Peru, Aspen and Leadville, Colorado, and the Comstock Lode, Georgetown, Nevada (USA).

**Uses**   The main ore for silver, which is extracted from "silver-bearing galena" chiefly by cupellation (a process of fusion which allows the light silver to float to the top where it can be skimmed off).

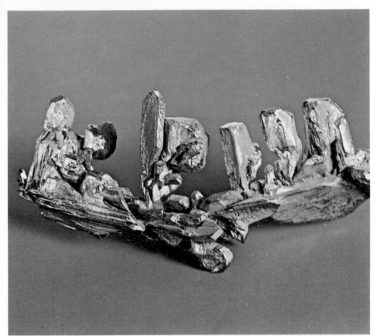

▲ Acanthite crystals (ca. ×2). Sardinia, Italy.

▲ Acanthite crystals (ca. ×1.5). Mexico.

## 16 PENTLANDITE

SULFIDES
(Fe,Ni)$_9$S$_8$ (Iron nickel sulfide)

**System** Isometric.
**Appearance** Massive. Usually with pyrrhotite as exsolution intergrowths. Color light bronze-yellow.
**Physical properties** Semi-hard (3.5–4), heavy, no cleavage, but uneven to conchoidal fracture. Opaque with metallic luster. Streak bronze-brown. Easily fused to a steel-gray bead.
**Environment** Occurs in mafic rocks, particularly norites, as a product of magmatic segregation. Besides pyrrhotite, also associated with cubanite, chalcopyrite and other nickel sulfides and arsenides.

**Occurrence** Pentlandite is the principal source of nickel at Sudbury, Ontario (Canada). Found in norite in the Bushveld Complex, Transvaal (South Africa) and in Norway. Not common in the USA but reported from Clark County, Nevada, and Yakobi Island, Alaska.
**Uses** An important ore of nickel.

## 17 SPHALERITE

SULFIDES
(Zn,Fe)S (Zinc iron sulfide)

**System** Isometric.
**Appearance** Tetrahedral or pseudo-octahedral crystals, often with rounded edges. Color varies from yellow or reddish-brown if nearly pure, blackish if iron is present (marmatite variety). Also pink, green or colorless. Aggregates of distorted crystals and pseudo-octahedral twins with striated faces are common. Banded concretionary masses (botryoidal) and compact cryptocrystalline masses.
**Physical properties** Semi-hard (3.5–4), heavy, very fragile, with perfect cleavage parallel to the twelve faces of the dodecahedron. Transparent or translucent with adamantine or resinous luster. Iron-rich varieties opaque with submetallic luster. Streak pale yellow or reddish. Soluble in hydrochloric acid, giving off hydrogen sulfide fumes. Infusible if pure, but becomes more fusible as the iron content increases. Varieties rich in manganese are triboluminescent. Others, especially the pale varieties, are fluorescent under ultraviolet light and x-rays. Wurtzite is a very rare high-temperature, hexagonal, polymorphic modification (over 1200°C; 2192°F).
**Environment** Mainly in pegmatitic-pneumatolytic veins and also in hydrothermal veins associated with galena, argentite, greenockite, chalcopyrite, barite and fluorite. In sedimentary deposits, perhaps of chemical origin, and in low- and high-

▲ Massive pentlandite with pyrrhotite (ca. ×1). Bohemia.

▲ Sphalerite crystals with galena (ca. ×1.5). Isère, France.

temperature replacement deposits (skarns). Also stable in metamorphic environments.

**Occurrence** Superb crystals are found in all sphalerite deposits of economic importance, particularly at Trepča (Yugoslavia), famous for its marmatites, and in the deposits at Kapnik (Hungary), Pribram (Czechoslovakia), Alston Moor (England), Joplin, Missouri (USA) and Santander (Spain). Small but fine transparent crystals of all colors found in the metamorphic dolomites of the Binnental (Switzerland) and in Italy in Carrara marble. Fine marmatites occur in Bottino (Tuscany, Italy). The world's main deposits are in the Tri-State mining district (Missouri, Kansas, Oklahoma, USA), at Sullivan (Canada), Broken Hill (Australia), Bleiberg (Austria), Trepča (Yugoslavia), Raibl (eastern Alps, Italy) and Sardinia (Montevecchio, Monteponi and Campo Pisano). Wurtzite is rare; splendid crystals are found at Butte, Montana (USA), Oruro and Potosí (Bolivia) and in Czechoslovakia.

**Uses** Sphalerite is the main ore for zinc, and usually also provides cadmium, gallium and indium as by-products. Zinc is one of the components of brass and other alloys and is used as sheet zinc and for galvanizing iron.

## 18  CHALCOPYRITE
SULFIDES
$CuFeS_2$ (Copper iron sulfide)

**System** Tetragonal.

**Appearance** Small, disphenoidal, pseudo-tetrahedral crystals uncommon. Generally in compact or microgranular masses, sometimes reniform or mammillary in appearance. Dark or brassy yellow in color, often with an iridescent film. Frequent lamellar twinning.

**Physical properties** Semi-hard (3.5–4), heavy, relatively fragile, no cleavage, conchoidal fracture. Opaque with metallic or semimetallic luster. Greenish-black streak. Burns in flame, coloring the flame green and giving off toxic, very irritating fumes. Softer than pyrite and fuses more readily.

4.2
4.3

**Environment** Typical of high-temperature, hydrothermal vein deposits, associated with pyrrhotite, sphalerite and pyrite or with nickel minerals. Found in mafic volcanic rocks which have undergone some metamorphism (porphyry copper ore), in contact metamorphic deposits (skarns) and in placers produced by the break-up of these rocks. Chalcopyrite also occurs by replacement of organic fragments like wood or remains of living organisms.

▲ Black sphalerite crystals (ca. ×1.5). Bottino, Italy.

▲ Chalcopyrite crystals with fluorite (ca. ×1). Isle of Man, Great Britain.

**Occurrence** The finest crystals, up to almost 2 cm (.75 in) long, come from Savoy (France). Sizable masses occur in the USA (Butte, Montana; Jerome, Arizona; Bingham, Utah), in Zaire (Katanga), Zambia, Canada (Sudbury), Chile, Cyprus and the USSR (Urals). In Europe there are important deposits at Sulitjelms (Norway), Falun (Sweden), Bor (Yugoslavia), Rio Tinto (Spain) and Rammelsberg and Mansfeld (Germany). The main deposits in Italy are at San Valentino in Predoi (Bolzano), Alagna Valsesia (Vercelli), Campiglia Marittima (Grosseto) and Funtana Raminosa (Sardinia). The Montecatini deposit in Val di Cecina (Pisa) is now worked out, but used to be the basis of one of Italy's main industries.

**Uses** One of the most important copper ores, yielding the by-products gold and silver. About 80 percent of the world's copper is derived from the treatment of chalcopyrite ore.

## 19 TETRAHEDRITE-TENNANTITE GROUP
SULFIDES
$(Cu,Fe)_{12}Sb_4S_{13} - (Cu,Fe)_{12}As_4S_{13}$ (Copper iron sulfides)

**System** Isometric.
**Appearance** Crystals of tetrahedral habit (often modified). Color varies from steel-gray to brown, depending on composition, occasionally purplish-red or dark blue. Usually blackish granular masses.
**Physical properties** A complex solid solution series exists among the following minerals: tetrahedrite $(Cu,Fe)_{12}Sb_4S_{13}$, tennantite $(Cu,Fe)_{12}As_4S_{13}$, goldfieldite $Cu_{12}(Sb,As)_4(Te,S)_{13}$ and freibergite $(Ag,Cu)_{12}(Sb,As)_4S_{13}$. The minerals in this group are semi-hard (3–4.5), heavy, fragile minerals with no cleavage and uneven fracture. Opaque with metallic luster, sometimes bright. Dark-gray streak. The minerals fuse easily and are soluble in nitric acid.
**Environment** In medium- or low-temperature hydrothermal veins, associated with copper, lead, zinc and silver min-

4.5
5.2

▲ Massive chalcopyrite (ca. ×0.5). Alagna Valsesia, Italy.

▲ Tennantite crystal (ca. ×1.5). Otavi, Namibia.

erals. Tetrahedrite and tennantite are good geological thermometers because they gradually become richer in silver and mercury as the temperature falls.

**Occurrence**  Splendid crystals at Botés and Kapnik (Romania), Boliden (Sweden), Pribram (Czechoslovakia), Tsumeb (Namibia) and Butte, Montana (USA). Freibergite is found at Freiberg (DDR) and Schwatz (Austria). Other important localities in Bolivia, Chile, Peru; Nevada, New Mexico, Arizona and California (USA). Occasionally found in the Sarrabus mines of Sardinia, in Tuscany and in the Trentino-Alto Adige region (Italy).

**Uses**  An important ore for copper and often for silver, mercury and antimony. In some places also contains tellurium or tin.

## 20  GREENOCKITE
SULFIDES
CdS (Cadmium sulfide)

**System**  Hexagonal.

**Appearance**  Rare, prismatic, hexagonal crystals, sometimes twinned. Generally as a yellowish film coating sphalerite or other zinc minerals.

**Physical properties**  Semi-hard (3–3.5), fragile, with perfect cleavage. Translucent with adamantine or resinous luster. Orange or brick-red streak. Infusible, soluble in hydrochloric acid, giving off unpleasant hydrogen sulfide fumes. When greenockite is rich in zinc it fluoresces orange-yellow.

**Environment**  A replacement mineral of cadium-rich sphalerites. Forms only on the surface or in parts of deposits affected by water.

**Occurrence**  Incrustations on sphalerite at Pribram (Czechoslovakia) and Joplin, Missouri (USA). Incrustations or small scattered granules on mammillary smithsonite, turning it bright yellow, at Gorno (Val Seriana, Bergamo, Italy) and Marion, Arkansas (USA). Minute crystals on prehnite, natrolite and in calcite veins at Renfrew (Scotland) and Paterson, New Jersey (USA).

**Uses**  An important ore for cadmium, a metal used in industry for electrical batteries, high-temperature alloys and electroplating for other metals.

▲ Tetrahedrite crystals (ca. ×2). Hungary.

▲ Greenockite (ca. ×2). Hungary.

## 21 ENARGITE
SULFIDES
Cu₃AsS₄ (Copper arsenic sulfide)

**System** Orthorhombic.

**Appearance** Rare, tabular or elongated crystals, with vertical striations. Mainly as lamellar black or iron-gray aggregates or granular masses.

**Physical properties** Soft (3), heavy, fragile, perfect cleavage. Opaque with metallic luster. Black streak. Soluble in nitric acid, with small flakes of sulfur appearing in the solution. Fuses easily.

**Environment** In medium-temperature hydrothermal veins, associated with copper minerals like bornite, covellite and tetrahedrite.

**Occurrence** Sizeable masses at Bor (Yugoslavia); Chuquicamata (Chile); Bingham and Tintic, Utah, and Butte, Montana (USA); in Peru, Mexico, Argentina and Namibia; on Luzon (Philippines). Found in Italy at Calabona (Sardinia).

**Uses** A fairly important ore of copper and arsenic.

---

## 22 PYRRHOTITE
SULFIDES
Fe₁₋ₓS (Iron sulfide)

**System** Hexagonal.

**Appearance** Crystals usually tabular with faces striated horizontally. Commonly in yellow-bronze, massive, granular aggregates, sometimes iridescent.

**Physical properties** Its composition, deficient in iron, is close to that of iron sulfide (FeS), so far found in its stoichiometric form only in meteorites (troilite). Semi-hard (3.5–4.5), heavy, fragile, with parting. Opaque with bright metallic luster. Fuses fairly easily to a black magnetic mass. Dissolves with difficulty, even in strong acids. Distinguished from pyrite by its ferromagnetism.

**Environment** In mafic and ultramafic extrusive rocks of early magmatic segregation, associated with pentlandite and other cobalt and platinum minerals, making it a valuable and important mineral (nickel-bearing pyrrhotite). In some high-grade metamorphic rocks, and in high-temperature hydrothermal veins, though never nickel-bearing under these conditions. On rare occasions also found in pegmatites and sedimentary deposits, associated with siderite.

**Occurrence** Fine crystals are found at Trepča (Yugoslavia), Kysbanya (Romania), Leoben (Austria) and Freiberg (DDR).

▲ Enargite crystals (ca. ×2). Peru.

▲ Pyrrhotite crystals (ca. ×1). Herja, Romania.

Large deposits in Canada: at Sudbury, where pyrrhotite is disseminated but associated with other minerals containing a high nickel content and rare metals; very large deposits in Manitoba near the new town of Thompson. Other deposits in Mexico, Bolivia, Brazil, Norway, Sweden and USA (Ducktown, Tennessee). Fine crystals occur at Standish, Maine, and Brewster, New York (USA). In Italy has been found and mined in the Sesia and Ossola valleys (Nibbio and Migiandone), on Elba and at Campiglia Marittima (Livorno) and Monteneve (Bolzano).

**Uses** Pyrrhotite masses which do not contain ores of other valuable metals are of no particular use. Nickel-bearing deposits are one of the main sources of nickel, cobalt and platinum.

## 23 NICKELINE
SULFIDES
NiAs (Nickel arsenide)

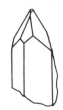

7.78

**System** Hexagonal.
**Appearance** Very rare stocky, tabular or pyramidal crystals. Generally occurs in compact pinkish-bronze masses, often with a dark or pale-green film when altered on the surface to annabergite.
**Physical properties** Hard (5–5.5), very heavy, fragile, no cleavage, but conchoidal fracture. Opaque with metallic luster. Brownish-black streak. Soluble in nitric acid, staining the solution green. Fuses easily, giving off fumes with a strong odor of garlic.
**Environment** In high-temperature hydrothermal veins, or disseminated in noritic gabbros, always associated with silver, nickel and cobalt sulfides.
**Occurrence** Rare small crystals at Richeldsdorf (Germany). Large masses at La Rioja (Argentina) and in Siegerland (Germany). Veins at Cobalt and Eldorado (Canada). Found rarely at Franklin, New Jersey (USA). Alternating concentric layers of nickeline and arsenopyrite at Natsume (Japan). Traces are found at Cortabbio (Val Sassina, Como) and Nieddoris (Sardinia) (Italy).
**Uses** The first mineral from which nickel was extracted and still used for this purpose when large masses are available.

▲ Massive pyrrhotite (ca. ×1). Trepča, Yugoslavia.

▲ Massive nickeline (ca. ×1.5). Mansfeld, Germany.

## 24 MILLERITE
SULFIDES
NiS (Nickel sulfide)

**System** Hexagonal.
**Appearance** Small, radiating acicular, brassy-yellow crystals. Often in hairlike tufts, felted, or in velvety crusts.
**Physical properties** Semi-hard (3–3.5), very heavy, with good cleavage. Elongated crystals are slightly elastic. Opaque with metallic luster. Greenish-black streak. Soluble in aqua regia. Fuses easily. Good conductor of electricity.
**Environment** In cavities in hematite or siderite masses or in marble and calcite veins. Also occurs as an alteration product of other nickel minerals.
**Occurrence** Small quantities have been found in several mines in Wales, the USA (Keokuk, Iowa; Antwerp, New York) and Germany (Harz). Also occurs in Arburese (Sardinia) and at Bombinia (Bologna) (Italy).
**Uses** Although it is the richest ore for nickel, it is too rare and scattered to be an important commercial ore.

---

## 25 GALENA
SULFIDES
PbS (Lead sulfide)

**System** Isometric.
**Appearance** Cubic, lead-gray crystals very frequent, cubo-octahedrons less common. Usually in compact granular masses with many shiny surfaces, where cleavage occurs, which blacken with time.
**Physical properties** Soft (2.5–2.8), very heavy, fragile with perfect cleavage parallel to the cube faces. Opaque with very bright metallic luster. Dark-gray streak. Fuses fairly easily, leaving a yellow stain (lead monoxide or litharge). Soluble when heated in hydrochloric acid, giving off hydrogen sulfide, recognizable by its characteristic bad smell. Dissolves in nitric acid, quickly producing small flakes of sulfur and a very fine white precipitate (lead sulfate, caused by rapid partial oxidation during dissolution).
**Environment** Typical hydrothermal mineral of medium-temperature deposits, associated with sphalerite and argentite, mainly with quartz and fluorite gangue. Also found in sedimentary and metamorphic deposits, where it probably derives from karst concentration of the disseminated mineral. The large

▲ Acicular millerite crystals (ca. ×2). Harz, Germany.

▲ Cubic galena crystals (ca. ×1). Joplin, Missouri.

▶ Cubic galena crystals (ca. ×1.5). Joplin, Missouri.

columnar replacement deposits in Cambrian and Ladinian limestones and dolomites are now thought to have developed in this way.

**Occurrence** The largest lead and zinc deposits are in the USA, in the Tri-State mining district (Missouri, Oklahoma and Kansas) near Joplin, in Australia (Broken Hill), England (Cumberland), Mexico (Santa Eulalia) and West Germany and the DDR (Andreasberg, Harz; Freiberg, Erzgebirge). There are also sizable deposits of silver-bearing galena at Leadville (Colorado), in the Coeur d'Alene district (Idaho) and at Tintic (Utah) in the USA. In Italy the Alpine deposits in Ladinian limestones and dolomites (including those at Raibl, Salafossa, Gorno and Dossena) are typical of high-temperature associations (sphalerite+galena+greenockite with quartz, calcite, fluorite and barite gangue). The deposits of Cambrian limestone in Sardinia (Monteponi and Montevecchio) are characteristic of lower-temperature deposits (galena+argentite+cerussite+sphalerite+phosgenite+anglesite with quartz, siderite, calcite and fluorite gangue). Large galena crystals have been found in all the places mentioned. The best known are from Joplin, Missouri (USA).

**Uses** The main ore of lead. Silver is a frequent by-product.

---

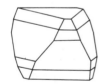

## 26 CINNABAR
SULFIDES
HgS (Mercury sulfide)

**System** Hexagonal.

**Appearance** Rhombohedral, tabular, crystals very rare. Normally occurs in scarlet microcrystalline or earthy masses. Also frequently in brightly colored earthy films or disseminated grains.

**Physical properties** Soft (2–2.5), very heavy, fragile, with perfect cleavage. Crystals translucent with adamantine luster. Opaque with dull appearance in fine-grained earthy aggregates. Always recognizable by its color and streak: bright scarlet. Does not alter on contact with air. Volatilizes if heated above 580°C (1076°F), depositing drops of mercury on cold surfaces nearby. Insoluble in acids but attacked by aqua regia and chlorine gas. Displays strong rotary polarization. Metacinnabar is a rare cubic dimorph of HgS found in the oxidation zone of mercury deposits in the form of black films.

8.10

**Environment** The mineral occurs in very low-temperature hydrothermal deposits as veins or impregnations and replacement deposits in sedimentary rocks associated with nearby igneous rocks. Also present as a sublimate in active craters and

▲ Cubo-octahedral galena crystals (ca. ×1.5). Aachen, Germany.

▲ Massive cinnabar (ca. ×1). Monte Amiata, Italy.

as a chemical deposit of hydrothermal, probably alkaline, springs. There are also placer-type concentrations produced by the erosion of ancient mercury-bearing rocks that were subsequently covered by layers of sediment.

**Occurrence** Found in masses and impregnations in lavas, cavernous limestone and clay at Almadén (Spain), Abbadia San Salvatore (Monte Amiata, Italy) and Idria (Yugoslavia). These are the three most important mercury mines in the world. There are other deposits at New Idria and New Almaden (USA), Nikotowa (USSR), Huancavelica (Peru), Kweichow and Hunan provinces (China) and in Algeria. Small, well-formed crystals are found in Pike County, Arkansas (USA). Metacinnabar has been found at Idria (Yugoslavia).

**Uses** The most important ore of mercury, as the native element is very rare. It was used in the past as the mineral pigment known as vermilion.

---

## 27 COVELLITE
SULFIDES
CuS (Copper sulfide)

**System** Hexagonal.

**Appearance** Flattened, hexagonal crystals rare. Usually platy or compact, indigo-blue masses, with strong iridescence.

**Physical properties** Very soft (1.5–2), very heavy, fragile with perfect cleavage, flexible in thin, black sheets. Opaque with submetallic luster and micaceous appearance. Fuses easily, giving a blue flame and sulfur dioxide fumes. Soluble in hydrochloric acid.

**Environment** Found in small hydrothermal veins associated with other copper minerals. Or produced by metamorphic dissociation in calcareous chalcanthite and chalcopyrite-bearing rocks. Sublimates directly from volcanic fumes on Vesuvius, where it was first identified.

**Occurrence** Beautiful iridescent crystals from Calabona (Sardinia) and Butte, Montana (USA). Also found in Italy in metamorphic dolomite at Crevola (Cal d'Ossola, Novara) and in Vesuvian lava. Large masses in Chile, Bolivia and Alaska.

**Uses** An excellent ore of copper when available in economic amounts.

4.59
4.76

▲ Cinnabar with gangue (ca. ×1). Monte Amiata, Italy.

▲ Iridescent covellite crystals (ca. ×1.5). Calabona, Sardinia.

## 28 STIBNITE

SULFIDES
Sb₂S₃ (Antimony sulfide)

**System** Orthorhombic.
**Appearance** Prismatic or acicular crystals, elongated and striated parallel to the long or *c*-axis. Sometimes curved. Steel-gray with iridescent film. Felted or compact masses also common, often altered to red kermesite or other yellow, powdery minerals ("antimony ochres").
**Physical properties** Soft (2), heavy. Perfect lengthwise cleavage. Sectile with subconchoidal fracture. Crystals flexible but inelastic. Opaque with bright metallic luster (slightly lighter than bismuthinite). Thin splinters fuse in the flame of a match (1 on the Kobell scale). In powder form, soluble in concentrated hydrochloric acid and decomposes readily in potassium hydroxide, staining the solution orange.
**Environment** In low-temperature hydrothermal veins, associated with silver, lead and mercury minerals. Also found as a chemical deposit from solutions of hot mineral springs.
**Occurrence** Splendid crystals up to 45 cm (18 in) long from the deposits at Ichinokawa on the island of Shikoku (Japan), though the deposit there is now worked out. Fine groups of crystals from Felsöbanya and Kapnik (Romania), Hollister County, California (USA) and Borneo. The largest deposits are now those of Hunan and Kwantung provinces (China), where the mineral is associated with cinnabar. In Italy fine crystals have been found at Rosia (Siena), Marciano and Niccioleta (Grosseto) and compact masses in the Arburese and Gerrei mines (Sardinia). Small amounts of stibnite occur in Val Carvagna, San Bartolomeo (Como) in quartz seams in schistose rocks. Deposits are also worked in Peru, Mexico, Bolivia and central France.
**Uses** The main ore of antimony, used for metal anti-friction alloys, metal type, shot and batteries, and in the manufacture of fireworks. The salts are used in the rubber and textile industries, in medicine and glass making.

▲ Acicular stibnite crystals (ca. ×1). Baia Sprie, Romania.

▶ Stibnite crystals (ca. ×2). Rosia, Italy.

▼ Kermesite with stibnite (ca. ×0.5). Tolfa, Italy.

## 29 BISMUTHINITE
SULFIDES
$Bi_2S_3$ (Bismuth sulfide)

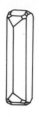

**System** Orthorhombic.
**Appearance** Prismatic, acicular, tin-gray crystals, with fine striations. In aggregates, granular and compact, or with a woven radiate structure.
**Physical properties** Soft (2), very heavy, with perfect cleavage. Crystals flexible but inelastic. Opaque with metallic luster. Does not fuse in a match flame (unlike stibnite, which it closely resembles). Soluble in nitric acid, causing delicate sulfur flakes to appear in the solution.
**Environment** In medium-temperature hydrothermal veins, associated with silver, tin and cobalt minerals.

**Occurrence** Sizable masses in Bolivia (Tasna, Chorole), Peru (Cerro de Pasco), Cornwall (England) and Australia. Fine crystals found at Haddam, Connecticut (USA) and in some pegmatites in Mexico and Canada. Found in aplitic veins in Val Masino (Sondrio) and more commonly on the island of Elba, at Brosso (Turin) and in Iglesiente (Sardinia) (Italy).
**Uses** The best ore of bismuth.

## 30 SYLVANITE
SULFIDES
$AgAuTe_4$ (Silver gold telluride)

**System** Monoclinic.
**Appearance** Stubby, prismatic, skeletal or arborescent crystals. Often occurs as branching incrustations which look rather like writing. Granular or bladed, silvery-white masses.
**Physical properties** Very soft (1.5–2), very heavy, with perfect cleavage. Opaque with silvery metallic luster. Color becomes yellow with increasing gold content. Soluble in nitric acid, leaving a yellow (gold) residue. Dissolves when heated in concentrated sulfuric acid, staining the solution dark red (tellurium). Fuses easily.
**Environment** In low-temperature hydrothermal veins, associated with calaverite and other tellurides, mainly in quartz gangue.

**Occurrence** Rather rare. Important masses at Kalgoorlie (Australia) and Cripple Creek, Colorado (USA). Also found in Transylvania (Romania).
**Uses** An industrial ore of gold, silver and tellurium.

▲ Bismuthinite crystals with hematite and siderite (ca. ×1). Brosso, Italy.

▲ Sylvanite crystals (ca. ×2). Săcărâmb, Romania.

**PYRITE**
SULFIDES
FeS$_2$ (Iron sulfide)

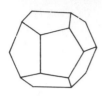

**System** Isometric.
**Appearance** Striated, cubic, octahedral or pyritohedral crystals, sometimes occurring as "iron cross" twins. Compact, granular aggregates. Concretions, mammillated and stalactitic nodules. Always fairly dark yellow, sometimes with an iridescent yellowish-brown film. Pseudomorphs after organic fossil remains frequent (e.g., pyritized ammonites).
**Physical properties** Dimorphic with marcasite. Hard (6–6.5), very heavy, very fragile, with poor cleavage. Opaque with very bright metallic luster. Greenish-black streak. Insoluble in hydrochloric acid, but powder dissolves in nitric acid. Gives off spark if struck with a hammer. If heated burns at a moderate temperature and fuses easily, giving off sulfur fumes and leaving a magnetic globule. The granular varieties eventually alter and crumble, producing first sulfides, then "limonite."
**Environment** Common in plutonic, volcanic, sedimentary and metamorphic rocks. Occurs in masses associated with chalcopyrite in magmatic segregation deposits in mafic rocks. Associated with sphalerite and galena in hydrothermal veins. On its own or associated with gold (auriferous pyrite) in hydrothermal, medium to low-temperature quartz veins. Disseminated in mafic lavas. Pyritized concretions formed by chemical deposition under the water. As a diagenetic mineral replaces fossils or forms nodules of various sizes. Stable in metamorphic environments up to the granulite facies, where it changes into pyrrhotite.
**Occurrence** Splendid crystals occur in deposits all over the world. The most important are Rio Tinto (Spain); Leadville and Gilman, Colorado, and Jerome, Arizona (USA); Rio Marina and Gavorrano (Tuscany, Italy); Sulitjelms (Norway); Falun (Sweden). There are significant deposits in Germany, Japan and the USSR. The striated cubes of Gavarrono (Grosseto, Italy) and Chester County, Pennsylvania (USA), the pyritohedra of Rio Marina (Livorno, Italy), the radiate disc forms of Sparta, Illinois (USA) and the pyritized fossils from various places in Italy, England and Germany are all well-known forms of pyrite.
**Uses** In the manufacture of sulfuric acid by the lead chamber method. Pellets of pressed dust are then treated to produce iron, gold, copper, cobalt, nickel, etc.

▲ Pyritohedral pyrite crystals (ca. ×1). Island of Elba, Italy.

▲ Cubic pyrite crystals (ca. ×2.5). Gavorrano, Italy.

## 32 HAUERITE

SULFIDES
MnS$_2$ (Manganese sulfide)

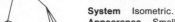

**System**  Isometric.
**Appearance**  Small octahedral or cubo-octahedral crystals, well-formed and blackish-brown in color.
**Physical properties**  Semi-hard (3.5–4.5), heavy, with perfect cleavage. Translucent or semi-opaque with subadamantine or submetallic luster. Pinkish-brown streak. Soluble in hydrochloric acid.
**Environment**  Forms in sedimentary environments by reduction of evaporites, then diagenetic recrystallization, associated with gypsum, calcite and native sulfur.
**Occurrence**  Splendid crystals up to 4 cm (1.06 in) long at Raddusa (Sicily) from sulfur-bearing clays. Associated with gypsum at Kalinka (Rutenia, USSR) and Big Hill, Texas (USA). Fairly common as concretions associated with salt domes in Louisiana and Texas (USA) and with ferro-manganese nodules deposited on the sea bed, particularly in the Pacific Ocean.
**Uses**  Purely of scientific interest.

3.5

## 33 COBALTITE

SULFIDES
CoAsS (Cobalt arsenic sulfide)

**System**  Orthorhombic.
**Appearance**  Pseudo-isometric cubes, pyritohedrons or combinations of these forms, sometimes striated. Color steel-gray often with a violet-to-purple tinge. Granular or compact masses.
**Physical properties**  Hard (5.5), very heavy, with perfect cleavage and uneven fracture. Opaque with metallic luster. Streak grayish-black. Soluble in nitric acid producing sulfur and arsenic oxide. Fuses forming a weakly magnetic globule.
**Environment**  Occurs in high temperature hydrothermal veins or disseminated in metamorphic rocks associated with other cobalt-nickel sulfides and arsenides.
**Occurrence**  Extremely sharp crystals up to 2.5 cm (1 in) occur in Sweden (Hankansbö and Tunaberg). Also found in fine crystals at Cobalt, Ontario (Canada).
**Uses**  An important ore of cobalt.

6.3

▲ Hauerite crystals (ca. ×2). Raddusa, Sicily, Italy.

▲ Pyritohedral crystal of cobaltite (ca. ×5). Tunaberg, Sweden.

## 34 MARCASITE
SULFIDES
FeS$_2$ (Iron sulfide)

**System** Orthorhombic.
**Appearance** Flattened prismatic crystals, often twinned in the form of ''spearheads'' and ''cockscombs.'' Radiate, concretionary, massive, nodular or stalactitic aggregates common. Always pale yellow with a slightly greenish tinge.
**Physical properties** Dimorphic with pyrite. Hard (6–6.5), heavy, fragile, with poor cleavage. Opaque with bright metallic luster on fresh surfaces. Greenish-black streak. Dissolves with difficulty in nitric acid, flakes of sulfur appearing in the solution. Oxidizes quickly in air, becoming covered with whitish patches of melanterite (FeSO$_4$ • 7H$_2$O), and then crumbles into dust. Fuses quite easily.
**Environment** In low-temperature hydrothermal veins, often with zinc and lead sulfides. Frequently in sedimentary environments as a chemical precipitate of primary acid waters in the reduction zone of deposits.

**Occurrence** Fine twinned crystals at Rammelsberg (Germany) and Karlovy Vary (Czechoslovakia). Sizable amounts in Romania and the Tri-State mining district (USA) and especially at Galena, Illinois (USA).
**Uses** Like pyrite, used for the extraction of sulfuric acid. Because of its bright luster it used to be cut for trinkets.

---

## 35 ARSENOPYRITE
SULFIDES
FeAsS (Iron arsenic sulfide)

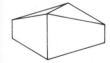

**System** Monoclinic.
**Appearance** Elongated, striated, prismatic crystals, pseudo-orthorhombic through twinning. Cruciform twins common. Granular masses of a silvery or whitish-gray color with pink tints.
**Physical properties** Hard (5.5–6), very heavy, fragile, with good cleavage. Opaque with metallic luster. Black streak. Produces sparks if struck with a hammer and gives off a strong odor of garlic. Soluble in nitric acid, yielding sulfur. Fuses easily and gives off white fumes smelling strongly of garlic.
**Environment** In sulfide deposits of early magmatic segregation (with gold and cobalt), pegmatite-pneumatolytic environments (with tin) and high-temperature hydrothermal veins (with gold, silver and nickel). Frequently in metamorphic deposits.

**Occurrence** The main deposits are at Boliden (Sweden), Freiberg (DDR), Cornwall (England), Sulitjelma (Norway) and Deloro (Canada). Fine crystals at Roxbury, Connecticut and Leadville, Colorado (USA) and in Valle Anzasca (Novara) and Val Sugana (Trento) (Italy).
**Uses** The principal ore of arsenic, with tin, gold, silver and cobalt as by-products.

▲ Marcasite rosettes (ca. ×1.5). Romania.

▲ Arsenopyrite crystals (ca. ×1). Hohenstein, Germany.

## 36 GLAUCODOT

SULFIDES
(Co,Fe)AsS (Cobalt iron arsenic sulfide)

**System** Orthorhombic.
**Appearance** Prismatic, elongated crystals. Granular or fibrous, radiate masses. White or grayish-white, often with pink patches of erythrite.
**Physical properties** Hard (5), very heavy, fragile with good cleavage. Opaque with metallic luster. Black streak. Soluble in nitric acid with separation of sulfur. Decomposes if heated, giving off white fumes smelling of garlic. Sometimes regarded as an intermediate member of the arsenopyrite-cobaltite series (FeAsS-CoAsS).
**Environment** In high-temperature pneumatolytic and hydrothermal veins, associated with pyrite, chalcopyrite, galena, etc. Also in greenschists derived from the metamorphism of lavas.
**Occurrence** Significant concentrations commonly in large crystals at Hakansbö (Sweden). Fine crystals in chlorite schists at Huesco (Chile). Also found in Norway, Romania, Tasmania and Cobalt, Ontario (Canada).
**Uses** An ore of cobalt, containing up to 25 percent Co.

6.04

## 37 MOLYBDENITE

SULFIDES
$MoS_2$ (Molybdenum sulfide)

**System** Hexagonal.
**Appearance** Tabular crystals with hexagonal outline. Bladed, foliated or finely interwoven masses more common. Bluish-gray color.
**Physical properties** Very soft (1–1.5), heavy, perfect cleavage, flexible plates. Greasy feel. Opaque with metallic luster. Bright blue-gray streak. Infusible, dissolves with difficulty. Leaves a grayish-green mark on paper.
**Environment** In pegmatites and very high-temperature pneumatolytic veins. Often found in cavities in granitic or dioritic rocks. Also occurs in contact metamorphic deposits associated with limestones (scheelite skarns).
**Occurrence** Disseminated but abundant in quartz veins traversing granite at Climax, Colorado (USA). Excellent crystals found at the Ogden mine, Edison, New Jersey (USA). Also found in Australia, Italy, Norway and Bolivia.
**Uses** The main ore of molybdenum, a metal used in many special alloys. The mineral is also used as a dry lubricant resistant to high temperature.

4.7

▲ Glaucodot crystals (ca. ×1.5). Hakansbö, Sweden.

▲ Molybdenite (ca. ×1). Québec, Canada.

▶ Molybdenite crystals (ca. ×1). Cuasso al Monte, Italy.

## 38 SKUTTERUDITE

SULFIDES

CoAs$_{2-3}$ (Cobalt arsenide)

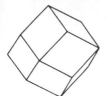

**System** Isometric.

**Appearance** Small, cubic, octahedral or sometimes pyritohedral, silvery-white, metallic-looking crystals, with iridescent film. Compact granular masses.

**Physical properties** Hard (6), very heavy, fragile, with perfect cleavage. Opaque with metallic luster. Black streak. Fuses fairly easily giving off garlic fumes. Soluble when heated in nitric acid, staining the solution pink. Smaltite is a low arsenic variety of skutterudite.

**Environment** In medium- or high-temperature hydrothermal veins, associated with arsenopyrite, native silver and bismuth, and with cobalt and nickel sulfides in calcite gangue.

**Occurrence** Superb crystals from Skutterud (Norway), Irhtem and Bou Azzer (Morocco) and Schneeberg, Saxony (DDR). Small crystals occur at Cobalt, Ontario (Canada) and Huelva (Spain). Found at Usseglio in Val di Lanzo (Turin) and in Iglesiente (Sardinia) (Italy).

**Uses** Fairly important cobalt, nickel and arsenic ore.

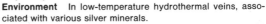

## 39 PROUSTITE

SULFIDES

Ag$_3$AsS$_3$ (Silver arsenic sulfide)

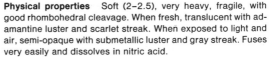

**System** Hexagonal.

**Appearance** Rarely in rhombohedral or scalenohedral crystals, characteristically striated, distorted and often twinned. Usually massive and shiny dark red.

**Physical properties** Soft (2–2.5), very heavy, fragile, with good rhombohedral cleavage. When fresh, translucent with adamantine luster and scarlet streak. When exposed to light and air, semi-opaque with submetallic luster and gray streak. Fuses very easily and dissolves in nitric acid.

**Environment** In low-temperature hydrothermal veins, associated with various silver minerals.

**Occurrence** Splendid translucent red crystals up to 15 cm (6 in) long found at Chañarcillo (Chile). Fairly rich deposits in the Erzgebirge (DDR), Harz (Germany), Cobalt, Ontario (Canada) and Mexico. Associated with silver-bearing galena in the Sarrabus deposits (Sardinia).

**Uses** A silver ore of limited importance, but prized by mineral collectors for its beauty.

▲ Skutterudite crystals (ca. ×1) Huelva, Spain.

▲ Proustite crystal (ca. ×1.5). Chañarcillo, Chile.

## 40 PYRARGYRITE
SULFIDES
Ag₃SbS₃ (Silver antimony sulfide)

**System** Hexagonal.
**Appearance** Prismatic crystals with different terminations (hemimorphic). Black with dark-red tints. Also in compact masses or disseminated grains.
**Physical properties** Soft (2.5), very heavy, fragile, with rhombohedral cleavage. Translucent with submetallic luster. Purplish-red streak. Fuses easily and dissolves in nitric acid.
**Environment** In low-temperature hydrothermal veins, usually associated with various silver minerals. An alteration product of argentite and silver-bearing galena in the cementation zone of deposits.
**Occurrence** More common than proustite but usually not found as well crystallized. Small crystals and disseminated masses occur in Harz (Germany) and Erzgebirge (DDR) and in the Sarrabus mines (Sardinia), Czechoslovakia, Guanajuato and Zacatecas (Mexico), Chile and Canada.
**Uses** A minor silver ore.

5.85

---

## 41 STEPHANITE
SULFIDES
Ag₅SbS₄ (Silver antimony sulfide)

**System** Orthorhombic.
**Appearance** Crystals prismatic to tabular, uncommon. Usually massive or disseminated. Color iron-black.
**Physical properties** Soft (2–2.5), very heavy, with poor cleavage and uneven to subconchoidal fracture. Streak iron-black. Opaque with metallic luster. Soluble in nitric acid, producing sulfur and arsenic oxide. Fuses easily.
**Environment** Occurs in low-temperature hydrothermal veins associated with native silver, tetrahedrite, polybasite, acanthite and proustite.
**Occurrence** Very fine crystal groups were found in Andreasberg and Freiberg, Saxony (Germany and DDR). Superb crystals recovered at Zacatecas and especially Arizpe, Sonora (Mexico). Stephanite was an important silver ore in the Comstock Lode, Virginia City, Nevada (USA) and is currently mined in Cobalt, Ontario (Canada).
**Uses** A minor ore of silver.

6.3

▲ Pyrargyrite crystals (ca. ×1.5) Andreasberg, Germany.

▲ Stephanite crystals on quartz in a cavity (ca. ×1.5). Mexico.

## 42 POLYBASITE

SULFIDES
$(Ag,Cu)_{16}Sb_2S_{11}$ (Silver copper antimony sulfide)

**System** Monoclinic.
**Appearance** Tabular pseudo-hexagonal crystals with bev-eled edges, often with triangular growth marks. Bladed or granular, iron-black masses.
**Physical properties** Soft (2–3), very heavy with poor cleav-age. Opaque with metallic luster. Black streak. Resembles hematite but much softer, and fuses easily.
**Environment** In medium- to low-temperature hydrothermal veins associated with proustite, pyrargyrite, tetrahedrite, etc.
**Occurrence** Beautiful crystals from Andreasberg (Germany), Pribram (Czechoslovakia) and Freiberg (DDR) and Arizpe, Sonora (Mexico). Masses in Chile, Bolivia and Colorado (USA). Found in some of the Sarrabus mines (Sardinia).
**Uses** An ore of silver.

6.0
6.2

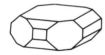

---

## 43 BOURNONITE

SULFIDES
$PbCuSbS_3$ (Lead copper antimony sulfide)

**System** Orthorhombic.
**Appearance** Stubby or tabular, prismatic crystals, with many faces, often in the shape of cog-wheels. Granular aggregates or disseminated grains. Dark gray or black.
**Physical properties** Soft (2.5–3), very heavy, fragile, with good cleavage. Opaque with metallic luster. Steel-gray streak. Fuses easily and dissolves in nitric acid, staining the solution green (copper).
**Environment** In medium-temperature hydrothermal veins as-sociated with galena, tetrahedrite, chalcopyrite, etc. Very common, especially in silver-bearing galena, silver and anti-mony deposits.
**Occurrence** Beautiful crystals in the Harz (Germany), in Cornwall (England), at Pribram (Czechoslovakia), in Bolivia and at Park City, Utah (USA). Has been found in Sarrabus (Sar-dinia), at Brosso and Giaveno (Turin) and in Val di Castello (Tuscany) (Italy).
**Uses** An ore of lead, copper and antimony.

5.7
5.9

▲ Polybasite crystals (ca. ×1). California.

▲ Bournonite crystals (ca. ×1.5). Horhausen, Germany.

## 44 BOULANGERITE

SULFIDES

$Pb_5Sb_4S_{11}$ (Lead antimony sulfide)

**System** Monoclinic

**Appearance** Very rarely in slender, acicular, opaque, gray crystals. Generally in fibrous, plumose masses, or small, fine tufts.

**Physical properties** Soft (2.5–3), very heavy, fragile, with good cleavage. Opaque with dullish metallic luster. Black streak. Fuses easily and dissolves if heated in strong acids.

**Environment** In vein deposits of lead, zinc and antimony, disseminated or in small economic masses.

**Occurrence** This mineral is probably not as rare as was previously thought. Economic concentrations at Molières (France). Beautiful crystals found at Pribram (Czechoslovakia), Boliden (Sweden), Claustal (Germany) and several mines in the Urals (USSR). Plumose masses found in Stevens County, Washington (USA). Also found at Bottino in the Apuanian Alps (Tuscany, Italy).

**Uses** A lead ore.

5.8
6.2

## 45 REALGAR

SULFIDES

AsS (Arsenic sulfide)

**System** Monoclinic.

**Appearance** Rarely in small, stubby, prismatic crystals. Usually in compact aggregates and scarlet films.

**Physical properties** Soft (1.5–2), fairly heavy, fragile with perfect cleavage. Semi-transparent with adamantine luster. Streak orange-yellow. Fuses easily, giving off arsenic fumes with a garlic smell. If exposed to light and air loses its color and slowly changes into powdered orpiment. Dissolves in aqua regia.

**Environment** A mineral deposited in hot springs and low-temperature hydrothermal veins, almost always associated with orpiment and antimony, silver, lead and tin minerals.

**Occurrence** Beautiful crystals found at Nagyag (Romania); in the Binnental (Switzerland); at Matra (Corsica) and in Manhattan, Nevada, and King County, Washington (USA). Compact masses in Hungary and Macedonia. Has been found in Val Malenco and Valfurva (Sondrio) and very occasionally in the cinnabar mines of Monte Amiata (Tuscany) (Italy).

**Uses** In the preparation of arsenious anhydride. In the manufacture of fireworks (giving a bright white color) and paint.

3.5
3.6

▲ Acicular boulangerite crystals (ca. ×1.5). Trepča, Yugoslavia.

▲ Massive realgar (ca. ×1). Valfurva, Italy.

## 46 ORPIMENT

SULFIDES
$As_2S_3$ (Arsenic sulfide)

**System** Monoclinic.
**Appearance** Small, prismatic, flattened crystals very rare. In crusts or bladed masses. Golden or orange-yellow.
**Physical properties** Soft (1.5–2), fairly heavy, fragile, with perfect cleavage. Translucent to transparent in very thin sections, with slightly greasy adamantine luster. Yellow streak. Fuses easily, giving off arsenic fumes with a strong garlic smell that conceals the sulfur dioxide fumes. Soluble in nitric acid and aqua regia with separation of sulfur, which floats to the surface of the solution. Tends to crumble if exposed to light and air.

3.48

**Environment** Deposited by hot springs, associated with realgar and sometimes cinnabar. Also found in metamorphic dolomites and in low-temperature hydrothermal veins associated with realgar and native arsenic.
**Occurrence** The largest masses are in Kurdistan (Turkey), Georgia (USSR), Persia and Manhattan, Nevada (USA). Occurs in the fumaroles of Vesuvius and the Campi Flegrei, at Amiata (Tuscany) and Ormea (Cuneo) (Italy).
**Uses** An ore of arsenic; used in the tanning of hides to remove hair.

---

## 47 HALITE (ROCK SALT)

HALIDES
NaCl (Sodium chloride)

**System** Isometric.
**Appearance** Cubic crystals, sometimes distorted with cavernous faces, "hopper" crystals. Colorless to white, yellow, red, brown and even black. Less frequently with irregular blue or purple patches. Groupings of divergent crystals, crusts and compact, white, opaque, microgranular masses.

2.1
2.2

**Physical properties** Soft (2.5), very light, fragile with perfect cubic cleavage. Transparent to translucent with vitreous luster. White streak. Very stable when pure, but if it contains impurities of calcium chloride or magnesium it becomes deliquescent. Dissolves readily in water, is an excellent conductor of heat and fuses easily, coloring the flame bright yellow and decrepitating. Moist to the touch and has a pleasant salty taste, not bitter like potassium and magnesium salts. A typical allochromatic mineral that changes color both because of chemical impurities or structural defects (turning black with bitumen or organic substances which discolor when heated, pink with minute algae, red with dusty hematite, yellowish with iron hydroxide, etc.) and when exposed to ultraviolet light (fluorescence), x-rays (turning blue then brown) or natural radioactivity (causing the common purple color).

**Environment** As a precipitate in sedimentary deposits

▲ Orpiment (ca. ×1). Karajdin, Turkey.

▲ Halite (rock salt) crystals (ca. ×1.5). Wieliczka, Poland.

caused by the evaporation of saline waters, usually saline lakes. Rock salt deposits are often interbedded with clay. As it is remarkably plastic and has a low density, rock salt in deep deposits tends to push its way upward toward the surface, forcing overlying sediments into domelike arches. Gypsum, anhydrite and sometimes native sulfur are concentrated at the top of these domes. Rock salt is also stable in a low-temperature metamorphic environment and is formed by direct sublimation from volcanic exhalations.

**Occurrence** There are large halite deposits at Stassfurt (Germany), Wieliczka (Poland), Cardona (Spain), Salzkammergut (Austria) and various places in the states of New York, Michigan and Ohio (USA). There are salt domes, intensively worked for sulfur, in Louisiana and Texas (USA). There is a small salt mine at Lungro (Cosenza, Italy), and magnificent crystals are found in Sicily. There are enormous expanses of salt, some covering hundreds of square kilometers, beneath the Mediterranean Sea, protected by a thick impervious layer of sediment.

**Uses** A vital ingredient in human and animal diet. Very important in the chemical industry for the preparation of soda, hydrochloric acid and sodium, and in the food-preserving industry. It is also used in scientific equipment for optical parts. Its by-products are potassium, magnesium, chlorine, bromine, iodine, etc.

---

## 48 SYLVITE
HALIDES
KCl (Potassium chloride)

**System** Isometric.

**Appearance** Cubic or cubo-octahedral crystals, colorless, white, bluish or yellowish-red. Often reddish because of disseminated hematite or purple due to exposure to natural radioactivity. Compact, whitish granular masses.

**Physical properties** Soft (2), very light, perfect cubic cleavage, fragile, plastic under prolonged unilateral pressure. Transparent with vitreous luster. White streak. Dissolves easily in water. Fuses readily coloring the flame purplish-red (this can be seen clearly through cobalt glass, which absorbs the yellow color of the sodium that substitutes for potassium in sylvite). It is not as deliquescent as halite and has a more bitter salty taste.

**Environment** In sedimentary evaporite deposits associated with halite, but much less frequent. Also a sublimation product of volcanic fumes.

**Occurrence** Sizable concentrations at Stassfurt (Germany), Kalusz (Poland), Cardona (Spain) and in Saskatchewan (Canada), the states of Utah, Texas and New Mexico at Carlsbad Caverns (USA). Found in the Ukraine (USSR) and in Italy at San Cataldo (Sicily) and on Etna and Vesuvius.

**Uses** An important industrial mineral used as a fertilizer, either directly or in the preparation of potassium salts.

▲ Violet halite (rock salt) crystals (ca. ×1). Calascibetta, Sicily.

▲ Sylvite crystals (ca. ×1). Vesuvius, Italy.

## 49 CHLORARGYRITE
HALIDES
AgCl (Silver chloride)

**System** Isometric.
**Appearance** Small, gray, cubic crystals, rarely in other associated forms. Generally in crusts or masses, gray or greenish in color, which tarnish to purplish-brown if exposed to light.
**Physical properties** Very soft (1–1.5), very heavy, sectile, malleable and ductile like wax. Opaque with waxy-to-submetallic luster. Insoluble in acids and water but dissolves easily in aqueous ammonia and fuses readily.
**Environment** In the oxidation zone of silver deposits as an alteration product from chlorine-rich solutions, associated with native silver, cerussite, etc.
**Occurrence** Large, mainly concretionary masses in the Atacama desert (Chile); Broken Hill (Australia), where it is associated with embolite Ag(Cl,Br); Potosí (Bolivia); Leadville and Treasure Hill, Colorado (USA), where blocks weighing several kilograms, some well crystallized, have been mined. Good individual crystals are found in the Poorman's Lode in Idaho (USA) and at Sarrabus (Sardinia).
**Uses** A silver-rich mineral (up to 75 percent Ag), but fairly rare.

---

## 50 FLUORITE
HALIDES
CaF$_2$ (Calcium fluoride)

**System** Isometric.
**Appearance** Cubes, octahedrons and dodecahedrons, other forms rarer. May be large. Penetration twins sometimes occur. Color extremely variable (allochromatic), colorless and completely transparent when pure, but may be yellow, green, blue, pink, purple or even black. Compact, banded and concretionary masses frequently found.
**Physical properties** Semi-hard (4), heavy, fragile, perfect octahedral cleavage. Transparent to translucent with vitreous luster. Insoluble in water, hydrochloric acid and nitric acid, but dissolves in concentrated sulfuric acid giving off fumes of hydrofluoric acid which attack glass. Decrepitates when heated and becomes phosphorescent. Fuses fairly easily, at least around the edges, turning the flame brick red. When exposed to ultraviolet rays often strongly fluorescent (usually blue or violet).
**Environment** In medium- and high-temperature hydrothermal veins associated with lead, zinc and silver sulfides and with barite and quartz. Rarer in pegmatites, where it is associated with cassiterite, topaz, tourmaline and apatite. Occurs as an accessory mineral in cavities of many felsic and

▲ Chlorargyrite crust (ca. ×1). Sonora, Mexico.

▲ Fluorite crystals (ca. ×1). Cumberland, England.

intermediate intrusives and as a sublimation product of volcanic rocks. It may also have a sedimentary origin, probably by deposition in enclosed basins from saline waters possibly of volcanic origin.

**Occurrence**  Splendid crystals are found in various localities. Colorless cleavage fragments of optical quality from Corvara in Val Sarentina (Bolzano, Italy), pink octahedrons from the St. Gotthard (Switzerland), green octahedrons from Cera (Brazil), Kongsberg (Norway) and Bancroft (Canada); green, white or violet cubes from Cumberland and Derbyshire (England); yellow cubes from Wölsenberg (Germany); purple crystals from Madoc (Canada) and various places in the Alps, especially near deposits of radioactive minerals (e.g., Valle Imagna, Bergamo, Italy). Significant masses worth mining are found in Ontario (Canada), Illinois and Kentucky (USA), England, the USSR, Mexico and Italy. There are large reserves of fluorite of potential industrial use in dried-up lake basins in southern Tuscany and Lazio (Italy).

**Uses**  In the production of hydrofluoric acid, which is vital in the pottery, optical and plastics industries, and in the metallurgical treatment of bauxite. Used as a flux in the metal industry. Very clear crystals are used as apochromatic lenses and as spectrographic prisms.

---

## 51  CRYOLITE
HALIDES
Na$_3$AlF (Sodium aluminum fluoride)

**System**  Monoclinic.

**Appearance**  Pseudo-cubic crystals, often in subparallel groupings, very rare. Granular, white or colorless masses common, associated with siderite.

**Physical properties**  Soft (2.5), light, fragile, has no cleavage but does exhibit parting. Transparent with vitreous or pearly luster. Almost disappears from sight when immersed in water, as the refractive index of cryolite (1.33) approaches that of water (1.00). White streak. Soluble in sulfuric acid, producing fumes of hydrofluoric acid that attack glass. Fuses readily, coloring the flame yellow (sodium) and giving a transparent globule which becomes opaque (glazed) when cool.

2.95
3.0

**Environment**  Always occurs in pegmatites, where it is probably a precipitate from fluoride-rich solutions.

**Occurrence**  The most important deposit is at Ivigtut (Greenland) where cryolite is associated with siderite, topaz, fluorite and other unusual fluorides, microcline, etc. Small amounts also occur in a topaz mine at Miask (Urals, USSR), at Pikes Peak, Colorado (USA), in Francon Quarry, Montreal (Québec, Canada) and in Nigeria.

**Uses**  Formerly as a flux in refining "bauxite," now replaced by synthetic products of fluorite. In pottery glazes and certain types of glass.

► Fluorite crystals (ca. ×1). Freiberg, DDR.

▼ Fluorite cleavage octahedron (ca. ×1). Cumberland, England.

▲ Cryolite crystals (ca. ×1.5). Ivigtut, Greenland.

## 52  CARNALLITE

HALIDES
$KMgCl_3 \cdot 6H_2O$ (Hydrated potassium magnesium chloride)

**System**  Orthorhombic.
**Appearance**  Crystals with pseudo-hexagonal outline rare. Compact granular masses, colorless or reddish because of tiny laminae of hematite scattered through the mineral.
**Physical properties**  Soft (2.5), no cleavage, but has conchoidal fracture. Transparent with vitreous or greasy luster. Extremely phosphorescent. As it is deliquescent it disintegrates quickly if exposed to air. Very soluble in water. Has a bitter, salty taste. Fuses easily, turning the flame violet (potassium).

1.6

**Environment**  One of the last minerals to precipitate in an evaporite deposit.
**Occurrence**  Large stratified or dome deposits associated with halite or kainite at Stassfurt (Germany), Solikamsk (Ukraine, USSR), in Iran, Mali, China, Tunisia, Texas and New Mexico (USA) and Spain. Sizable masses are found at San Cataldo and Pasquasia (Sicily).

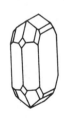

**Uses**  As a potassium fertilizer. In the chemical industry for the extraction of magnesium, potassium, chlorine, bromine, cesium, etc.

---

## 53  ATACAMITE

HALIDES
$Cu_2Cl(OH)_3$ (Hydrous copper chloride)

**System**  Orthorhombic.
**Appearance**  Green, prismatic, vertically striated crystals. Fibrous, acicular, lamellar, sometimes concretionary aggregates.
**Physical properties**  Semi-hard (3–3.5), heavy, perfect cleavage. Translucent with vitreous luster. When heated first decrepitates; then fuses fairly easily turning the flame blue, then green without hydrochloric acid (unlike brochantite). Soluble in hydrochloric acid without effervescence (unlike malachite).

3.76

**Environment**  An alteration product of copper sulfides in surface oxidation zones as in desert conditions. Also occurs as a sublimation product of volcanic exhalations.
**Occurrence**  Large dark-green crystals at Ravensthorpe and Wallaroo (Australia). Sizable masses in the Atacama desert (Chile), in Australia, Peru, Bolivia, Mexico, Namibia, the USA (Bisbee and Jerome, Arizona; Tintic, Utah) and in Russia (Kazakhstan and the Urals). In Italy it occurs as a crust on Vesuvian lava and as an alteration product of copper minerals in the Nurra mines (Sardinia).

**Uses**  A secondary copper mineral. It is only worked in the Chilean deposits, which are large enough to make it economically worthwhile.

▲ Carnallite (ca. ×1). Stassfurt, DDR.

▲ Radiating atacamite crystals (ca. ×1.5). Copiapo, Chile.

## 54 BOLEITE
HALIDES
$Pb_9Ag_3Cu_8Cl_{21}(OH)_{16} \cdot H_2O$ (Hydrated lead copper chloride)

**System** Tetragonal.
**Appearance** Twinned crystals with three interpenetrant individuals, giving a pseudo-cubic or pseudo-octahedral appearance. Dark blue color. Often intergrown with other halides.
**Physical properties** Semi-hard (3.5), very heavy, perfect cleavage. Translucent with vitreous luster. Light-blue streak with grayish tinge. Soluble in nitric acid but not in water. Fuses readily with decrepitation. Related to pseudoboleite which often occurs as epitaxial overgrowths on boleite.
**Environment** A surface alteration mineral in sulfide deposits, generally in residual clays in the leached zone, in kidney-shaped formations or isolated crystals.
**Occurrence** Found at Boleo (Baja California, Mexico), where cubes 2 cm (0.8 in) on edge have occurred, in copper mines of Chile, at Broken Hill (Australia) and at the Mammoth mine, Tiger, Arizona (USA).
**Uses** Purely of scientific interest; highly prized by collectors.

## 55 CUPRITE
OXIDES AND HYDROXIDES
$Cu_2O$ (Copper oxide)

**System** Isometric.
**Appearance** Octahedral, cubic or dodecahedral crystals, or a combination of the above forms. Dark ruby-red color, sometimes altered on the surface to green malachite. Aggregates of fine, acicular, hairlike crystals (chalcotrichite variety) occur.
**Physical properties** Semi-hard (3.5–4), very heavy, fragile, poor cleavage. Translucent with adamantine luster, altering to semi-opaque if exposed to air. Bright red streak. Soluble in concentrated acids. Fuses readily, turning the flame green (copper).
**Environment** Found in the oxidation zone of copper deposits associated with azurite, tenorite, malachite and native copper.
**Occurrence** Fine octahedral crystals at Chessy (France), Redruth and Liskeard (Cornwall, England), Bisbee and Morenci, Arizona (USA), Corocoro (Bolivia), Chuquicamata (Chile), Onganja (Namibia); deposits in the Urals and Altai (USSR) and on Sardinia. Chalcotrichite is found at Morenci, Arizona (USA), Monsol (France), in Chile and at Libiola (Liguria, Italy).
**Uses** Important ore of copper. Some flawless, transparent crystals have been cut as gems.

▲ Pseudo-cubic boleite crystals (ca. ×1). Boleo, Mexico.

▲ Cuprite crystals (ca. ×1.5). Cornwall, England.

## 56 ZINCITE
OXIDES AND HYDROXIDES
(Zn,Mn)O (Zinc manganese oxide)

**System** Hexagonal.
**Appearance** Crystals very rare. Usually massive, orange to dark red.
**Physical properties** Semi-hard (4–4.5), very heavy, perfect cleavage. Translucent with subadamantine luster. Orange-yellow streak. Infusible. Soluble in hydrochloric acid.
**Environment** In the contact metamorphic deposit at Franklin and Ogdensburg, New Jersey (USA), associated with white or pink calcite, green willemite, gray tephroite and black franklinite. Rare in other deposits.
**Occurrence** Besides New Jersey, traces have been found in the Bottino mine (Tuscany, Italy) and possibly in the Riso mine (Bergamo, Italy), in Poland, Spain and Australia.
**Uses** An important zinc mineral in New Jersey, whose crystals are eagerly sought by mineralogists and collectors.

5.4
5.7

## 57 SPINEL
OXIDES AND HYDROXIDES (Spinel Group)
MgAl₂O₄ (Magnesium aluminum oxide)

**System** Isometric.
**Appearance** Small, perfect octahedrons, frequently twinned. Also common in aggregates of rounded grains. Colorless, white, pink, light blue and black.
**Physical properties** Very hard (8), heavy, poor cleavage. Transparent to nearly opaque with vitreous luster. White streak. Infusible and insoluble. Several varieties of spinel are known. Color is related to composition: pleonaste is greenish, iron-bearing; picotite is blackish or brown, chromium- and iron-bearing; transparent varieties of various colors are called gem spinels.
**Environment** In contact metamorphic rocks, especially in marly dolomites rich in magnesium and aluminum. In granulites and in high-pressure ultramafic rocks. Common in many alluvial and marine placers.
**Occurrence** Plentiful in the riverine sands of the Indian subcontinent, Thailand, Malagasy Republic, Sri Lanka and Afghanistan. Fine crystals at Monte Somma–Vesuvius and the volcanoes of Lazio (Italy). Large crystals found in Orange County, New York, and Sterling Hill mine, Ogdensburg, New Jersey (USA). Widespread as an accessory mineral of rocks.
**Uses** "Ruby spinel" is a semi-precious gemstone.

3.5
4.1

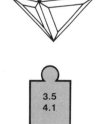

▲ Zincite (ca. ×1.5). Franklin, New Jersey.

► Gem spinel crystals (ca. ×1.5). India.

▼ Spinel (ca. ×1). India.

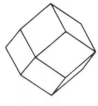

## 58 MAGNETITE

OXIDES AND HYDROXIDES (Spinel Group)
$Fe^{+2}Fe^{+3}_2O_4$ (Iron oxide)

**System** Isometric.

**Appearance** Black, shiny, perfect octahedrons or dodecahedrons with striated faces. Iron-black, compact and granular masses with bluish iridescence.

**Physical properties** Hard (5.5–6.5), very heavy, no cleavage, but exhibits octahedral parting. Opaque with metallic luster. Black streak. Strongly magnetic and sometimes acts as a natural magnet (lodestone). Infusible. Dissolves very slowly in concentrated hydrochloric acid. May contain titanium or chromium in solid solution.

**Environment** Common in a great variety of rocks. Frequent in mafic and ultramafic extrusive rocks as a differentiation product during the magmatic stage. Rather rare in pegmatites and hydrothermal veins. Large masses found in detrital sedimentary rocks (alluvial and marine sands) and in dune deposits in desert conditions. In a metamorphic environment it is formed by reduction of hematite derived from the dissociation of sulfides and iron silicates. Plentiful in contact metasomatic conditions (skarns). Intimately associated with corundum and a few other minerals in natural emery.

**Occurrence** Huge deposits are found in the Kiruna district (northern Sweden), where it is associated with apatite; the Bushveld Complex (South Africa); the Adirondacks, New York (USA); Iron springs, Utah (USA); Iron Mountains, Wyoming (USA); the Urals (Gora Blagodat and Magnitnaya, USSR). In Italy a deposit is worked at Cogne (Val d'Aosta). Beautiful crystals are found in the chlorite schists of Val Malenco (Sondrio) and Val di Vizze (Bolzano, Italy); in the Binnental (Switzerland) and Pfitschtal (Austrian Tyrol); Brewster, New York (USA); the effusions of Vesuvius and the volcanoes of Lazio (Italy). Small quantities of magnetite are also found at Traversella (Turin, Italy) where marvelous crystals occur, in Nurra (Sardinia) and at Capo Calamita (island of Elba). Found as lodestone at Magnet Cove, Arkansas (USA).

**Uses** The richest and most important ore of iron. Vanadium and phosphorus are often extracted from the slag. Swedish magnetite, associated with iron silicates, produces a very hard silicon steel.

▲ Octahedral magnetite crystals (ca. ×1.5). Binnental, Switzerland.

▲ Dodecahedral magnetite crystals (ca. ×1.5). Traversella, Italy.

## 59 FRANKLINITE

OXIDES AND HYDROXIDES (Spinel Group)
$(Zn,Mn^{+2},Fe^{+2})(Fe^{+3},Mn^{+3})_2O_4$ (Zinc manganese iron oxide)

**System** Isometric.

**Appearance** Octahedral or dodecahedral crystals. Massive black aggregates with metallic appearance and reddish tints.

**Physical properties** Hard (5.5–6.5), very heavy, fragile, no cleavage but good conchoidal fracture. Opaque with metallic luster. Reddish-brown streak. Weakly magnetic, becoming strongly magnetic if heated in a reducing flame. Infusible. Soluble in hydrochloric acid.

**Environment** In contact metasomatic conditions in crystalline dolomites, associated with red zincite, green willemite, magnetite, rhodonite and garnet.

5.0
5.2

**Occurrence** Mainly at Franklin and Sterling Hill, Ogdensburg, New Jersey (USA), where octahedral crystals measuring almost 30 cm (12 in) on edge have been found, as well as compact masses.

**Uses** In the past used for the extraction of zinc and manganese, which were then alloyed with iron.

---

## 60 CHROMITE

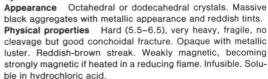

OXIDES AND HYDROXIDES (Spinel Group)
$FeCr_2O_4$ (Iron chromium oxide)

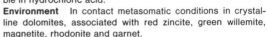

**System** Isometric.

**Appearance** Rare, small, black, octahedral crystals. Usually in compact, granular masses or in disseminated grains.

**Physical properties** Hard (5.5), heavy, no cleavage, but has even fracture. Opaque with submetallic luster. Dark-brown streak. Weakly magnetic. Infusible and insoluble in acids (unlike magnetite, which dissolves slowly). Chromite forms a solid solution series with magnesiochromite $MgCr_2O_4$, spinel $MgAl_2O_4$ and other spinels.

**Environment** Exclusively in mafic and ultramafic rocks as a crystal accumate in the early stages of magmatic crystallization. Occurs in peridotite and serpentinite, and concentrated in placers.

4.5
4.8

**Occurrence** Large deposits in Turkey, South Africa, Rhodesia, the USSR (Urals), the Philippines, Albania and Cuba. Small lenses occur in the Italian alps. The best crystals come from Namibia. Found in the Stillwater Complex, Montana, and serpentine in Texas, Pennsylvania and various localities in California (USA).

**Uses** The main ore for chromium, important in steel alloys, stainless steel, chrome-plating and used in the manufacture of some paints, in leather tanning and in mordants.

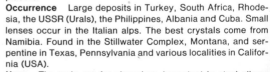

▲ Franklinite (ca. ×1). Franklin, New Jersey.

▲ Chromite in peridotite (ca. ×1). Makri, Turkey.

## 61 HAUSMANNITE
OXIDES AND HYDROXIDES
$Mn^{+2}Mn^{+3}_2O_4$ (Manganese oxide)

**System** Tetragonal.

**Appearance** Pseudo-octahedral, dipyramidal crystals, often with repeated (cruciform) twinning. Blackish, compact or granular aggregates.

**Physical properties** Hard (5–5.5), heavy, fragile with good cleavage. Transparent in thin section with submetallic luster. Reddish-brown streak. Not magnetic. Infusible, but soluble in concentrated hydrochloric acid.

**Environment** In contact metasomatic rocks (skarns) or metamorphic rocks by dehydration of manganese hydroxides under reducing conditions.

4.7
4.8

**Occurrence** Sizable masses at Sapalsk (USSR) in contact with metamorphosed limestones. Also considerable amounts at Långban (Sweden), Ilfeld (Germany), in Val d'Avers (Grisons, Switzerland) and in Bulgaria. Found intimately mixed with psilomelane at Batesville, Arkansas (USA).

**Uses** An excellent manganese ore, not very common.

## 62 CHRYSOBERYL
OXIDES AND HYDROXIDES
$BeAl_2O_4$ (Beryllium aluminum oxide)

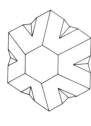

**System** Orthorhombic.

**Appearance** Prismatic, tabular crystals often twinned forming pseudo-hexagonal crystals. Colorless, green, yellow, gray or brown. Alexandrite is red in artificial light, dark green in natural light. "Cat's eye" caused by microscopic inclusions of rutile.

**Physical properties** Very hard (8.5), heavy, fragile, fairly good cleavage. Transparent to translucent with subadamantine to silky luster. Insoluble, infusible.

**Environment** In pegmatites and aplites or in mica schists, especially in contact with granitic intrusions. Easy to find in clastic alluvial or marine deposits.

3.7

**Occurrence** Alexandrite is found at Takowaja in the Urals (USSR) in crystals 1–10 cm long (0.5–5 in), and in Sri Lanka. Cat's eyes are found in alluvial deposits in Brazil and Sri Lanka. Beautiful transparent or translucent yellow-green twins occur at Espirito Santo (Brazil). Gray chrysoberyl is found in many pegmatites including Haddam, Connecticut (USA), Norway, and in pegmatites at Olgiasca (Lake Como, Italy).

**Uses** The colorless varieties, alexandrite and honey-yellow cat's eyes are highly prized gemstones.

▲ Hausmannite (ca. ×1.5). Ilfeld, DDR.

▶ Cyclic twin of chrysoberyl (ca. ×1). Urals, USSR.

▼ Cyclic twin of chrysoberyl (ca. ×1). Brazil.

## 63　BIXBYITE
OXIDES AND HYDROXIDES
(Mn,Fe)$_2$O$_3$ (Manganese iron oxide)

**System**　Isometric.

**Appearance**　Cubic crystals often modified by trapezohedrons. rare. Penetration twins noted. Usually massive. Color black.

**Physical properties**　Hard (6–6.5), heavy, indistinct cleavage and uneven fracture. Streak black. Opaque with metallic to submetallic luster. Dissolves with difficulty in nitric acid evolving chlorine. Fusible.

**Environment**　Occurs in cavities in highly altered rhyolitic rock.

**Occurrence**　Sharp crystals up to 1.2 cm (0.5 in) occur associated with topaz and pseudobrookite in rhyolitic cavities in the Thomas Range, Utah (USA). Large well-formed crystals also found at Postmasburg (South Africa). Massive bixbyite occurs at Långban (Sweden), Sitapár (India) and Ribes (Spain).

**Uses**　A minor ore of manganese.

---

## 64　CORUNDUM
OXIDES AND HYDROXIDES
Al$_2$O$_3$ (Aluminum oxide)

**System**　Hexagonal.

**Appearance**　Hexagonal, barrel-shaped crystals with either tapered or stubby terminations. Variable in color, sometimes spotted. Usually gray or brown, semi-opaque and granular (with magnetite as emery), but sometimes transparent, and colored red (ruby variety), blue, yellow, green, purple and colorless (sapphire).

**Physical properties**　Very hard (9), heavy, no cleavage but exhibits nearly rectangular parting. Transparent to translucent or semi-opaque, with adamantine luster. Infusible and insoluble. Some varieties are fluorescent yellow in ultraviolet light. Others display asterism caused by crystallographically oriented inclusions of rutile (star sapphire and ruby). The color of grayish varieties is due to a partial alteration to margarite or zoisite caused by hydrothermal solutions.

**Environment**　An accessory mineral of igneous rocks undersaturated with respect to silica (syenites and nepheline syenite pegmatites) and in high-grade metamorphic rocks, poor in silica and rich in aluminum (marbles, some mica schists,

▲ Bixbyite cube on topaz (ca. ×18). Thomas Range, Utah.

▲ Corundum (var. ruby) (ca. ×1.5). India.

granulites). Also found in eclogites and occasionally in roding-ites. Sedimentary concentrations are frequent in alluvial and marine sands.

**Occurrence** Enormous gray crystals (up to 170 kg; 375 lb) have been found in mica schists in Malagasy Republic and pegmatites in South Africa. The largest rubies, up to 6 cm (2.36 in) long, come from the marbles of Mogok (Burma), alluvial deposits in the Malay Peninsula and Sri Lanka and granulites in Tanzania and Brazil. Sapphires are commonest in Sri Lanka, but also occur in Burma, Thailand, Kashmir (India), Australia, Afghanistan and the United States (Helena and Yogo Gulch, Montana). There are large emery deposits on Naxos and Samos (Greece), at Smyrna (Turkey), in Australia and in Chester County, Pennsylvania, and Peekskill, New York (USA). In Italy large crystals of gray corundum are found in the plumasites of Val Sabbiola (Vercelli) and in various other places in Val Sesia and Biellese (Piedmont).

**Uses** Emery and unattractive varieties are used as abrasives. The clear, colored varieties are highly regarded gemstones. Synthetic corundum, easily manufactured by the Verneuil process, has replaced natural corundum in mechanical precision engineering ("rubies" for watches and other instruments).

---

## 65 HEMATITE
OXIDES AND HYDROXIDES
$Fe_2O_3$ (Iron oxide)

**System** Hexagonal.
**Appearance** Stubby, black, rhombohedral crystals, often with individual crystals arranged like the petals of a rose ("iron rose"). More commonly in massive, granular masses, compact, sometimes with iridescent surfaces, soft and earthy (red ochre). Often oolitic, botryoidal or concretionary in appearance. Often colors both rocks and minerals shades of red and reddish-brown.

**Physical properties** Hard (5.5–6.5), very heavy, fragile, no cleavage. Opaque with metallic luster, blood-red tints in thin section. Streak dark-cherry red, making it easy to distinguish among hematite, magnetite and ilmenite. Dissolves slowly when heated in concentrated hydrochloric acid. Infusible. Becomes magnetic if heated in reducing flame.

**Environment** A common accessory mineral of many igneous rocks, especially lavas, because it forms under oxidizing conditions, as compared to magnetite. Rare in plutonic rocks but common in pegmatites and hydrothermal veins. Much hematite is formed under sedimentary conditions through diagenesis of "limonite," retaining its concretionary and oolitic forms. Re-

▲ Corundum (ca. ×1). India.

▲ Ruby crystals in marble (ca. ×1). India.

▲ Iridescent hematite crystals (ca. ×1). Island of Elba, Italy.

mains stable in a low-temperature metamorphic environment where it often replaces magnetite (martite variety). Also occurs as a sublimation product of volcanic exhalations.

**Occurrence** The main deposits are sedimentary in origin, formed by the action of meteoric waters on sedimentary formations rich in iron carbonate and silicate. The largest deposits are at Lake Superior (USA), in Québec (Canada), Venezuela, Brazil and Angola. There are oolitic deposits in Tennessee (USA), the Ukraine (USSR) and Canada. Beautiful crystals are found at Rio Marina on the island of Elba, at Bahia (Brazil) and in Cumberland (England). Iron roses are typical of the St. Gotthard region (Switzerland), Minas Gerais (Brazil), where they sometimes reach 15 cm (6 in) in diameter, and the Binnental (Switzerland). Beautiful martites have been found at Twin Peaks, Utah (USA), and in the deposits of Lake Superior (USA) and Nova Scotia (Canada).

**Uses** The most important iron ore. Red ochre is also used as a pigment and a polishing powder.

## 66 ILMENITE
OXIDES AND HYDROXIDES
$FeTiO_3$ (Iron titanium oxide)

**System** Hexagonal.

**Appearance** Very flat, tabular, black or dark brown, rhombohedral crystals. Also in compact or granular aggregates.

**Physical properties** Hard (5–6), heavy, no cleavage, but fairly good pseudo-rhombohedral fracture. Opaque, even in thin section, with submetallic luster. Black to brownish-red streak. Soluble in powder form in concentrated hydrochloric acid. Infusible. Tends to become magnetic if heated and is sometimes weakly magnetic when cool. Physical characteristics vary according to the amount of magnesium present in solid solution. Ilmenite forms a solid solution series with geikielite ($MgTiO_3$).

**Environment** A common accessory in plutonic rocks, as a high-temperature segregation product. Also occurs in pegmatites and nepheline syenites. Large concentrations are found in sands, especially marine sands. Also found in metamorphic rocks such as gneiss, chlorite schist, etc.

**Occurrence** Large crystals are found in diorite at Kragerö (Norway). Beautiful, small shiny crystals in Val Dévero (Novara, Italy) and the St. Gotthard region (Switzerland). Crystals to 2.5 cm (1 in) found in Orange and Warwick counties, New York (USA). Huge deposits in Norway, India, Brazil, Canada, Florida (USA) and the USSR.

**Uses** A major ore of titanium.

► Hematite crystal (ca. ×0.5).
Island of Elba, Italy.

▼ Hematite rosette (ca. ×1).
St. Gotthard, Switzerland.

▲ Ilmenite crystals (ca. ×1.5). Froland, Norway.

## 67 PEROVSKITE

OXIDES AND HYDROXIDES
$CaTiO_3$ (Calcium titanium oxide)

**System** Orthorhombic.

**Appearance** Pseudo-cubic crystals, striated parallel to the edges, reddish-brown, yellow or blackish-gray in color. Kidney-shaped aggregates and masses.

**Physical properties** Hard (5.5), heavy, no cleavage, conchoidal fracture. Translucent with adamantine to submetallic luster. Pale yellow streak. Infusible and soluble only in heated sulfuric acid. Some varieties contain alkaline metals and rare earths (loparite, knopite).

**Environment** Accessory mineral in undersaturated igneous rocks (melilite and nepheline basalts, nepheline syenites, carbonatites), in ultramafic rocks, including those which have undergone metamorphism, and in some contact metamorphosed marbles.

**Occurrence** Splendid crystals in carbonatite at Alnö (Sweden) and Oka (Canada), in skarns at Magnet Cove, Arkansas (USA) and in the Urals (USSR). In Italy, found in serpentine in Val Malenco (Sondrio), at Emarese (Val d'Aoste) and in the volcanic effusions of Monte Somma.

**Uses** Varieties rich in niobium, cerium and rare earth elements are ores for these minerals.

## 68 STIBICONITE

OXIDES AND HYDROXIDES
$SbSb_2O_6(OH)$ (Hydrous antimony oxide)

**System** Isometric.

**Appearance** Compact or massive; botyroidal, or forming crusts. Color yellow, yellow-white or orange, brown or black due to impurities.

**Physical properties** Hard (4–5.5), very heavy, no cleavage, but uneven fracture. Streak yellow-white. Opaque with pearly to earthy luster. Gives off water in closed tube and fuses with difficulty to a gray slag with white coating on charcoal block.

**Environment** Occurs as an alteration product of stibnite in low-temperature hydrothermal veins or hot springs.

**Occurrence** Superb pseudomorphs of stibiconite after stibnite found at San Luis Potosí (Mexico) associated with cervantite. In the USA in several counties in California, especially Kern County, and in Shoshone County, Idaho; Unionville, Nevada; and Garfield County, Utah. Also occurs in Ouro (Bolivia) and various deposits in Peru, the UK, Spain, western Australia and Sarawak (Borneo).

**Uses** Of interest to scientists and collectors.

▲ Perovskite crystal in serpentine (ca. ×2). Val Malenco, Italy. *Inset,* perovskite (ca. ×1). Ural Mountains, USSR.

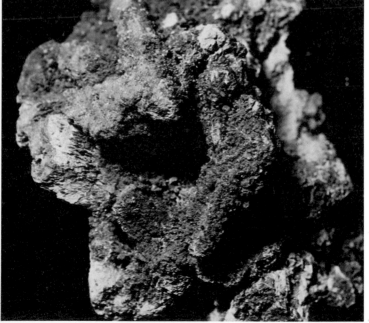

▲ Brown stibiconite crust on quartz (ca. ×2). New Mexico.

## 69 BETAFITE

OXIDES AND HYDROXIDES
$(Ca,Na,U)_2(Nb,Ta)_2O_6(OH)$ (Hydrous uranium titanium niobium oxide)

**System** Isometric.
**Appearance** Octahedral or dodecahedral crystals, sometimes elongated. Pitch-black, usually coated with a greenish-brown or yellow surface alteration.
**Physical properties** Hard (5), heavy, no cleavage but good conchoidal fracture. Transparent with vitreous luster. Opaque with earthy luster when metamict, as is often the case. Dissolves easily in acid but fuses with difficulty. Radioactive.
**Environment** A common accessory mineral in many granitic pegmatites rich in rare earth elements. Also occurs in metamorphosed limestones in contact with pegmatites.
**Occurrence** Occurs in its classic form in pegmatites of the Malagasy Republic, where betafite specimens weighing up to 104 kg (231 lb) have been found, and where several varieties have been identified. Unaltered crystals have also been found in Norway and in pegmatite at Sludianka, near Lake Baikal, Siberia (USSR). Found occasionally in pegmatites in Val d'Ossola (Italy).
**Uses** An important ore of thorium, uranium, rare earth elements and niobium.

## 70 MICROLITE

OXIDES AND HYDROXIDES
$(Na,Ca)_2Ta_2O_6(O,OH,F)$ (Hydrous sodium calcium tantalum oxide)

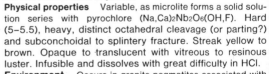

**System** Isometric.
**Appearance** Octahedral (often highly modified) crystals, rare. Commonly massive or disseminated. Color usually brown, but can be colorless or yellow.
**Physical properties** Variable, as microlite forms a solid solution series with pyrochlore $(Na,Ca)_2Nb_2O_6(OH,F)$. Hard (5–5.5), heavy, distinct octahedral cleavage (or parting?) and subconchoidal to splintery fracture. Streak yellow to brown. Opaque to translucent with vitreous to resinous luster. Infusible and dissolves with great difficulty in HCl.
**Environment** Occurs in granite pegmatites associated with lepidolite, spodumene and tantalite.
**Occurrence** Very large sharp octahedral crystals to 6.5 cm (2.5 in) found at Amelia Court House, Virginia (USA). In numerous pegmatites in Maine (Newry) and Connecticut (Haddam Neck, Middletown). Very fine crystals in albite are being produced in Virgem da Lapa, Minas Gerais, Brazil, and in the Malagasy Republic.
**Uses** Of interest to mineralogists and collectors.

▲ Betafite crystals (ca. ×2). Malagasy Republic.

▲ Cubo-octahedron of microlite in albite (ca. ×6). Amelia Court House, Virginia.

## 71 RUTILE

OXIDES AND HYDROXIDES
α-TiO₂ (Titanium oxide)

**System** Tetragonal.
**Appearance** Elongated, prismatic crystals, often striated. Sometimes very slender crystals occur as inclusions in other minerals (those found in quartz are known as "maiden hair") or criss-crossing at 60° (sagenite). Yellow, red, brown or black. Elbow- and heart-shaped (geniculated) twins common.

**Physical properties** Hard (6–6.5), heavy, fragile, with perfect cleavage. Normally opaque or translucent with metallic luster. Rarely transparent with adamantine luster. Brown streak. Infusible, insoluble in acid.
**Environment** A very common accessory mineral in intrusive and metamorphic rocks or in quartz veins running through them. Occurs in concentrations with other heavy minerals in sands derived from these rocks.
**Occurrence** Splendid crystals are found in Alpine fissures (St. Gotthard, Tavetschtal, Cavradi and Castione, Switzerland) and in pyrophyllite in Graves Mountain, Georgia (USA). Common in apatite veins in Norway and in hornblende-bearing rocks in Nelson County, Virginia (USA). Also found in Australia, Italy, Mexico and Brazil.
**Uses** Important commercial ore of titanium.

## 72 CASSITERITE

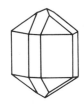

OXIDES AND HYDROXIDES
SnO₂ (Tin oxide)

**System** Tetragonal.
**Appearance** Stubby, prismatic crystals, brown to near black in color, often twinned (elbow-shaped). Also found in black granular banded fibrous masses which look like wood, and in scattered granules.
**Physical properties** Hard (6–7), heavy, fragile with imperfect cleavage and conchoidal fracture. Transparent to translucent with adamantine luster. White streak. Infusible and insoluble in acids.

**Environment** Typical of pegmatites and greisens. However, the largest deposits are sedimentary and occur in river and marine placers.
**Occurrence** Fine crystals are found in the Erzegebirge (DDR) and Cornwall (England), but cassiterite for industry comes from ore deposits in Malaysia, Sumatra, the USSR, China and Bolivia. In the USA it is found in pegmatite veins in Oxford County, Maine, and Custer, South Dakota. Also found on Elba, and near Campiglia Marittima (Livorno) and in pegmatite at Piona (Lake Como) (Italy).
**Uses** An important ore of tin.

▲ Reticular network of rutile crystals (ca. ×1). Ticino, Switzerland.

▶ Rutile (ca. ×1). Graves Mountain, Georgia.

▼ Twinned cassiterite crystals (ca. ×1), Germany, and (*right*) cassiterite (ca. ×1), Zaire.

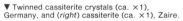

## 73 PYROLUSITE
OXIDES AND HYDROXIDES
MnO$_2$ (Manganese oxide)

**System** Tetragonal.
**Appearance** Very rare, black, prismatic crystals (polianite). Usually fibrous, dendritic, concretionary aggregates or earthy masses, associated with other manganese oxides, often non-crystalline (*wad*).
**Physical properties** Hard (6–6.5), heavy, poor cleavage. Opaque with submetallic luster. Blue-black streak. Infusible. Soluble in HCl, giving off Cl. Earthy varieties are very soft and greasy, leaving a distinct black mark when rubbed on a surface. It becomes hard when heated, giving off various gases.
**Environment** Usually sedimentary, as a chemical precipitate in lakes, lagoons or in stagnant bodies of water. Also formed by surface alteration of Mn minerals.
**Occurrence** Pyrolusite has been found at Platten (Bavaria, Germany), in Czechoslovakia and in Cornwall (England). Enormous deposits of *wad* and compact pyrolusite occur in the USSR (Chiatura, Georgia), in India (the Deccan), Brazil, Ghana and South Africa. Found in Italy on the island of San Pietro (Sardinia). In the USA mined at Batesville, Arkansas.
**Uses** An important ore of manganese.

---

## 74 ROMANECHITE (PSILOMELANE)
OXIDES AND HYDROXIDES
BaMn$^{+2}$Mn$^{+4}_8$O$_{16}$(OH)$_4$ (Hydrous barium manganese oxide)

**System** Orthorhombic
**Appearance** Botryoidal, stalactitic and dendritic black aggregates. Earthy masses. Never in crystals.
**Physical properties** Semi-hard (5–7), heavy, cryptocrystalline, often with greasy feel. Opaque with submetallic luster. Black or brown streak. Infusible. Soluble in hydrochloric acid, giving off chlorine. The nature of romanechite (psilomelane) is not quite clear. It is probably partly amorphous, containing alkaline and alkaline earth metals (calcium, potassium, barium) in nonstoichiometric proportions.
**Environment** Usually sedimentary. Forms in the oxidation zone of manganese oxide and silicate deposits.
**Occurrence** In large deposits at Chiatura (Georgia, USSR) and Nikopol (Ukraine, USSR). Dendritic psilomelane typically forms on the bedding or fracture planes of many limestones.
**Uses** An important manganese ore when available in sufficient quantity.

▲ Pyrolusite (ca. ×1.5). Nova Scotia, Canada.

▲ Dendritic psilomelane (ca. ×1). Solenhofen, Germany.

## 75 ANATASE
OXIDES AND HYDROXIDES
$\beta$-TiO$_2$ (Titanium oxide)

**System** Tetragonal.
**Appearance** Very sharp, dipyramidal crystals, more rarely pseudo-octahedral or tabular (octahedrite). Metallic-black, honey yellow or sapphire-blue in color.
**Physical properties** Hard (5.5–6), heavy, perfect cleavage. Transparent or translucent with adamantine to submetallic luster. Light-yellow streak. Infusible and insoluble.
**Environment** In low-temperature veins and Alpine fissures, associated with quartz, titanite, adularia, etc. Common in alluvial placers and strata of humus decomposition as an early oxide of titanium, formed under metastable sedimentary conditions.

**Occurrence** Beautiful crystals on quartz are found in the Binnental and Tavetschtal (Switzerland) and in Italy in Val Devero, near Domodossola and in Valtellina (Sondrio). Fine specimens occur in the Dauphiné (France) and in Norway. Small blue crystals at Gunnison, Colorado, and Sommerville, Massachusetts (USA). Occurs in diamond- and gold-bearing placers in Brazil, South Africa and the southern Urals (USSR).
**Uses** Occasionally as an ore of titanium.

---

## 76 BROOKITE
OXIDES AND HYDROXIDES
$\gamma$-TiO$_2$ (Titanium oxide)

**System** Orthorhombic.
**Appearance** Tabular or lamellar crystals, sometimes striated vertically, brown to black, sometimes banded.
**Physical properties** Hard (5.5–6), heavy, fragile, imperfect cleavage. Transparent or translucent with adamantine or submetallic luster. Light yellow or light grayish-brown streak. Infusible and insoluble.
**Environment** In veins and Alpine fissures running through gneisses and granites. Sometimes in contact metamorphic rocks. Common as a diagenetic mineral in sedimentary rocks.

**Occurrence** Splendid crystals occur at Bourg d'Oisans (Dauphiné, France), in Grisons and Uri cantons (Switzerland) and in the Tyrol and the Untersulzbachtal (Salzburg, Austria). In Italy found at Beura (Val d'Ossola) and the Piatta Grande near Sondalo (Valtellina, Sondrio). Large crystals are found in skarns at Magnet Cove, Arkansas (USA) and in the Virgental (Austria). Common in alluvial placers in the Urals (USSR) and Brazil.
**Uses** Brookite deposits are rarely large enough to be commercially exploited as a source of titanium.

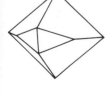

▲ Anatase crystal on quartz (ca. ×1.5). Dauphiné, France.

▲ Brookite twin (ca. ×2). Magnet Cove, Arkansas.

## 77 COLUMBITE-TANTALITE GROUP
OXIDES AND HYDROXIDES
(Fe,Mn)(Nb,Ta)₂O₆ (Iron manganese niobium tanta-
lum oxides)

**System** Orthorhombic.

**Appearance** Prismatic, stubby or tabular crystals, often in
heart-shaped twins, striated, black with metallic appearance
and sometimes coated with an iridescent film. Massive aggregates. Epitaxial overgrowths of columbite and samarskite
occur.

**Physical properties** A complex solid solution series exists
between ferrocolumbite, manganocolumbite (niobium rich) and
ferrotantalite and manganotantalite (tantalum rich). Pure end
members are very rare. Hard (6), very heavy, fragile, with distinct cleavage. Opaque with submetallic luster and reddish
tints. Black or dark-red streak. Insoluble and nearly infusible.

**Environment** In granitic pegmatites rich in lithium silicates
and phosphates, associated with spodumene, lepidolite, beryl,
etc. Often concentrated in sands.

**Occurrence** Economic deposits occur in western Australia,
Zaire (Katanga), Brazil, USSR (Ilmen Mountains), Norway,
Canada and the Malagasy Republic. Fine crystals found at
Londonderry (Australia), in pegmatites in South Dakota and
New England (USA). Found in pegmatites also in Italy.

**Uses** The principal ore of niobium and tantalum, used in metallurgy for heat-resistant alloys, rust-proofing (stainless steel)
and electromagnetic superconductors.

## 78 FERGUSONITE
OXIDES AND HYDROXIDES
YNbO₄ (Yttrium niobium oxide)

**System** Tetragonal.

**Appearance** Prismatic to pyramidal crystals rare. Usually in
irregular masses or grains. Color gray, yellow or brown on
fresh surfaces, dark brown on altered.

**Physical properties** Hard (5.5–6.5), very heavy, indistinct
cleavage and subconchoidal fracture. Streak pale brown.
Opaque (transparent in splinters) with dull (vitreous on broken
surfaces) luster. Streak brown, yellow-brown or black. Infusible. Dissolves completely in strong hydrofluoric acid and partially in sulfuric and hydrochloric acids.

**Environment** Occurs in granite pegmatites rich in rare earths
associated with zircon, monazite, gadolinite and euxenite.

**Occurrence** Found in several localities in Norway, Ilmen
Mountains (USSR) and the Malagasy Republic (especially near
Ambatouohangy). In the USA at Baringer Hill, Llano County,
Texas; and at Amelia Court House, Virginia.

**Uses** Source of the rare earth yttrium.

▲ Columbite crystal (ca. ×1). Raahe, Finland.

▲ Massive fergusonite (ca. ×1.5). Froland, Norway.

## 79 SAMARSKITE

OXIDES AND HYDROXIDES

(Y,Ce,U,Ca,Pb) (Nb,Ta,Ti,Sn)$_2$O$_6$ (Complex rare earth niobium tantalum oxide)

**System** Monoclinic.

**Appearance** Crystals are rectangular prisms, striated, velvety-black with brown surface alteration. Sometimes in epitaxial overgrowth with columbite-tantalite group minerals. Compact masses.

5.69

**Physical properties** Hard (5–6), very heavy, very poor cleavage, conchoidal fracture. Translucent or opaque with vitreous to resinous luster. Dark reddish-brown streak. Very radioactive. Soluble when heated in acids. If heated loses its color, decrepitates and breaks, fusing only at the edge of the mass.

**Environment** An accessory of many granitic pegmatites rich in rare earth elements and niobium. Common in heavy mineral sands.

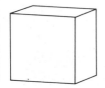

**Occurrence** In pegmatites at Miask (Urals, USSR) and in the sands of the Caspian and Black seas. In pegmatites in southern Norway and Switzerland, Borneo, India, Zaire and Brazil. In the USA found in Mitchell County, North Carolina.

**Uses** An important ore of niobium, tantalum, yttrium and uranium (some varieties contain up to 23 percent U).

---

## 80 THORIANITE

OXIDES AND HYDROXIDES

(Th,U)O$_2$ (Thorium uranium oxide)

**System** Isometric.

**Appearance** Cubic, black or dark-gray crystals, often modified by octahedral faces. Granular masses.

**Physical properties** Hard (6.5), very heavy, fragile, poor cleavage. Semi-opaque with dull to submetallic luster. Greenish-gray streak. Strongly radioactive. Insoluble in hydrochloric acid but dissolves slowly in nitric acid. Infusible.

9.7
9.8

**Environment** Accessory mineral of granitic pegmatites and contact metamorphosed limestones. Usually concentrated in sands in the form of rounded grains.

**Occurrence** In marine and alluvial sands in Sri Lanka, Malagasy Republic and the rivers of Siberia (USSR). In the USA it has been found in a contact marble with serpentine at Easton, Pennsylvania.

**Uses** One of the most important ores of thorium and uranium.

▲ Massive samarskite (ca. ×1.5). Kragerö, Norway.

▲ Thorianite cubes (ca. ×2). Malagasy Republic.

**URANINITE**
OXIDES AND HYDROXIDES
$UO_2$ (Uranium oxide)

7.5
9.7

**System** Isometric.
**Appearance** Occasionally in black, cubic, cubo-octahedral or more rarely dodecahedral crystals, sometimes of considerable size. Usually in granular masses or aggregates, botryoidal, reniform, colloform and banded. Pitch black (pitchblende variety) often associated with colorful secondary uranium minerals (e.g., gummite).
**Physical properties** Variable, depending on the degree of radioactive decay of uranium into lead, or on the presence of thorium and rare-earth elements which sometimes are present in sizable amounts. Well crystallized uraninite (mainly with tetravalent uranium) is hard (5.5), very heavy, fragile and without distinct cleavage. Opaque with submetallic luster. Shiny black streak. Highly radioactive. Depending on the amount of alteration, dissolves fairly easily in nitric, sulfuric and hydrofluoric acids but not in hydrochloric acid. Infusible. Pitchblende, in which part of the uranium is hexavalent, is semi-hard (4), light and fragile. Opaque with characteristic pitchlike appearance. Again highly radioactive.
**Environment** In pegmatites and medium- and high-temperature hydrothermal veins. In sedimentary clastic deposits (conglomerate matrix, sometimes gold-bearing) and associated with urano-organic complexes.
**Occurrence** Impressively large, beautiful crystals are found in Canada (Wilberforce, Ontario), where there are also large deposits in the Great Bear Lake region. The rich mineral deposit at Shinkolobwe, near Lumumbashi (Katanga), is very important. Here uraninite is associated with cobalt and copper minerals, with secondary molybdenum, tungsten, gold and platinum minerals. Other notable localities are Yachymov (formerly Joachimsthal, Czechoslovakia) and the Colorado Plateau region of the USA (Colorado, Arizona, New Mexico and Utah). In Italy uraninite has been found in the pegmatites at Piona (Lake Como), Montescheno (Novara), Novazza (Gergamo) and Lurisia (Cuneo). There are important deposits in Witwatersrand (South Africa), Portugal and France. The Cornish deposits (in England) are now worked out and are of purely scientific interest.
**Uses** In this mineral, or more specifically in a large piece of pitchblende from Yachymov, the Curies in 1898 first identified polonium and radium, and helium, already known to exist in the Sun. Uraninite is now universally recognized as an important source of uranium, a vitally important mineral because of its energy potential through nuclear fission. Pitchblende also provides radium, used as a source of radioactivity in various branches of industry and medicine, and other uranium compounds used as pigments for porcelain and in photography.

▲ Uraninite in pegmatite (ca. ×1). Montescheno, Italy.

▶ Uraninite crystal (ca. ×1). Czechoslovakia.

▼ Uraninite (var. pitchblende) (ca. ×1). Canada.

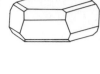

## 82  BRUCITE
OXIDES AND HYDROXIDES
Mg(OH)$_2$ (Magnesium hydroxide)

**System**  Hexagonal.

**Appearance**  Lamellar, hexagonal crystals. Colorless, green, blue or less frequently pinkish-yellow, altered to brown in manganese-bearing varieties. Foliated, scaly or finely granular masses. A white variety (nemalite) occurs in fibers up to 2 m (6.5 ft) long.

**Physical properties**  Soft (2.5), light, perfect cleavage into flexible but inelastic sheets or thin fibers. Transparent or translucent with pearly luster on cleavage faces, otherwise waxy or silky. White streak. Infusible. Dissolves easily in cold diluted acids, without effervescence.

**Environment**  A metamorphic mineral, found in many low-temperature serpentinized rocks, chlorite schists and talc schists. In contact metamorphism of dolomites it often forms at the expense of periclase (MgO) as an alteration product in the final or descending stages of metamorphism. Also found in low-temperature hydrothermal veins and in fissures in carbonate rocks where hot mineral solutions (thermal waters) were able to percolate through them. Occurs less frequently under very alkaline conditions.

**Occurrence**  Nemalite occurs mainly on fracture planes in serpentinized peridotites at Asbestos and Thetford Mine (Québec, Canada) and in the asbestos mines of Bajénov (USSR). Lamellar brucite has been found at Brewster, New York, in Texas and in Lancaster County, Pennsylvania (USA), where it is associated with hydromagnesite, and in the Shetland Islands (Scotland). In Italy, fine white brucite is found at Cogne and Emarese (Val d'Aosta) and in veins several centimeters long at Carro (Passo del Bracco, La Spezia). Pale-blue specimens occur in Val d'Astico (Vicenza) and in veins in Valle Serra (Trento). Disseminated, replacing periclase in calcite-brucite rocks at Predazzo (Trento) and Capo Teulada (Sardinia). Also noted in dolomite effusions of Vesuvius.

**Uses**  Used extensively for magnesia refractories or as a source of metallic magnesium and its salts.

▲ Lamellar brucite (ca. ×1). Valle Serra, Italy.

▲ Brucite (var. nemalite) (ca. ×1.5). Asbestos, Québec, Canada.

## 83   "BAUXITE"

OXIDES AND HYDROXIDES

$\alpha$-FeO(Oh) (Hydrous aluminum oxides)

2.3
2.7

**System**   None (an aggregate).

**Appearance**   Bauxite is a general term for a rock composed of hydrated aluminum oxides rarely found in distinct, separate crystals (monoclinic gibbsite Al(OH)$_3$, orthorhombic boehmite AlO(OH) and orthorhombic diaspore, AlO(OH)), and amorphous aluminous substances ("cliachite" Al$_2$O$_3$·H$_2$O). Occurs in earthy, clayey or pisolitic masses, white or yellowish when pure, reddish or brown when containing impurities of iron oxides or bitumen.

**Physical properties**   Soft (1–3), light, fragile with earthy fracture. Opaque. Insoluble and infusible. The component minerals have the following physical properties. Gibbsite is semi-hard (2.5–3), light, with perfect cleavage. Transparent with vitreous or pearly luster in crystals, semi-opaque when massive. When wet, smells strongly of clay or fresh earth. Almost infusible and insoluble in acids. Boehmite is semi-hard (2.5–3), very heavy, with perfect cleavage. Translucent with vitreous or pearly luster. Insoluble and infusible. Flakes and turns white when heated. Diaspore is hard (6.5–7), heavy, fragile, with good cleavage. Translucent with vitreous to pearly luster. Infusible and insoluble.

**Environment**   Residual sedimentary materials, formed by selective concentration of alumina after the dissolution of carbonate and silicate rocks in subtropical regions. Boehmite is also found in cavities in nepheline syenite pegmatites as a product of hydrothermal alteration of nepheline and feldspar. Gibbsite also occurs in low-grade metamorphic rocks (talc schists) associated with serpentine and magnetite. Diaspore is common with corundum and emery in metamorphic rocks and in bauxite deposits.

**Occurrence**   There are large bauxite deposits in Surinam, Jamaica, Ghana, Indonesia, the USSR, Yugoslavia, France, Hungary, the USA, and Italy. Gibbsite crystals have been found in talc schists at Slatoust (Urals, USSR), Richmond, Massachusetts (USA) and Caldas (Brazil). Boehmite is especially frequent in light-colored bauxite in Ariège (France), Arkansas (USA), Ayrshire (Scotland) and the Urals (USSR). Diaspore is common in many bauxite deposits in Italy and France, in emery at Chester, Massachusetts (USA) and on Naxos (Greece).

**Uses**   The main ore of aluminum, extracted by electrolytic treatment in a cryolite bath. Also used in the production of synthetic corundum (Al$_2$O$_3$) and aluminous refractories.

▲ ''Bauxite'' (ca. ×0.5). Matese, Italy.

▲ ''Bauxite'' (ca. ×0.5). San Giovanni Rotondo, Italy.

## 84 DIASPORE
OXIDES AND HYDROXIDES
AlO(OH) (Aluminum hydroxide)

**System** Orthorhombic.
**Appearance** Tabular or acicular crystals rare. Usually in foliated or stalactitic, pale-pink, gray or greenish aggregates. Often finely disseminated.
**Physical properties** Hard (6.5–7), heavy, very fragile, perfect cleavage. Transparent or translucent with vitreous to pearly luster on cleavage faces. White streak. Infusible and insoluble even in strong acids.
**Environment** Associated with boehmite and gibbsite as a constituent of some "bauxites." Stable under metamorphic conditions and sometimes associated with corundum and margarite in emery deposits in the amphibolite facies. Sometimes found in marble.
**Occurrence** Fine plates occur in emery at Chester, Massachusetts (USA) and on Naxos (Greece), and in chlorite schists at Mramorskoi (Urals, USSR). Small colorless crystals found in dolomite at Campolongo (Switzerland). Common in many bauxites in France, Hungary, USA, Canada and Italy (San Giovanni Rotondo, Foggia).
**Uses** A constituent of bauxites, an important source of aluminum.

## 85 GOETHITE
OXIDES AND HYDROXIDES
$\alpha$-FeOOH (Iron hydroxide)

**System** Orthorhombic.
**Appearance** Prismatic crystals with vertical striations rare. Usually in tabular, acicular, felted, botryoidal, stalactitic, oolitic or pisolitic brown aggregates. Also occurs in porous, amorphous, earthy masses. Pseudomorphs after pyrite common.
**Physical properties** Hard (5–5.5), heavy, has perfect cleavage, making it seem very soft. Feels greasy. Translucent in thin sheets, with silky luster, opaque when massive. Brownish-yellow streak. Dissolves slowly in hydrochloric acid. Almost infusible. Becomes magnetic with prolonged heating.
**Environment** An important constituent of "limonite" and ochres. "Limonite" is a rock formed under oxidizing conditions at the expense of iron-bearing minerals or precipitated directly in enclosed basins (bog iron). Very rare as a primary mineral in some pegmatites.
**Occurrence** Large deposits in Cuba, Alsace-Lorraine (France), Westphalia (Germany) and Labrador (Canada). Crystals found at Pribram (Czechoslovakia), in Cornwall (England), the USSR and near Pikes Peak, Colorado (USA). Also common in the hematite deposits of Lake Superior (USA).
**Uses** An important iron ore. Some ochres are used as pigments.

▲ Diaspore crystals (ca. ×1.5). Chester, Massachusetts.

▲ Radiating goethite balls (ca. ×1). Thuringia, Germany.

## 86 MANGANITE
OXIDES AND HYDROXIDES
MnO(OH) (Hydrated manganese oxide)

**System** Monoclinic.
**Appearance** Elongated, prismatic, black crystals, deeply striated and often grouped in bundles. Often twinned. Microcrystalline, concretionary and oolitic masses.
**Physical properties** Semi-hard (4), heavy, perfect cleavage. Semi-opaque with submetallic luster and reddish-brown tints, dark at the edges. Dark-brown streak. Infusible, soluble in hydrochloric acid giving off chlorine.

**Environment** In low-temperature hydrothermal veins associated with calcite and barite. In diagenitic sedimentary deposits associated with romanechite (psilomelane) and pyrolusite. Often partly altered to pyrolusite while keeping its own external form.
**Occurrence** Fine crystals 7–8 cm (2.8–3.2 in) long found at Ilfeld, Harz (DDR), Ilmenau (Thuringia) and Negaunee, Michigan (USA). Compact masses occur in many sedimentary manganese deposits, including Nikopol (Ukraine, USSR) and the island of San Pietro (Sardinia).
**Uses** A useful ore of manganese, only rarely found in economically exploitable amounts.

---

## 87 LEPIDOCROCITE
OXIDES AND HYDROXIDES
γ-FeO(OH) (Hydrous iron oxide)

**System** Orthorhombic.
**Appearance** Rarely as flattened, platy crystals often in rosette groupings, blood red in color. Frequently as fibrous brown aggregates.
**Physical properties** Hard (5–5.5), heavy, perfect cleavage. Transparent in thin sheets. Translucent with subadamantine luster. Bright orange streak. Dissolves slowly in hydrochloric acid and quickly in nitric acid. Infusible. Becomes strongly magnetic when heated.

**Environment** A constituent of many laterites, with the commoner goethite. Seems to form as an alteration product of pyrite.
**Occurrence** Beautiful crystals from Hesse and Westphalia in Germany (Eisenfeld, Herdorf, Rossbach). Also found in the USSR, as iron deposits at Lake Superior and Easton, Pennsylvania (USA), in India (Bihar) and Japan.
**Uses** An iron ore.

▲ Manganite crystals (ca. ×1). Ilfeld, Germany.

▲ Lepidocrocite (ca. ×1). Westerwald, Germany.

**"LIMONITE"**
OXIDES AND HYDROXIDES
(Hydrated iron oxides)

**System**    None (amorphous aggregate).

**Appearance**    "Limonite" is a field term used to describe a rock made up of a mixture of mainly amorphous mineral-like substances. The basic constituent is microcrystalline or cryptocrystalline goethite, together with lepidocrocite and amorphous iron hydroxides (some as gels). The term "limonite," like "ochre," is often applied generally to any iron hydroxide which cannot be defined compositionally or mineralogically without elaborate tests. Generally occurs as botryoidal, stalactitic, oolitic or pisolitic, colloform, earthy or porous masses, or in the form of a crust, yellowish brown when loose, blackish and iridescent when more coherent. Frequently occurs as a pseudomorph after iron minerals (pyrite, etc.) and organic remains (shells, etc.).

**Physical properties**    Very varied, depending on the constituents and the type of aggregation. Frequently semi-hard (5–5.5) and fragile, disintegrates easily. Translucent or semi-opaque with vitreous to earthy luster. Pale-brown streak (a characteristic which distinguishes it from hematite). Dissolves very slowly in acid and fuses with difficulty. If heated in air it alters to hematite and becomes magnetic.

**Environment**    A secondary mineral in surface oxidation zones of iron deposits or a residual mineral after the dissolution of carbonate and silicate rocks ("laterite") under tropical or subtropical conditions. Sizable masses form in the littoral zone of lake or marine basins as a result of the flocculation of iron hydroxides caused by electrolytic or bacterial action. Often occurs in oxidized lava crusts, in fractures of many intrusive rocks and in Alpine quartz veins.

**Occurrence**    There are large lateritic deposits in Cuba (Mayari and Moa districts), Venezuela, Brazil, Angola, Zaire, Canada and India. The Nurra iron-bearing deposit in Sardinia is made up of pisolitic limonite, and fine iridescent botryoidal masses are found on the island of Elba. There are oolitic sedimentary deposits in Alsace-Lorraine and Luxembourg.

**Uses**    A rather insignificant iron mineral, not much used in the modern iron and steel industry as it often contains impurities of phosphorus. Earthy varieties ("yellow ochre") are used as a coloring agent and in the manufacture of noncaustic modeling clay.

▲ Iridescent and botryoidal "limonite" (ca. ×0.5). Herdorf, Germany.

▲ "Limonite" (yellow ochre) (ca. ×1). Elba, Italy.

## 89 MAGNESITE

MgCO₃ (Magnesium carbonate)

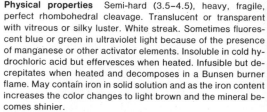

**System** Hexagonal.

**Appearance** Rarely in rhombohedral, white crystals. More often in compact and microcrystalline, white, yellow or gray, porcelain-like masses with conchoidal fracture. Granular and fibrous aggregates.

**Physical properties** Semi-hard (3.5–4.5), heavy, fragile, perfect rhombohedral cleavage. Translucent or transparent with vitreous or silky luster. White streak. Sometimes fluorescent blue or green in ultraviolet light because of the presence of manganese or other activator elements. Insoluble in cold hydrochloric acid but effervesces when heated. Infusible but decrepitates when heated and decomposes in a Bunsen burner flame. May contain iron in solid solution and as the iron content increases the color changes to light brown and the mineral becomes shinier.

**Environment** Formed by alternation of ultramafic rocks (peridotites and serpentinites) through the action of waters containing carbonic acid. As a diagenetic mineral replaces calcite and dolomite. Also occurs in hydrothermal veins, pegmatites and occasionally intrusive rocks (carbonatites and sagvandites).

**Occurrence** There are large deposits of compact magnesite in Styria (Austria), Manchuria (China), Silesia (Poland), the Urals (USSR), the Coast Range of California (USA), Madras (India) and the island of Euboea (Greece). Beautiful clear crystals, 6×10 cm (2.36×4 in), are found in pegmatite at Bom Jesus near Bahia (Brazil). In Italy, magnesite occurs in Piedmont at Monte Calvo (Val di Susa), Baldissero Canavese, Castellamonte and Prali (Val Germanasca), and is also found in Tuscany, on Elba and in Van Solda (Bolzano). Very beautiful crystals occur in Alpine fissures (St. Gotthard, Switzerland; Val di Vizze and Val d'Ultimo, Alto Adige, Italy; the Zillertal, Austria), and at Brosso and Traversella (Piedmont, Italy) and Kranbath (Austria).

**Uses** An important ore of magnesium and its salts. Used in the manufacture of basic refractories capable of withstanding extremely high temperatures and for special types of cement and powders used in the paper, rubber and pharmaceutical industries.

▲ Rhombohedral magnesite crystals (ca. ×1.5). Styria, Austria.

▲ Ferroan-magnesite crystals (ca. ×1). Traversella, Italy.

## 90 SMITHSONITE
NITRATES, CARBONATES, BORATES
$ZnCO_3$ (Zinc carbonate)

**System** Hexagonal.

**Appearance** Rhombohedral or scalenohedral crystals usually with curved faces, uncommon. More often found in botryoidal, mammillated, reniform or stalactitic aggregates, often porous and concretionary. A typical allochromatic mineral, especially the mammillated varieties: white when pure, pale blue or green when copper impurities (probably malachite) are present; turning bright yellow with cadmium, pink or violet with cobalt or manganese and brown with tiny particles of iron hydroxides.

**Physical properties** Hard (5.5), heavy, fragile with perfect rhombohedral cleavage (though this is rarely seen, as in masses conchoidal fracture is more common). Translucent with vitreous to greasy luster. Streak always white. Many varieties show pinkish fluorescence if exposed to ultraviolet light. Soluble in cold concentrated hydrochloric acid, giving off carbon dioxide. Infusible.

**Environment** Typical sedimentary precipitate produced by the action of waters rich in zinc sulfate on carbonate rocks. Characteristically found in the oxidation zone of sulfide deposits (zinc, lead and copper). Generally associated with hemimorphite, cerussite, malachite, anglesite, pyromorphite, etc.

**Occurrence** Small, colorless, rhombohedral crystals found at Tsumeb (Namibia) and Broken Hill (Zambia). Blue-green botryoidal masses at Magdalena and Kelly, New Mexico (USA), Santander (Spain) and Laurium (Greece). Stalactitic yellow masses at Marion, Arkansas (USA) and in Sardinia (mainly at Masua, near Iglesias, and at Monteponi). Masses and crusts of various colors in Italy in the Bergamo deposits (Gorno and Valle del Riso) and the Alps (Raibl, Udine). Sizable masses worked commercially at Leadville, Colorado (USA), in the USSR (southern Kazakhstan) and Turkey.

**Uses** Zinc is extracted from this mineral. When it shows good coloring or attractive banding it is polished and used as an ornamental stone. Of interest to collectors and scientists concerned with the study of mineral deposits.

▲ Concretionary smithsonite (ca. ×1). Sardinia, Italy.

▲ Botryoidal smithsonite (ca. ×1). Sardinia, Italy.

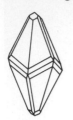

## 91 SIDERITE

NITRATES, CARBONATES, BORATES
$FeCO_3$ (Iron carbonate)

**System** Hexagonal.

**Appearance** Rhombohedral crystals with curved, striated faces. Sometimes with tabular habit and in saddle-shaped aggregates. Botryoidal, compact, oolitic, concretionary masses common, varying in color from pale yellow to dark brown or almost black when a large amount of manganese is present.

**Physical properties** Semi-hard (3.5–4), heavy, fragile, perfect rhombohedral cleavage. Transparent or translucent with bright vitreous luster. White streak. Alters readily on the surface, becoming coated by brown goethite. Practically infusible. Decomposes at a moderate temperature, turning brown and becoming magnetic. Insoluble in cold hydrochloric acid, but dissolves in hot HCl with distinct effervescence.

3.7
3.9

**Environment** Very common in medium- to low-temperature hydrothermal veins, associated with fluorite, barite, galena, sphalerite, etc., and occasionally with cryolite as well. Deposits of oolitic sedimentary siderite are even more common—formed in an environment of continental waters with low oxygen content, associated with clay and carbonaceous material.

**Occurrence** Fine crystals are found in Styria (Austria), at Redruth and Camborne, Cornwall (England), in Brazil, in carbonatites at Mont-Saint-Hilaire (Canada) and cryolite veins in Ivigtut (Greenland) and finally at Brosso and Traversella (Piedmont, Italy). There are large sedimentary deposits in England, Romania, Germany, Lorraine (France), Lancaster County, Pennsylvania (USA), Scotland and Nurra (Sardinia), where the siderite is clearly oolitic. Large masses found in veins with sphalerite and pyrite at Roxbury, Connecticut (USA). The large amounts of manganese-bearing siderite in the Triassic formation of Servino in the Pre-Alps and the Alps near Bergamo and Brescia in Italy are probably also sedimentary in origin. The deposits in Spain (Bilbao), Algeria, Tunisia, central Germany, Austria (Hüttenberg and Herzberg) and the DDR (Harz and Freiberg) occur in veins. Siderite veins are found throughout the crystalline basement of the Orobian Alps.

**Uses** An important iron mineral (containing about 48 percent Fe) because it is free of sulfur and phosphorus and is sometimes rich in manganese. It is also of interest to collectors and scientists concerned with the study of ore deposits.

▲ Rhombohedral siderite crystals (ca. ×1). Isère, France.

▲ Mamillary siderite (ca. ×1). Baia Sprie, Romania.

3.3
3.6

## 92 RHODOCHROSITE

NITRATES, CARBONATES, BORATES
$MnCO_3$ (Manganese carbonate)

**System**  Hexagonal.

**Appearance**  Rhombohedral or scalenohedral crystals, pink in color, rare. Granular, concretionary, mammillated, reniform or stalactic masses common.

**Physical properties**  Semi-hard (3.5–4.5), heavy, fragile with perfect rhombohedral cleavage. Translucent with vitreous to pearly luster. Pale, slightly pinkish streak. When exposed to air becomes covered with a dark film through alteration (oxidation) of manganese. Infusible. Blackens gradually when heated. Insoluble in cold hydrochloric acid (unlike pink calcite) but becomes soluble when heated (unlike rhodonite).

**Environment**  In medium-temperature hydrothermal veins associated with copper, silver and lead sulfides and other manganese minerals. Much rarer in pegmatites. Fairly common as a sedimentary mineral in the oxidation zone of sulfide deposits or as a precipitate from hydrothermal solutions. Stable with rhodonite, tephroite and manganese oxides under a wide range of metamorphic conditions, with complete dissociation in the amphibolite facies.

**Occurrence**  Splendid rhombohedral crystals measuring up to 5 cm (2 in) on edge are found in many localities in Colorado (USA) and in silver-bearing veins in Transylvania (Romania) and Freiberg (DDR) and in groups of perfect blood-red scalenohedra near Hotazel (South Africa). There are sizable crystals at Butte, Montana (USA), Huelva (Spain), Ariège (France) and Chiatura (USSR). Radiate concretionary aggregates, often with fine color banding, come from Catamarca (Argentina), Magdalena (Sonora, Mexico), Kapnik (Romania) and various places in Germany. In Italy found at St. Marcel (Val d'Aosta), Ulzio (Piedmont), Val Malenco (Sondrio), Elba (Livorno), in the Gambatesa deposit (Sestri Levante) and associated with manganese oxides on the island of San Pietro (Sardinia).

**Uses**  An ore of manganese when available in large enough masses. Concretionary masses are polished and used mainly as an ornamental stone. Sometimes rhodochrosite is used as a semiprecious gem, faceted or cut as a cabochon. Also of interest to scientists and collectors.

▲ Rhombohedral rhodochrosite crystals (ca. ×1). Colorado.

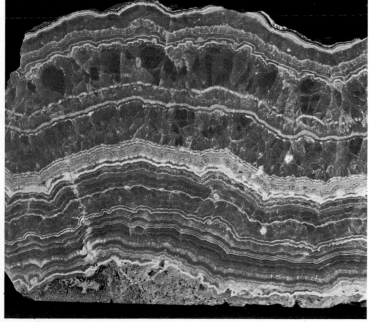

▲ Banded rhodochrosite (ca. ×1). Argentina.

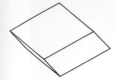

## 93 CALCITE
NITRATES, CARBONATES, BORATES
CaCO₃ (Calcium carbonate)

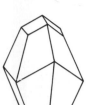

**System** Hexagonal.

**Appearance** Rhombohedral, scalenohedral or prismatic crystals, often intergrown or twinned. Variable color (colorless, white, pink, green, yellow, etc.). Occurs in masses: compact, microcrystalline (limestones), saccharoidal (marbles), fibrous (alabasters), concretionary (stalactites and stalagmites), pulverulent, oolitic, etc.

2.71

**Physical properties** Semi-hard (3), light, with perfect rhombohedral cleavage. Transparent with vitreous to iridescent pearly luster depending on direction (highly birefringent), translucent or opaque. White streak. Some varieties are fluorescent in ultraviolet light (red, yellow, pink or blue) and often visibly thermoluminescent. Soluble in cold, dilute hydrochloric acid, with brisk effervescence. Infusible but dissociates at a high temperature in a Bunsen burner flame, giving lime (CaO) and carbon dioxide. May contain various cations in solid solution—such as iron, manganese, zinc, strontium, magnesium—giving the corresponding varieties ferroan calcite, magnanoan calcite, strontian calcite and magnesian calcite, intermediate between calcite and the corresponding carbonates. Polymorphs of CaCO₃ include aragonite (orthorhombic), vaterite (hexagonal) and possibly elaterite (a high temperature and pressure phase). When heated these phases alter to the more stable calcite.

**Environment** A typical sedimentary mineral formed by chemical precipitation through the evaporation of solutions rich in calcium bicarbonate (e.g., stalactites and travertines) or by extraction through the action of marine and freshwater organisms (e.g., organogenetic limestones). Remains stable under metamorphic conditions up to the highest grades, simply recrystallizing and increasing in grain size as long as the pressure of carbon dioxide remains high (marbles). If the pressure is reduced it dissociates, resulting in complex calcium silicates (e.g., in contact metamorphism). Primary calcites of igneous origin (carbonatites) are rare, though calcite is certainly present in magma since late volcanic solutions may deposit it in vacuoles in lava. Also found in low-temperature hydrothermal veins associated with sulfides. Calcareous rocks constitute 4 percent, by weight, of the earth's crust and cover 40 percent of its surface.

**Occurrence** Large, very clear, rhombohedra, formerly used to make polarizing prisms, occur in basalt cavities in Iceland (Iceland spar), the Harz (Germany), the Erzgebirge in Bohemia and Saxony (Czechoslovakia and DDR) and in Colorado (USA). Pointed scalenohedra, sometimes pink or green, are found in Cumberland (England), and golden-yellow crystals up to 1 m (3.3 ft) long in Joplin, Missouri, and Pitcher, Oklahoma (USA). Other localities famous for their beautiful crystals are Rossie, New York, Bergen Hill and Paterson, New Jersey (USA), Kapnik (Hungary), Freiberg (DDR), St. Andreasberg and Bräunsdorf (Germany), Guanajuanato (Mexico), Rhisnes (Belgium) and Kara Dag (Crimea, USSR). Italy is famous for translucent rhombohedra from Porretta (Moxena), Sarrabus (Sardinia), the basalts of the Lessini mountains (Vicenza) and

▲ Rhombohedral calcite crystals (ca. ×1.5). Ontario, Canada.

▶ Scalenohedral calcite crystal (ca. ×0.5). Val di Fassa, Italy.

▼ Calcite scalenohedron modified by rhombohedron (ca. ×0.5). Iceland.

large scalenohedra from Passo Molignon (Val di Fassa, Trento). Onyx is a colored variety of microcrystalline calcite from Mexico and Pakistan. Collectors are familiar with the calcite crystals containing dendritic copper from the Keweenaw Peninsula, Michigan (USA) and the crystals covered with sand from Fontainebleau (France).

**Uses**  Clear crystals were formerly used to make polarizing prisms (the Nicol prism) for petrological microscopes. Compact masses have a variety of uses, especially in building (cement, lime, structural and ornamental stone), metallurgy (as flux and slag), the manufacture of fertilizers (for soil enrichment) and the chemical industry (in caustic soda, calcium chloride, liquid carbon dioxide, etc.). Other uses include marble for sculpture, lithographic stones for printing and loose earthy masses as powders for polishing and as fillers in the rubber and paint industries.

▲ Rhombohedral calcite crystal (ca. ×1). Northern Apennines, Italy.

▲ Calcite crystals (ca. ×1). Sardinia, Italy.

▲ Calcite crystals (ca. ×1.5). Cumberland, England.

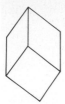

## 94 DOLOMITE
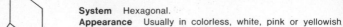
NITRATES, CARBONATES, BORATES
CaMg(CO₃)₂ (Calcium magnesium carbonate)

**System**  Hexagonal.

**Appearance**  Usually in colorless, white, pink or yellowish rhombohedral crystals. Aggregates of crystals with curved "saddle" faces frequent. Compact, saccharoidal, sometimes cavernous (dolomite) masses.

**Physical properties**  Semi-hard (3.5–4), not very heavy, fragile, perfect rhombohedral cleavage. Transparent or translucent with vitreous or sometimes pearly luster. White streak. Infusible, decomposes like all carbonates. In fragments dissolves with difficulty in cold, dilute hydrochloric acid, but when heated dissolves quickly with effervescence. The magnesium may be partially or totally replaced by other cations, as dolomite forms a solid solution series with ankerite Ca(Fe,Mg,Mn)(CO₃)₂ and the very rare kutnohorite Ca(Mn,Mg,Fe)(CO₃)₂.

2.85
2.95

**Environment**  A basic constituent of sedimentary carbonate rocks (dolomites and dolomitic limestones) formed under diagenetic conditions by the action of sea water on calcareous mud or by organogenetic formation. Also found in low-temperature hydrothermal veins and in metamorphic environments, either as dolomitic marbles or associated with talc schists and serpentinite (metamorphosed ultramafic rock).

**Occurrence**  The finest crystals come from Brosso and Traversella (Piedmont, Italy) where ankerite specimens have also been found, the Binnental (Switzerland), the Freiberg and Schneeberg mines (DDR), Cornwall (England), Joplin (Missouri) and various other places in the USA (Iowa, New York, Vermont, New Jersey, Michigan), St. Eustache (Québec, Canada), Navarra (Spain), Bahia (Brazil) and Guanajuanato (Mexico). Beautiful ankerite crystals also occur in porphyry cavities at Cuasso al Monte (Varese, Italy) and in the Binnental (Switzerland). Pink, massive kutnohorite is found in the Kutnohora locality (Bohemia, Czechoslovakia), at Franklin, New Jersey (USA), Providencia (Mexico) and Nagano (Japan).

**Uses**  Dolomitic rock has many uses in industry: in building (structural and ornamental stone, for special cements), metallurgy (basic refractories, for the extraction of metallic magnesium, fluxes and slag in the iron and steel industry), and the chemical industry (in preparing magnesium salts). It is also of particular interest to scientists and collectors.

▲ Dolomite crystals (ca. ×1). Traversella, Italy.

▲ "Saddle-shaped" dolomite crystals (ca. ×1). Val di Fassa, Italy.

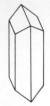

## 95 ARAGONITE
NITRATES, CARBONATES, BORATES
CaCO₃ (Calcium carbonate)

2.95

**System** Orthorhombic.

**Appearance** Small, elongated, prismatic crystals. Often in radiating groups or polysynthetically twinned to resemble hexagonal prisms in which three individuals can be identified, since the base faces are striated in three different directions and a v-shaped reentrant angle occurs where each individual crystal meets its adjacent neighbor. Concretionary, pisolitic, stalactitic, fibrous and radiating, or coralloid ( *flos ferri*, "flowers of iron") masses. Colorless white, yellow, reddish and various other colors. Often forms the skeleton of a number of marine organisms, living or recently fossilized.

**Physical properties** Semi-hard (3.5–4), heavy, fragile, indistinct prismatic cleavage. Transparent or translucent with vitreous luster. White streak. Dissolves readily in cold dilute hydrochloric acid with strong effervescence. When heated, converts to calcite at medium temperature, decrepitating loudly without fusing. Some varieties are fluorescent blue, pink and yellow under ultraviolet light. X-rays color it yellow or brown. Powder boiled in a solution of cobalt nitrate turns purple whereas calcite stays white.

**Environment** A high-pressure polymorph of calcite, only stable in metamorphic rocks formed at high pressure and low temperature (glaucophane schists). However, forms easily in a sedimentary environment, under metastable conditions, helped by biological action or chemical precipitation from solutions lightly charged with ions like strontium, lead and zinc. Found in deposits of hot springs (travertine), alteration zones of sulfide deposits, biogenetic deposits (lumachelles), and evaporite deposits (sulfur-bearing gypsum formation).

**Occurrence** Beautiful pseudo-hexagonal, gray or reddish sixlings come from Molina de Aragon (Spain) and Fort Collins, Colorado (USA). Fine orthorhombic crystals from the Sicilian sulfur mines (Agrigento), Monte Somma (Vesuvius, Italy), Bastennes (France) and Alston Moor (Cumberland, England). *Flos ferri* is typical of the siderite mines of Styria (Austria) but is also found in Arizona, New Mexico (USA) and Mexico. Some onyx and alabaster varieties are formed from aragonite (Mexico, Pakistan). In Italy, also occurs in cavities in lava on Etna and in the Alpine lithoclases of Val Malenco (Sondrio) and Monte Ramazzo (Liguria).

**Uses** Apart from the alabaster varieties, which are cut into semitransparent slabs as ornamental stones, the mineral is of no practical use and is of interest only to scientists and collectors.

▲ Cyclic twin of aragonite (ca. ×1). Molina de Aragon, Spain:

▶ Acicular aragonite crystals (ca. ×1). Vesuvius, Italy.

▼ Coralloid aragonite (ca. ×1). Styria, Austria.

## 96 STRONTIANITE
NITRATES, CARBONATES, BORATES
SrCO₃ (Strontium carbonate)

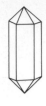

**System** Orthorhombic.
**Appearance** Needle-like (acicular) crystals, generally in bundles or sheaves. Colorless or white, sometimes pink, gray or green. Microgranular, compact or fibrous masses.
**Physical properties** Semi-hard (3.5–4), heavy, fragile, good prismatic cleavage. Transparent or translucent with glassy luster. White streak. May be fluorescent blue under ultraviolet light and even thermoluminescent. Soluble in dilute hydrochloric acid. Fuses slowly, turning the flame bright red.
**Environment** Typical of low-temperature hydrothermal veins, associated with calcite, celestite and sulfides. Also found in geodes and concretionary masses in clays and limestones.
**Occurrence** Large economic deposits in Germany (Westphalia, Harz), Spain, Mexico and the USA (Strontium Hills, California). Fine crystals at Strontian (Argyll, Scotland), Leogang (Austria), Braundsdorf (DDR). In Italy, small amounts in the barite mines of Varesotto (Vignazza).
**Uses** An ore for strontium, used in making fireworks (produces a purplish-red flame), in the sugar industry and for special types of glass.

---

## 97 WITHERITE
NITRATES, CARBONATES, BORATES
BaCO₃ (Barium carbonate)

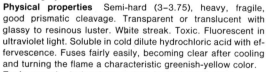

**System** Orthorhombic.
**Appearance** Trilling crystals resembling hexagonal pyramids, often scepter-shaped or globular with striations parallel to the base. Fibrous, mammillated and granular; colorless, white or gray masses.
**Physical properties** Semi-hard (3–3.75), heavy, fragile, good prismatic cleavage. Transparent or translucent with glassy to resinous luster. White streak. Toxic. Fluorescent in ultraviolet light. Soluble in cold dilute hydrochloric acid with effervescence. Fuses fairly easily, becoming clear after cooling and turning the flame a characteristic greenish-yellow color.
**Environment** In low-temperature hydrothermal veins, associated with barite and galena.
**Occurrence** Beautiful crystals at Alston Moor (Cumberland) and Hexham (Northumberland) (England), and Rosiclaire, Illinois (USA). Compact masses occur at El Portal, California (USA) and in Siberia (USSR). In Italy associated with barite in the Donori vein (Sardinia).
**Uses** A barium ore, when available in large enough amounts, and also used in special types of glass.

▲ Strontianite crystals (ca. ×1.5). Westphalia, Germany.

▲ Witherite crystals (ca. ×1.5). Northumberland, England.

## 98 CERUSSITE
NITRATES, CARBONATES, BORATES
PbCO$_3$ (Lead carbonate)

**System** Orthorhombic.
**Appearance** Colorless or white crystals with grayish tints, elongated and generally twinned to form a reticulated network with 60° angles. Stubby, tabular crystals in star- or heart-shaped twins. Rare compact, fibrous, stalactitic or even earthy masses.
**Physical properties** Semi-hard (3–3.5), very heavy, very fragile, prismatic cleavage. Transparent or translucent with adamantine luster. White streak. Bright blue-green fluorescence in ultraviolet light. When heated turns brown, then fuses easily. Insoluble in hydrochloric acid, but dissolves in nitric acid, with strong effervescence, distinguishing it from anglesite (PbSO$_4$).
**Environment** Occurs in the oxidation zone of lead deposits, produced by chemical alteration of galena through the action of waters rich in carbonic acid. Associated with primary minerals like galena and sphalerite and secondary minerals like anglesite, smithsonite, pyromorphite and goethite.
**Occurrence** Very large, clear crystals at Tsumeb (Namibia) and Dona Ana, New Mexico (USA). Splendid crystals found at Broken Hill (Australia); Rhodesia; Phoenixville, Pennsylvania, and Leadville, Colorado (USA); Nassau and Bohemia (Germany and Czechoslovakia); Tunisia; Siberia (USSR); Mezica (Yugoslavia) and Mechernich (Germany). In Italy commonly found as small, very clear crystals in close bundles at Montevecchio and Monteponi (Sardinia). Sizable deposits are worked in the USSR (southern Kazakhstan, Altai).
**Uses** An ore for lead and to a lesser extent silver. Important for the study of deposits and of interest to collectors.

▲ Cerussite crystals (ca. ×1). Unknown locality.

▲ Cerussite crystals (ca. ×1). Italy.

▲ Cerussite crystals (ca. ×1). Montevecchio, Sardinia, Italy.

## 99 AZURITE
NITRATES, CARBONATES, BORATES
$Cu_3(CO_3)_2(OH)_2$ (Hydrous copper carbonate)

**System** Monoclinic.

**Appearance** Elongated or tabular prismatic crystals, often highly modified. Striated. Often intergrown or grouped in radiating aggregates. Also frequently occurs as a film or reniform, earthy, granular or concretionary masses. Pseudomorphs of azurite after other minerals also common.

**Physical properties** Semi-hard (3.5–4), heavy, fragile, fairly good cleavage. Transparent or translucent with vitreous luster tending to adamantine. Pale-blue streak. Soluble in ammonia and effervesces in dilute acids. Fuses easily, first turning black through water loss when heated.

**Environment** A secondary copper mineral in sulfide deposits associated with carbonate rocks. Forms at lower temperatures than malachite, which often replaces it pseudomorphically through hydration. Also found in sandstones as an impregnation caused by carbonatic water coming into contact with water rich in copper sulfates. Normally associated with malachite, "limonite," calcite, chalcocite, chrysocolla and other secondary copper minerals.

**Occurrence** Masses are rare. The finest crystals are from Chessy, near Lyons (France), Tsumeb (Namibia), Calabona and Rosas (Sardinia), Laurium (Greece) and many localities in Arizona (Bisbee, Morenci) (USA). Magnificent groups of crystals, sometimes almost completely replaced by malachite, have been found at Broken Hill (New South Wales, Australia). Also found in Chile, Australia, Siberia, Persia, Romania and Mexico.

**Uses** A valuable ornamental stone. Small pieces are cut as cabochons. Rarely faceted into beautiful blue gemstones that are rather fragile and easily scratched. In the past it was crushed to make a pigment (mountain blue or Armenia stone) that eventually becomes greenish as it alters to malachite. Also a copper ore of secondary importance and of interest to scientists and collectors.

▲ Azurite crystals (ca. ×1). Chessy, France.

▲ Prismatic azurite crystals (ca. ×1). Calabona, Sardinia.

## 100 MALACHITE

$Cu_2(CO_3)(OH)_2$ (Hydrous copper carbonate)

**System** Monoclinic.

**Appearance** Acicular crystals common. Sometimes in fibrous, radiating aggregates. Commonly occurs as a green film on other copper minerals and as botryoidal or reniform masses with concretionary, banded structure and emerald-green color. Pseudomorphs of malachite after azurite are also common.

**Physical properties** Semi-hard (3.5–4), heavy, fragile, good cleavage. Semi-opaque or translucent with vitreous to silky luster. Light green streak. Soluble in concentrated acids with effervescence. In hydrochloric acid it turns the solution green. When heated thoroughly loses water and turns black, then fuses fairly easily, coloring the flame green (copper).

**Environment** Typically found in the oxidation zone of copper deposits, produced by reaction of sulfides with carbonate gangue. Sometimes occurs in large masses with a core of azurite. Also found disseminated in sandstones deposited by meteoric waters.

**Occurrence** During the last century enormous blocks weighing many kilos were extracted from iron-rich clay in mines in the Urals (Mednoroudianskoy and Goumechevskoy, USSR). Large masses are still found in the Urals, Katanga (Zaire), Zambia, Chile, New South Wales and southern Australia, the Tintic district, Utah, and Bisbee, Arizona (USA). Beautiful crystals occur at Betzdorf (Germany), Chessy (France), Rezbanya (Romania), Tsumeb (Namibia) and above all in mines in Arizona (Copper Queen mine near Bisbee, Morenci). Fairly rare in Italy, with compact masses on Elba and acicular crystals in Sardinia (Arenas, Sa Duchessa, Campo Posano).

**Uses** A very valuable decorative stone used in polished slabs for small tables, boxes and ornaments. The magnificent columns of St. Isaac's Cathedral and the hall facings of the Winter Palace (Leningrad, USSR) are made of malachite. In the past it was crushed and used as an inorganic pigment (mountain green). It is also a copper ore and of interest to scientists and collectors.

▲ Acicular malachite crystals (ca. ×1). Herdorf, Germany.

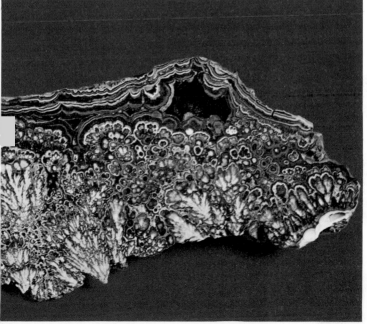

▲ Banded malachite (ca. ×1). Zambia.

## 101 ROSASITE

NITRATES, CARBONATES, BORATES
(Cu,Zn)$_2$(CO$_3$)(OH)$_2$ (Hydrous copper zinc carbonate)

**System** Monoclinic.
**Appearance** Groups of acicular radiating crystals (spherules or balls) encrusting "limonite" common. Color green to bluish-green.
**Physical properties** Semi-hard (4.5), heavy, with right-angle cleavage. Gives off water when heated in closed tube, fusible and soluble in most acids.
**Environment** A secondary mineral found in the oxidation zone of zinc-copper-lead deposits associated with other copper carbonates.
**Occurrence** Occurs in fine mammillary crusts at Mapimi, Durango (Mexico), Kelly, New Mexico (USA) and Tsumeb (Namibia).
**Uses** Minor ore of copper.

---

## 102 HYDROZINCITE

NITRATES, CARBONATES, BORATES
Zn$_5$(CO$_3$)$_2$(OH)$_6$ (Hydrous zinc carbonate)

**System** Monoclinic.
**Appearance** Hardly ever found as crystals. Earthy, compact, white or gray masses, or colloform crusts with concentric, fibrous structures, or stalactites.
**Physical properties** Soft (2–2.5), heavy, perfect cleavage. Translucent with glassy luster. White streak. When heated decomposes into yellow zincite. Soluble in hydrochloric acid. Characteristic pale-blue fluorescence in ultraviolet light.
**Environment** Secondary alteration of zinc minerals under oxidizing conditions. Found mainly in deposits associated with smithsonite.
**Occurrence** Large masses found at Santander (Spain), in Carinthia (Austria), Algeria, and Nevada and New Mexico (USA). In Italy common in the Gorno deposit (Val Seriana, Bergamo) and also found at Raibl (Udine), in Val Parina and Val del Riso (Bergamo) and on Sardinia.
**Uses** An important ore for zinc, when available in economic quantities.

▲ Green balls of rosasite with calcite on limonite (ca. ×1). Mapimi, Mexico.

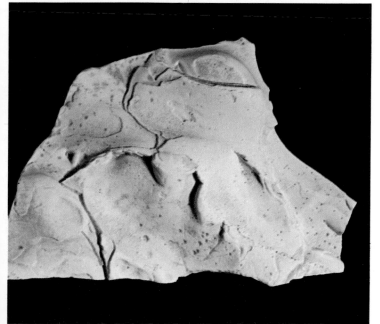

▲ Hydrozincite crust (ca. ×1). Monteponi, Sardinia, Italy.

## 103 AURICHALCITE
NITRATES, CARBONATES, BORATES
$(Zn,Cu)_5(CO_3)_2(OH)_6$ (Hydrous zinc copper carbonate)

**System** Monoclinic.
**Appearance** Very rarely in good isolated crystals. Generally in pearly, light-blue crusts composed of mammillary balls of small, slender, acicular crystals on a matrix of "limonite."
**Physical properties** Very soft (2–2.5), heavy, perfect cleavage into thin, flexible, translucent, pearly plates. Very light blue-green streak. Soluble in dilute hydrochloric acid with effervescence. When heated turns the flame green (copper) but does not fuse.
**Environment** Occurs in the oxidation zone of zinc and copper sulfide deposits, where it is deposited by circulating carbonate-rich solutions.
**Occurrence** Fine specimens from Laurium (Greece), Chessy (France), Tsumeb (Namibia), Leadhills (Scotland), Romania, Mapimi (Mexico) and Stockton, Utah and Kelly, New Mexico (USA). Found at Monteponi and Rosas (Sardinia) and Oneta (Bergamo) (Italy).
**Uses** Of interest to scientists and collectors.

## 104 PHOSGENITE
NITRATES, CARBONATES, BORATES
$Pb_2(CO_3)Cl_2$ (Lead carbonate chloride)

**System** Tetragonal.
**Appearance** Prismatic, stubby or tabular crystals, striated lengthwise, often with pointed terminations. Colorless to amber yellow.
**Physical properties** Soft (2.75–3), very heavy, good cleavage. Transparent or translucent with greasy subadamantine luster. White streak. Soluble in cold distilled water and briskly effervescent in dilute nitric acid. Fuses readily. Often has beautiful yellow fluorescence in ultraviolet light.
**Environment** Alteration product of galena (lead slags) in contact with sea water or brackish continental waters. Also forms under hydrothermal conditions.
**Occurrence** Splendid and enormous crystals weighing up to 45 kg (99 lb) are found at Monteponi and Montevecchio (Sardinia) in a vein with silver-bearing galena. Also found at Laurium (Greece), Tsumeb (Namibia), Matlock (England), Tainowitz (Poland) and Custer County, Colorado (USA).
**Uses** Of interest to scientists and collectors.

▲ Aurichalcite (ca. ×1). Durango, Mexico.

▲ Phosgenite crystal (ca. ×0.5). Monteponi, Sardinia, Italy.

▲ Phosgenite crystal (ca. ×1). Monteponi, Sardinia, Italy.

## 105 TRONA

NITRATES, CARBONATES, BORATES
Na₃(CO₃)(HCO₃)·2H₂O (Hydrated sodium carbonate)

**System**  Monoclinic.

**Appearance**  Prismatic, tabular, colorless crystals. Earthy crusts or white or yellowish-white efflorescences.

**Physical properties**  Soft to semi-hard (2.5–3), light, fragile with perfect cleavage. Transparent to translucent with dull, earthy luster. White streak. Soluble in hydrochloric acid. When heated it loses water and then carbon dioxide, fusing only at a high temperature. Does not alter with exposure to air.

**Environment**  Typical mineral of evaporite deposits in arid regions (salt lakes).

2.17

**Occurrence**  Large economic deposits at Lake Magadi (Kenya) and Wyoming (USA). Also in Egypt, Tripolitania, Libya, Venezuela, in the deserts of central Asia and various places in Africa, associated with other evaporite minerals. Fine crystals at Borax, California, and Soda Lake, Nevada (USA). Has been found in cavities in lava on Vesuvius (Italy).

**Uses**  An ore of sodium.

## 106 HYDROMAGNESITE

NITRATES, CARBONATES, BORATES
Mg₅(CO₃)₄(OH)₂·4H₂O (Hydrated magnesium carbonate)

**System**  Monoclinic.

**Appearance**  Small, white, acicular crystals rare. Generally in small, radiating mammillary balls or earthy crusts.

**Physical properties**  Semi-hard (3.5), light, fragile, with perfect cleavage. Transparent with glassy to silky luster. Colorless or white streak. Soluble with effervescence in dilute acids. When heated loses water and carbon dioxide and turns white but does not fuse.

**Environment**  Very low-temperature hydrothermal mineral, occurring in fractures in serpentinites and vacuoles of mafic lava. Also found as an alteration product of brucite in contact dolomitic marbles.

2.24

**Occurrence**  In serpentine in the Alps (Kraubath, Austria; Val Malenco, Val d'Aosta, Italy) and in Moravia, Serbia, Cuba, California, Nevada, New York (USA). Found in dolomitic blocks from Monte Somma (Campania) and tufa from Ariccia (Roma) and in the contact metamorphic aureole at Predazzo (Trent) (Italy).

**Uses**  Of interest to scientists and collectors.

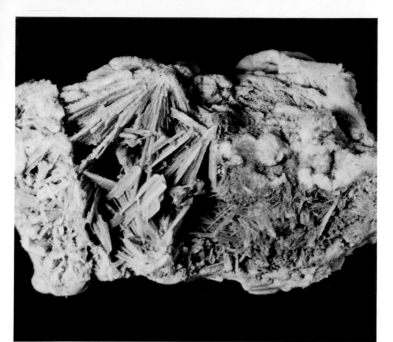

▲ Trona crystals (ca. ×1.5). Egypt.

▲ Hydromagnesite balls (ca. ×1.5). Emarese, Italy.

## 107 ARTINITE

NITRATES, CARBONATES, BORATES
$Mg_2(CO_3)(OH)_2 \cdot 3H_2O$ (Hydrated magnesium carbonate)

**System** Monoclinic.
**Appearance** Slender white or gray needles. Generally in round ball-like aggregates or crusts and small veins.
**Physical properties** Soft (2), light, fragile, with perfect cleavage. Transparent with silky luster. White streak. Dissolves easily in cold dilute acids, with effervescence. When heated turns white, losing water and carbon dioxide, but does not fuse.
**Environment** A very low-temperature hydrothermal mineral, in fractures of serpentinites, often associated with hydromagnesite and brucite.
**Occurrence** Classic specimens are found in Italy, in Val Malenco (Val Brutta, Val Lanterna, Torre S. Maria) and many other areas in the Italian Alps (Emarese, Cogne, Lanzo, Baldissero), and from Monte Ramazzo (Sestri Levante). Also found in Austria (Kraubath), Yugoslavia and the USA (Hoboken, New Jersey, and Staten Island, New York), always in serpentinized peridotites.
**Uses** Of interest to scientists and collectors.

## 108 KERNITE

NITRATES, CARBONATES, BORATES
$Na_2B_4O_7 \cdot 4H_2O$ (Hydrated sodium borate)

**System** Monoclinic.
**Appearance** Slightly elongated, occasionally wedge-shape heavily striated crystals. Also in cleavable masses.
**Physical properties** Soft (2.5), very light, with prismatic cleavage (perfect one direction). Streak white. Opaque to translucent with vitreous to satiny luster. Soluble in cold water and acids. Fuses easily swelling to a cauliflower mass.
**Environment** Formed in playa lake basins supplied with boron from thermal springs emanating from underlying volcanic rock.
**Occurrence** The large crystals and cleavage fragments are found in Kern County, California, where crystals to 8.75 m (28.66 ft) have been recorded.
**Uses** A principal ore of boron. Used in the manufacture of heat-resistant glass and soaps, enamel and fertilizers.

▲ Radiating artinite balls (ca. ×1.5). Val Malenco, Italy.

▲ Cleavage prism of kernite (ca. ×1). Kern County, California.

## 109 BORAX

NITRATES, CARBONATES, BORATES
Na$_2$B$_4$O$_7$·1OH$_2$O (Hydrated sodium borate)

**System** Monoclinic.
**Appearance** Prismatic, stubby crystals, rather flat and striated vertically. Colorless, white, yellowish or bluish. Earthy masses. Crusts.
**Physical properties** Soft (2–2.5), very light, perfect cleavage, conchoidal fracture. Transparent to translucent with greasy luster. Opaque white streak. Sweetish taste that sets the teeth on edge. Soluble in water and fuses easily into transparent beads, coloring the flame yellow (sodium). Desiccates, losing water and turning into a white powder (kernite).

1.74

**Environment** Occurs in the waters of saline lakes and in salt pans in desert regions. In both cases derived from solar evaporation of saline waters.
**Occurrence** Large amounts on the shores of the lakes of central Asia (Tibet, Lop Nor, Kashmir) and the western USA, where it is common in Death Valley (California). Worked industrially in deposits in Searles Lake (San Bernardino County, California). Enormous crystals occur at Clear Lake and Boron, California.
**Uses** The principal source of boric acid.

---

## 110 ULEXITE

NITRATES, CARBONATES, BORATES
NaCaB$_5$O$_9$·8H$_2$O (Hydrated sodium calcium borate)

**System** Triclinic.
**Appearance** Very rarely in distinct crystals. Light, spongy, rounded masses, composed of aggregates of white, silky, hairlike fibers.
**Physical properties** Extremely soft (1), light and fragile, with perfect prismatic cleavage, though this can only be observed under the microscope. Transparent with silky luster, often chatoyant. Soluble in hot water. When heated, first swells and turns white, then fuses easily into a bubble of transparent glass, coloring the flame yellow. When surfaces of acicular crystals or ordinary fibrous fragments are polished across the fiber ends, the material shows *fiber-optics,* transmitting light up each fiber as it appears at the opposite end, earning them the name "television stones."

2.0

**Environment** Produced by evaporitic precipitation in lake basins in arid regions, associated with borax and saltpeter.
**Occurrence** In crystals at Boron, California (USA). Large expanses occur in the Mojave desert (California, USA) and Chile (Atacama). Also found in Argentina and Peru. Small amounts occur in Italy as products of the borax-bearing fumaroles at Larderello.
**Uses** An ore of boron.

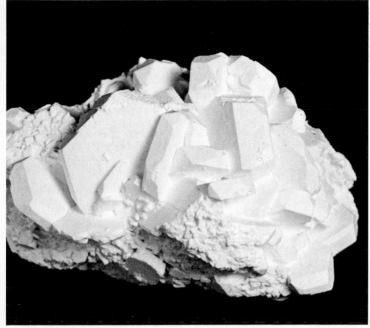

▲ Borax crystals (ca. ×1). Turkey.

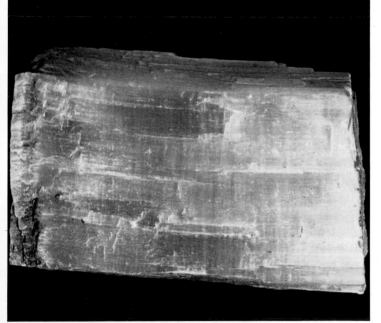

▲ Ulexite cleavage fragment (ca. ×1). Turkey.

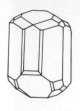

## 111 COLEMANITE
NITRATES, CARBONATES, BORATES
$Ca_2B_6O_{11} \cdot 5H_2O$ (Hydrated calcium borate)

**System** Monoclinic.
**Appearance** Splendid, clear, dipyramidal, prismatic crystals. White, less frequently tinted pink or red by impurities. Also granular or massive.
**Physical properties** Semi-hard (4–4.5), light, perfect cleavage. Transparent or translucent with subadamantine to vitreous luster. White streak. Insoluble in water, soluble in hot hydrochloric acid (as the solution cools, needles of boric acid are precipitated). Fuses easily, decrepitating and coloring the flame green (boron).

**Environment** In continental evaporite basins, probably fed by thermal springs.
**Occurrence** There are large mines in Death Valley and other places in California (USA), Biğadiç (Turkey), Chile, Kazakhstan (USSR) and Argentina.
**Uses** One of the main ores for boron and its salts, used in the chemical, pharmaceutical and cosmetics industries and in the manufacture of steel paints, heat-resistant glass and fluxes for welding. Boron is increasingly important as rocket fuel and as an ingredient in very strong alloys.

## 112 GLAUBERITE
SULFATES, CHROMATES, MOLYBDATES, TUNGSTATES
$Na_2Ca(SO_4)_2$ (Sodium calcium sulfate)

**System** Monoclinic.
**Appearance** White, yellowish and brick-red crystals. Tabular, prismatic or dipyramidal habit, sometimes with striated faces and rounded edges. Compact masses and crusts.
**Physical properties** Soft (2.5–3), light, perfect cleavage. Transparent with vitreous to greasy luster. White streak. Soluble in hydrochloric acid. Fuses very easily, decrepitating when heated. Alters to gypsum when exposed to air.
**Environment** A sedimentary mineral formed by evaporation of saline or meteoric water in sands and desert conglomerates or nitrate deposits. Has also been found in lava as a sublimation product of fumaroles.
**Occurrence** In crystals at Vallarubia (Spain), in Lorraine (France) and in the salt lakes of California and Arizona (USA). Sizable masses in the Triassic evaporites of the Salzburg region (Austria) and Permian evaporites in Ruthenia, central Russia, Texas and New Mexico (USA). Beautiful sublimations, associated with thenardite and sassolite, occur in the lavas of Lipari and Vulcano (Messina, Sicily).
**Uses** For the extraction of Glauber's salt ($Na_2SO_4 \cdot 10H_2O$). Used as a mordant to fix dyes and as a medicine.

▲ Colemanite crystals (ca. ×1). Turkey.

▲ Glauberite crystals (ca. ×1.5). Camp Verde, Arizona.

## 113 ANHYDRITE
SULFATES, CHROMATES, MOLYBDATES, TUNGSTATES
CaSO₄ (Calcium sulfate)

**System** Orthorhombic.
**Appearance** Rarely in stocky or tabular prismatic crystals. Usually in compact, saccharoidal, fibrous masses that break into small cubes. White, gray or reddish.
**Physical properties** Semi-hard (3–3.5), heavy, good cleavage parallel to three pinacoids yielding rectangular fragments. Translucent or transparent with vitreous to pearly luster. White streak. Almost insoluble in hydrochloric acid and in sulfuric acid, except when powdered. Fuses easily, turning the flame brick red. Alters to gypsum when attacked by watery solutions.
**Environment** In chemical sedimentary deposits of the evaporite type, formed in a hot climate. In metamorphosed evaporites (resulting from the dehydration of gypsum).
**Occurrence** Large masses in the apex of salt domes in Louisiana and Texas (USA). In stratified salt deposits in Poland, Germany, Austria, France and India. Rectangular quartz molds after anhydrite are found in basalt at Paterson, New Jersey (USA). Purple crystals at Bex (Switzerland). Found at Costa Volpino (Bergamo, Italy).
**Uses** For the production of sulfuric acid, as a filler in paper and as an ornamental stone.

## 114 CELESTITE
SULFATES, CHROMATES, MOLYBDATES, TUNGSTATES
SrSO₄ (Strontium sulfate)

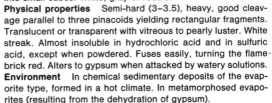

**System** Orthorhombic.
**Appearance** Prismatic or more rarely tabular crystals. Colorless, milky white, pale blue or yellowish. Radiating or parallel, massive, granular and concretionary aggregates.
**Physical properties** Semi-hard (3–3.5), heavy, fragile with perfect cleavage parallel to the base. Transparent to translucent with vitreous to pearly luster. Becomes fluorescent and sometimes thermoluminescent when impurities of organic substances are present. Slightly soluble in water and acids. Fuses easily, decrepitating, coloring the flame crimson (strontium) and forming a white bead. Easily distinguished from barite by its lower specific gravity and the flame test.
**Environment** A primary hydrothermal mineral in cavities of volcanic rocks as a late infilling produced by the action of solutions. Also in veins, associated mainly with galena, sphalerite and other sulfides. More often found disseminated in carbon-

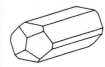

▲ Anhydrite crystals (ca. ×1.5). Hanover, Germany.

▲ Celestite crystals with sulfur crystals (ca. ×1). Sicily, Italy.

ate-rich rocks or in evaporite deposits, associated with other minerals like gypsum, aragonite and sulfur, and chlorides.

**Occurrence** Splendid pale-blue crystals found in basalt cavities at Montecchio Maggiore (Vicenza, Italy), and in pegmatites in Bohemia (Herrengrund) and Malagasy Republic. Blue crystals of sedimentary origin found in nodules and geodes in marl at Bristol (England) and in the gypsum and sulfur-bearing formation at Agrigento (Sicily) and Perticara (Romagna, Italy). Exceptional crystals found at Put-in-Bay and Strontian Island on Lake Erie (USA) and at Dornburg (DDR). Stratified deposits of economic size occur in England (Bristol and elsewhere in Gloucestershire), Tunisia and the USSR (Volga, Turkestan). The largest crystals—white, semi-opaque, 50–75 cm (20–30 in) long and weighing 2–3 kg (4.5–6.66 lb), come from Put-in-Bay, Ohio (USA).

**Uses** The main ore for strontium and its salts, used in making fireworks and flares (giving a crimson flame). Also used in nuclear industry, in the manufacture of rubber, paint and electrical batteries, in the refining of beet sugar and in the preparation of iridescent glass and porcelain. Much sought after by museums and collectors.

## 115 BARITE
SULFATES, CHROMATES, MOLYBDATES, TUNGSTATES
$BaSO_4$ (Barium sulfate)

**System** Orthorhombic.

**Appearance** Tabular crystals. Colorless, yellow, red, green or sometimes black because of inclusions of bituminous matter. Sometimes groups of plates form fan shapes (''cockscomb'' crystals) or rosettes (''desert roses''). Compact, granular, massive, earthy and stalactitic masses.

**Physical properties** Semi-hard (2.5–3.5), very heavy, though not a metallic mineral (an important identifying feature). Fragile with perfect prismatic cleavage. Transparent or translucent with vitreous luster. White streak. Insoluble in acids. Fuses with difficulty, decrepitating and turning the flame yellow-green. Some varieties are fluorescent in ultraviolet light.

**Environment** Common gangue mineral in medium- and low-temperature hydrothermal veins, associated with lead, silver and antimony sulfides, and in replacement veins and cavities in limestones and dolomites (recently redefined as residual deposits of the Karst type). Also deposited by hot springs. The petrifaction of some fossils into barite is probably achieved by

4.48

▲ Blue celestite crystals (ca. ×1), Malagasy Republic.

▲ Lamellar barite with pyrite (ca. ×1). Germany.

► Barite crystal (ca. ×1). Unknown locality.

the same conditions. Occurs in massive form in iron- and manganese-bearing jaspers and rarely in crystals in cavities in basaltic rocks.

**Occurrence** Enormous crystals (up to 1 m [3.3 ft] long) are found in England (Cumberland, Cornwall, Derbyshire), in Czechoslovakia (Pribram), in Transylvania and Felsöbanya (Romania) and the DDR (Freiberg), always occurring in hydrothermal veins. The same conditions are found in Italy in mines at Storo (Trento), Cortabbio (Como) and Guspini and Fluminimaggiori (Sardinia). Sedimentary barite in the form of globular, radiating or massive concretions, often thermoluminescent, occurs at Monte Paderno (Bologna, Italy). In the USA fine crystals are found at Cheshire, Connecticut, Sterling, Colorado, and Elk Creek, South Dakota. Desert roses are common in the Sahara and Oklahoma (USA). Fine clear crystals are found in sandstone at Calafuria (Livorno) and in the Appenines near Piacenza (Italy).

**Uses** The main barium ore. Also used as a heavy additive in mud when drilling for oil, in the paper and rubber industries, in radiography (''barium milk-shake''), as a screen for intense radiation mixed with cement mortar. Used to make an expensive white pigment.

---

## 116  ANGLESITE
SULFATES, CHROMATES, MOLYBDATES, TUNGSTATES
$PbSO_4$ (Lead sulfate)

**System** Orthorhombic.
**Appearance** Prismatic crystals with stocky or tabular habit, rarely dipyramidal. Colorless or white, gray, green, brown or black because of inclusions of unaltered galena. Occasionally emerald-green or purple. Powdery crusts. Nodular and reniform, granular aggregates in concentric layers, sometimes containing a core of galena.
**Physical properties** Soft (2.75–3), very heavy and fragile, with perfect cleavage. Translucent or transparent with adamantine to greasy luster. Colorless streak. Often fluorescent (yellow) in ultraviolet light. Fuses very easily, decrepitating when heated. Dissolves slowly in nitric acid, unlike cerussite ($PbCO_3$), which effervesces briskly.
**Environment** A typical alteration product of galena in the oxidation zone of lead deposits, associated with galena, cerussite, sphalerite, smithsonite, hemimorphite, iron oxides

▲ Barite (ca. ×1). Baia Sprie, Romania.

▲ Colorless anglesite crystal (ca. ×1.5). Monteponi, Sardinia, Italy.

and sometimes phosgenite. Rarely occurs as a result of direct sublimation of volcanic fumes.

**Occurrence** The finest crystals come from Tsumeb (Namibia), Musen (Westphalia, Germany) and the Wheatley mine near Phoenixville, Pennsylvania (USA). Fairly good specimens come from the island of Anglesey in Wales, where the mineral was first identified (hence the name). Fairly large masses found in Germany, at Montevecchio (Sardinia), in Spain, New Caledonia, Mexico, Tunisia and at Bisbee, Arizona, and Tintic, Utah (USA). The small, transparent, green crystals growing on galena from Montevecchio and the splendid, prismatic, blackish, white and yellow crystals from Monteponi, Sardinia, are famous but rare. Distinctive crystals covered with sulfur come from Los Lamentos (Chihuahua, Mexico).

**Uses** An ore of lead sent to concentration plants along with other oxidized minerals. Also of interest to mineralogists and collectors and in the study of ore deposits.

---

## 117 BROCHANTITE
SULFATES, CHROMATES, MOLYBDATES, TUNGSTATES
$Cu_4(SO_4)(OH)_6$ (Hydrous copper sulfate)

**System** Monoclinic.
**Appearance** Small, prismatic, stubby or acicular, striated crystals. Bright green. Fibrous, felted or granular crusts and nodules. Less frequently massive.
**Physical properties** Semi-hard (3.5–4), heavy, perfect cleavage. Translucent or transparent with vitreous to pearly luster. Light green streak. Soluble in very dilute acids and fuses fairly easily, with decrepitation.

**Environment** A secondary mineral found in the oxidation zone of copper deposits, especially in arid climates.

3.97

**Occurrence** Very common in mines in Chuquicamata, Chile, Blanchard Mine, New Mexico, and Bisbee, Arizona (USA). However, the finest crystalline aggregates come from Ain Barbar (Algeria), Tsumeb (Namibia), Rezbanya (Romania) and Italy, from the Rosas, Sa Duchessa and Funtana Raminosa mines in Sardinia.
**Uses** An important ore of copper.

▲ Green anglesite crystal (ca. ×1). Monteponi, Sardinia, Italy.

▲ Brochantite crystals (ca. ×1.5). Tsumeb, Namibia.

## 118 LINARITE

SULFATES, CHROMATES, MOLYBDATES, TUNGSTATES
$PbCu(SO_4)(OH)_2$ (Hydrous lead copper sulfate)

**System** Monoclinic.
**Appearance** Small, bright-blue acicular or tabular crystals, often in groups of crystals or crusts.
**Physical properties** Soft (2.5), very heavy, perfect cleavage. Translucent, with adamantine or subadamantine luster. Light-blue streak. Soluble in dilute nitric acid. Fuses, decrepitating and turning black.

5.35

**Environment** In the oxidation zone at the upper portion of lead and copper deposits, associated with other secondary minerals.
**Occurrence** The finest crystals, over 10 cm (4 in) long, come from Mammoth mine, Tiger, Arizona (USA). Also found in fine groups at Linares (Spain), Tsumeb (Namibia), in the Tintic district (Utah, USA) and Butte district (Montana, USA), at Cerro Gordo (California, USA) and Serra de Capitillas (Argentina). Found in the Arenas, Rosas and San Giovanni mines in Sardinia.
**Uses** Of interest mainly to mineralogists and collectors.

## 119 ALUNITE

SULFATES, CHROMATES, MOLYBDATES, TUNGSTATES
$KAl_3(SO_4)_2(OH)_6$ (Hydrous potassium aluminum sulfate)

**System** Hexagonal.
**Appearance** Small, white, pseudo-cubic crystals rare. Usually occurs in granular or compact white, yellowish, gray or red masses.
**Physical properties** Semi-hard (3.5–4), fairly light, poor cleavage, with conchoidal or irregular fracture. Translucent or transparent with vitreous luster. White streak. Soluble in sulfuric acid. Infusible but decrepitates when heated and turns the flame purplish-red (potassium).

2.75

**Environment** An alteration product of rocks rich in alkalic feldspar (pegmatites, syenites, trachytes), through the action of sulfate-rich water produced by the alteration of pyrite. Sometimes occurs in veins in schistose rocks and in fumarole products.
**Occurrence** Large masses in lava in Tolfa, Lazio and Tuscany (Italy), and in Hungary, France, the Greek archipelago, Australia, several localities in the USA (Goldfield, Nevada, and Rosita Hills, Colorado) and the USSR.
**Uses** For the production of alum and some potash fertilizers.

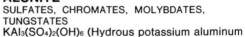

▲ Crystalline linarite (ca. ×1.5). Las Condes, Chile.

▲ Massive alunite (ca. ×1). Azerbaijan, USSR.

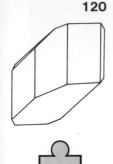

## 120 CHALCANTHITE
SULFATES, CHROMATES, MOLYBDATES, TUNGSTATES
$CuSO_4 \cdot 5H_2O$ (Hydrated copper sulfate)

**System** Triclinic.
**Appearance** Crystals, short tabular or prismatic, rare. Parallel groups of curved crystals ("ram's horn") are sometimes found. Color sky-blue.
**Physical properties** Soft (2.5), light, with indistinct cleavage and conchoidal fracture. Streak colorless. Transparent to translucent with vitreous luster. Soluble in water and, though infusible, will give off water when heated.
**Environment** A secondary mineral found in the oxidized zone of copper deposits associated with other hydrated copper-iron sulfates. Has been known to crystallize as stalactites and crusts on mine timbers when it is dissolved in mine waters.
**Occurrence** The largest deposit of chalcanthite occurs in Chuquicamata (Chile). Found in many copper deposits in the southwestern USA.
**Uses** A major ore of copper in Chile.

2.3

## 121 EPSOMITE
SULFATES, CHROMATES, MOLYBDATES, TUNGSTATES
$MgSO_4 \cdot 7H_2O$ (Hydrated magnesium sulfate)

**System** Orthorhombic.
**Appearance** Small, acicular, colorless to white crystals. Usually in crusts, efflorescences and felty-looking stalactites.
**Physical properties** Soft (2.0–2.5), very light, very fragile, with perfect cleavage. Transparent or translucent with vitreous or silky luster in crystals, earthy when massive. Tastes bitter when dissolved in water (Epsom salts). Fuses easily.
**Environment** A precipitate of hot thermal solutions, fumaroles, and saline waters. Common on the walls of sulfide (pyrite) mines as a deposition by cold sulfatic waters.
**Occurrence** The crusts produced by thermal waters at Epsom (England) are famous. Large crystals measuring 2–3 m (6.66–10 ft) have been found near the salt lakes of the Kruger Mountains (Washington) and Carlsbad, New Mexico (USA). Found in efflorescences at Stassfurt (Germany), in many mines in arid regions (Arizona, [USA], Chile, Tunisia) and in Italy in Valle Antrona, on Elba, in the Lake Como area and at Altaville Irpina (Campania).
**Uses** In the preparation of pharmaceutical products, as a mordant for tanning hides and dyeing cloth. In the paper and sugar industries.

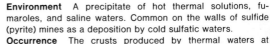

1.67

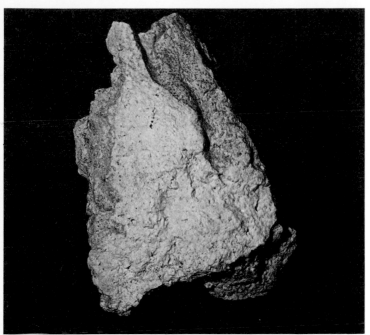

▲ Massive chalcanthite (ca. ×1.5). Boccheggiano, Italy.

▲ Epsomite crystals (ca. ×2). Switzerland.

## 122 GYPSUM

SULFATES, CHROMATES, MOLYBDATES, TUNGSTATES

CaSO₄·2H₂O (Hydrated calcium sulfate)

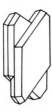

2.35

**System**  Monoclinic.

**Appearance**  Clear, tabular crystals up to 1 m (3.3 ft) long, often in swallowtail or spearhead twins. Transparent crystals and cleavage fragments (selenite). Fibrous aggregates of elongated satiny crystals (satin spar). Granular and compact, waxy-looking masses, sometimes banded (alabaster). White, gray, yellowish or brown. Also occurs in rosette-shaped aggregates, often incorporating grains of sand, of a reddish color, known as "desert roses."

**Physical properties**  Soft (2) and light. Perfect cleavage into slightly flexible but inelastic plates and very fine flakes. Transparent, with vitreous or silky luster, often pearly on cleavage faces. Sometimes fluorescent in ultraviolet light. Soluble in hydrochloric acid and hot water. Fuses fairly easily, turning cloudy and opaque in the flame, with loss of water.

**Environment**  A typical sedimentary evaporite mineral. Forms through direct precipitation from saline waters or through alteration of anhydrite. May also form by direct sublimation from fumaroles or precipitate from hot volcanic springs. Also diagenetic as concretionary blocks in clay and marl.

**Occurrence**  Giant crystals are found in clay near Bologna and Pavia in Italy and in the sulfur mines of Sicily, at Chihuahua (Mexico), in Chile and in Utah (USA). Splendid desert roses occur in Tunisia, Morocco and Arizona and New Mexico (USA). Stratified deposits are worked intensively in the Paris basin (France), in Nova Scotia (Canada), at Volterra (Pisa, Italy) and in many parts of the USA (New York, Michigan, Iowa, Texas, California) and in the USSR (western Urals, northern Caucasus, Uzbekistan).

**Uses**  In the making of plaster of Paris for construction, as a retarder in Portland cement, as a flux for pottery and as a fertilizer. Some varieties of alabaster are used in interior decorating and for sculpture.

▲ Gypsum crystals (ca. ×1). Sicily, Italy.

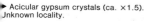

► Acicular gypsum crystals (ca. ×1.5). Unknown locality.

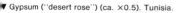

▼ Gypsum (''desert rose'') (ca. ×0.5). Tunisia.

## 123 CYANOTRICHITE

SULFATES, CHROMATES, MOLYBDATES, TUNGSTATES

$Cu_4Al_2(SO_4)(OH)_{12} \cdot 2H_2O$ (Hydrated copper aluminum sulfate)

**System** Orthorhombic.

**Appearance** Stubby, wooly needles, arranged in ball-like aggregates or in radiating groups of sky-blue needles.

**Physical properties** Very soft (1), light, very fragile, with good cleavage. Translucent or transparent, with silky luster. Pale-blue streak. Soluble in acids, fusible with decrepitation.

**Environment** A rare alteration mineral in the oxidation zone of copper deposits.

**Occurrence** Crusts on sandstone near the Cap Garonne (France) and in several mines in Nemaqualand (South Africa) and the USSR (Nizhin Tagilsk, Berkara). Also found in Romania, Scotland, the Grandview mine, Coconino County, Arizona (USA) and at Laurium (Greece). Has been found at Traversella (Piedmont, Italy) and on Elba.

**Uses** Of interest to scientists and collectors.

## 124 COPIAPITE

SULFATES, CHROMATES, MOLYBDATES, TUNGSTATES

$Fe^{+2}Fe_4^{+3}(SO_4)_6(OH)_2 \cdot 20H_2O$ (Hydrated iron sulfate)

**System** Triclinic.

**Appearance** Small, tabular crystals varying in color from olive-green to yellow or orange. Platy crusts, sometimes powdery.

**Physical properties** Soft (2.5), very light, perfect cleavage. Translucent with pearly luster. Dissolves easily in water, giving a yellow solution. Fuses, losing much water between 70° and 300°C (158°–572°F). Other members of the copiapite group include: copper-bearing (cuprocopiapite), calcium (calciocopiapite), aluminum (aluminocopiapite) and magnesium (magnesiocopiapite).

**Environment** A secondary mineral that occurs in the upper portions of sulfide deposits, associated with other sulfates.

**Occurrence** Very common in the Atacama desert (Copiapò, Chile). Also found in mines in the Harz (Germany), Cyprus, Sweden (Falun) and in the USA (Comstock, Nevada, Bisbee, Arizona, and Knoxville, California). Found on Elba (Vigneria and Capo Calamità).

**Uses** Of interest to scientists and collectors.

▲ Acicular cyanotrichite crystals (ca. ×1). Garonne, France.

▲ Copiapite (ca. ×2). Copiapo, Chile.

## 125 CROCOITE
SULFATES, CHROMATES, MOLYBDATES, TUNGSTATES
$PbCrO_4$ (Lead chromate)

**System** Monoclinic.
**Appearance** Prismatic, orange-red crystals, sometimes striated. Crystals often hollow. Large groups of acicular crystals common.
**Physical properties** Soft (2.5–3), very heavy, fragile, with poor cleavage. Translucent with slightly greasy adamantine luster. Orange-yellow streak. Fuses easily. Dissolves in strong acids, giving off chlorine and leaving a residue. When exposed to light loses some of its luster.
**Environment** A secondary mineral deposited from hydrothermal solutions containing chromic acid that had attacked other lead minerals.
**Occurrence** Fine crystals are found at Dundas (Tasmania), Goyabeira and Congonhas de Campo (Brazil), Labo (Philippines), the central Urals area (USSR), Nontron (France) and Lancaster County, Pennsylvania (USA).
**Uses** Of interest only to scientists and collectors because of its rarity. The element chromium was first extracted from this mineral.

## 126 SCHEELITE
SULFATES, CHROMATES, MOLYBDATES, TUNGSTATES
$CaWO_4$ (Calcium tungstate)

**System** Tetragonal.
**Appearance** Dipyramidal, pseudo-octahedral crystals, sometimes tabular, striated on some faces. Yellow, green or reddish-gray. Rarely in granular masses, white or yellowish (especially when molybdenum bearing).
**Physical properties** Semi-hard (4.5–5), very heavy, fragile, good cleavage. Translucent or transparent with vitreous to adamantine luster. White streak. Fluorescent in short-wave ultraviolet light, giving off a pale-blue light which turns yellow in the molybdenum-rich member of the scheelite group (powellite, $CaMO_4$). Soluble in acids, fuses with difficulty.
**Environment** In high-temperature pegmatitic and hydrothermal veins. In Alpine fissures and medium-grade metamorphic rocks, including some contact metamorphic rocks.
**Occurrence** Splendid crystals weighing 0.5 kg (1 lb) are found in Brazil, but the most important economic masses occur in Bolivia, Burma, Malaysia, Japan, China, the USA and Austria. Fine crystals up to 7.5 cm (3 in) are found in Tong Wha (Korea). Fine crystals are found at Traversella (Turin), in Val di Fiemme and Valsugana (Trento) (Italy) and in various localities in Sarrabus and Gerrei in Sardinia. Fine pseudomorphs of ferberite-heubnerite after scheelite have been found at Trumbull, Connecticut (USA).
**Uses** An important ore of tungsten.

▲ Crocoite crystal (ca. ×1). Dundas, Tasmania.

▲ Scheelite crystal (ca. ×1). Traversella, Italy.

## 127 FERBERITE — HEUBNERITE
SULFATES, CHROMATES, MOLYBDATES, TUNGSTATES
$FeWO_4$—$MnWO_4$ (Iron manganese tungstates)

**System** Monoclinic.
**Appearance** Tabular crystals with vertically striated faces or lamellar crystals, reddish-brown or blackish-brown. Brown-black granular masses.

**Physical properties** Ferberite ($FeWO_4$) and heubnerite ($MnWO_4$) form a solid solution series. "Wolframite" is said to be intermediate in composition between iron-rich ferberite and manganese-rich heubnerite. The use of this term should be discouraged. Both are semi-hard (5–5.5), very heavy, fragile with perfect cleavage. Semi-opaque or translucent with sub-metallic to resinous luster and black to reddish-brown streak. Fuses slowly, forming a magnetic globule. Insoluble in acid.

**Environment** In pegmatites and medium- to high-temperature hydrothermal veins. In altered pneumatolytic zones (greisen) and concentrated in alluvial placers.

**Occurrence** Large deposits in southern China (Nan-ling), Malaysia, Burma, Bolivia, Canada and Australia. In Europe found in Cornwall (England), the Erzgebirge in Bohemia and Saxony (Czechoslovakia and DDR), in Portugal and in Spain. In Italy found at Traversella (Piedmont) and in Sardinia. In the USA it is mined in Colorado (Boulder).
**Uses** The main ore of tungsten. Used as a filament in light-bulbs and as a carbide in drilling equipment.

---

## 128 WULFENITE
SULFATES, CHROMATES, MOLYBDATES, TUNGSTATES
$PbMoO_4$ (Lead molybdate)

**System** Tetragonal.
**Appearance** Stubby, pyramidal crystals, often tabular and with a square outline. Massive, granular or earthy aggregates, yellowish or reddish-orange in color.

**Physical properties** Soft (2.75–3), very heavy, fairly good cleavage. Transparent or translucent with resinous or adamantine luster. White streak. Dissolves slowly in acids and fuses easily.
**Environment** A secondary mineral occurring in the oxidation zone of lead deposits, often as pseudomorphs after cerussite.

**Occurrence** The most famous crystals are yellow crystals from Bleiberg (Austria); gray from Pribram (Czechoslovakia); reddish from Rézbanya (Romania), Red Cloud, Arizona (USA) and the Congo; and above all the colorless crystals from Tsumeb (Namibia) and Phoenixville, Pennsylvania (USA). Other deposits found in Morocco, Algeria, Mexico, Yugoslavia, Australia (Queensland) and the USA (Tiger, Arizona, Organ Mountains, New Mexico). Wulfenite has been found in Italy in Val Seriana (Bergamo) and at Raibl (Udine).
**Uses** A secondary ore of molybdenum. Much sought after by collectors.

▲ Ferberite—Heubnerite crystals (ca. ×1.5). Zinnwald (Cinovec), Czechoslovakia.

▲ Wulfenite crystals (ca. ×1). Yugoslavia.

## 129 PURPURITE

PHOSPHATES, ARSENATES, VANADATES
$(Mn^{+3}, Fe^{+3})PO_4$ (Manganese iron phosphate)

**System** Orthorhombic.

**Appearance** Very small crystals of a purplish-red color, usually covered with a brown or black surface alteration. Dark compact masses.

**Physical properties** Semi-hard (4–4.5), heavy, fragile, with good cleavage. Translucent to opaque with satiny or submetallic luster. Deep-red streak. Fuses easily and dissolves in hydrochloric acid.

**Environment** An alteration product of lithiophilite that occurs in pegmatites.

**Occurrence** Fairly large masses found in pegmatites at Mangualde (Portugal) Wodgina (Australia) and Chanteloupe (France). In the USA found with the related mineral heterosite $(Fe^{+3}, Mn^{+3})PO_4$ in Black Hills, South Dakota, at Pala, California, and several other places

**Uses** Of interest to scientists and collectors.

## 130 MONAZITE

PHOSPHATES, ARSENATES, VANADATES
$(Ce, La, Nd, Th)PO_4$ (Rare earth phosphate)

**System** Monoclinic.

**Appearance** Stubby to tabular prismatic crystals, yellow to brownish-red in color. Generally in disseminated granules.

**Physical properties** Hard (5–5.5), very heavy, good cleavage and conchoidal fracture. Translucent with vitreous to resinous (waxy) luster, depending on the degree of metamict alteration caused by the radioactive decay of the thorium usually present in the mineral (up to 12 percent $ThO_2$). Insoluble and infusible, turning gray when heated.

**Environment** A common accessory in granites, syenites and gneisses. Large crystals weighing several kilograms have been found in some pegmatites. Concentrated in fluvial and marine placers (monazitic sands).

**Occurrence** Fine crystals in the pegmatites of Amelia Court House, Virginia, and Alexander County, North Carolina (USA) and Malagasy Republic. Exploitable sedimentary deposits in beach sands in Brazil, India, Sri Lanka, Australia and Florida (USA). Found in pegmatite at Piona (Como, Italy).

**Uses** The main ore for cerium and thorium.

▲ Massive purpurite (ca. ×1). Mangualde, Portugal.

▲ Monazite crystals (ca. ×1). Malagasy Republic.

## 131 HERDERITE
PHOSPHATES, ARSENATES, VANADATES
CaBe(PO₄)F (Calcium beryllium phosphate)

**System**  Monoclinic.
**Appearance**  Stout pseudo-orthorhombic prismatic crystals rare. Also found as fibrous botryoidal aggregates. Color yellow, greenish-white and purple.
**Physical properties**  Hard (5–5.5), light, with poor cleavage and subconchoidal fracture. Transparent to translucent with vitreous luster. Soluble in most acids. Fuses with difficulty.
**Environment**  Occurs as a late-stage hydrothermal product in granite pegmatites associated with microcline.
**Occurrence**  The finest herderite crystals in the world, to 15 cm (6 in), occur at Virgem da Lapa pegmatite, Minas Gerais (Brazil). Well-formed crystals are also found at Newry, Hebron, Paris, Auburn and Poland, Maine, and North Groton, New Hampshire (USA).
**Uses**  Of interest to mineralogists and collectors.

---

## 132 AMBLYGONITE
PHOSPHATES, ARSENATES, VANADATES
(Li,Na)Al(PO₄)(F,OH) (Hydrous lithium aluminum phosphate)

**System**  Triclinic.
**Appearance**  Rough, nearly equant crystals common. Also compact or cleavable masses. Color white, yellow, pink, green, blue or rarely colorless.
**Physical properties**  Hard (5.5–6), light, with excellent prismatic cleavage and uneven fracture. Streak white. Transparent to translucent with vitreous or pearly luster. Soluble with difficulty in acids. Fuses easily with intumescence, coloring the flame red (lithium).
**Environment**  Occurs in lithium- and phosphate-rich granite pegmatites. Associated with spodumene, apatite, lepidolite and tourmaline.
**Occurrence**  Giant crystals weighing several metric tons have been mined near Custer, South Dakota. Found in the pegmatites of Pala, California, Taos County, New Mexico, and Yauapa County, Arizona. Transparent crystals suitable for faceting found at Newry, Maine (USA) and Minas Gerais (Brazil).
**Uses**  Clear crystals or fragments faceted.

▲ Large herderite crystal (ca. ×0.5). Minas Gerais, Brazil.

▲ Amblygonite with muscovite (ca. ×0.5). Newry, Maine.

## 133 OLIVENITE
PHOSPHATES, ARSENATES, VANADATES
$Cu_2(AsO_4)(OH)$ (Hydrous copper arsenate)

**System** Orthorhombic.
**Appearance** Small, prismatic crystals, acicular or tabular, olive-green to brownish-yellow. Often in reniform aggregates and fibrous nodules. More rarely gray or white.
**Physical properties** Soft (3), heavy, poor cleavage with conchoidal fracture. Transparent to translucent with subadamantine luster, becoming silky in fibrous varieties. Yellow streak. Soluble in acids and fuses easily, giving off an odor of garlic.
**Environment** A secondary mineral found in the oxidation zone of copper sulfide deposits associated with arsenopyrite, azurite and malachite.
**Occurrence** Splendid crystals at Tsumeb (Namibia). Also in crystal groups from Cornwall (England), at Tintic, Utah (USA), in Zinnwald (Saxony, DDR) and at Cap-Garonne (France).
**Uses** Of scientific interest and much in demand by mineral collectors.

## 134 ADAMITE
PHOSPHATES, ARSENATES, VANADATES
$Zn_2(AsO_4)(OH)$ (Hydrous zinc arsenate)

**System** Orthorhombic.
**Appearance** Elongated, fairly small crystals, usually mamillary or ball-like aggregates. Bright green or colorless, pink, yellow and violet.
**Physical properties** Semi-hard (3.5), heavy, fragile, with good cleavage. Transparent to translucent with vitreous luster. White streak. Dissolves easily in dilute acids. Fuses, decrepitating and turning white. Sometimes fluorescent lemon-yellow in ultraviolet rays.
**Occurrence** Classic specimens from Tsumeb (Namibia) and Laurium (Greece), and splendid crystals from deposits at Mapimi (Mexico), Tintic, Utah (USA), and Chañarcillo (Chile). Pink crystals (containing traces of cobalt) and green crystals (containing traces of copper) have been found together at Cap-Garonne (France). Very rare in the Mt. Valerio mine (Tuscany, Italy).
**Uses** Of purely scientific interest and much sought after by mineral collectors.

▲ Olivenite crystals (ca. ×1.5). Cornwall, England.

▲ Botryoidal adamite (ca. ×1.5). Laurium, Greece.

## 135 TARBUTTITE

PHOSPHATES, ARSENATES, VANADATES
$Zn_2(PO_4)(OH)$ (Hydrous zinc phosphate)

**System** Triclinic.

**Appearance** Prismatic, stubby crystals, striated, colorless or shades of white, pink, yellow and green. Crusts and aggregates.

**Physical properties** Semi-hard (3.7), heavy, perfect cleavage. Transparent or translucent with vitreous or pearly luster. White streak. Soluble and easily fusible.

**Environment** In the oxidation zone of zinc deposits, as a result of the percolation of phosphatic solutions over existing zinc carbonates.

**Occurrence** So far only found in the Broken Hill deposit in Zambia.

**Uses** Of interest to mineralogists and collectors.

4.1

## 136 LAZULITE

PHOSPHATES, ARSENATES, VANADATES
$MgAl_2(PO_4)_2(OH)_2$ (Hydrous magnesium aluminum phosphate)

**System** Monoclinic.

**Appearance** Pointed, pseudo-dipyramidal, bright-blue crystals. Microgranular masses.

**Physical properties** Hard (5-6), medium heavy, fragile with indistinct prismatic cleavage. Translucent with vitreous luster. Infusible, discolors and breaks into small fragments when heated. Dissolves in strong hot acids with difficulty.

**Environment** In hypersilicic rocks, igneous (pegmatites, quartz veins), where it is associated with andalusite and rutile, or metamorphic rocks (quartzites), where it occurs with quartz, corundum, kyanite, sillimanite, garnet and sapphirine.

**Occurrence** Beautiful crystals up to 5 cm (2 in) long are found in the quartzites of Graves Mountain, Georgia (USA); at Zermatt (Switzerland), Werfen and Vorai (Austria); and Horrsjoberg (Sweden). Scorzalite, the iron analog of lazulite, is found mainly in pegmatites in Brazil.

**Uses** An attractive ornamental stone of minor importance.

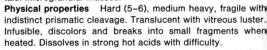

3.0
3.1

▲ Tarbuttite crystals (ca. ×1.5). Broken Hill, Zambia.

▲ Lazulite crystal (ca. ×2). Graves Mountain, Georgia.

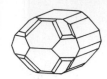

## 137 CONICHALCITE

PHOSPHATES, ARSENATES, VANADATES
CaCu(AsO₄)(OH) (Hydrous calcium copper arsenate)

**System** Orthorhombic.
**Appearance** Microscopic, short, prismatic crystals in botryoidal or reniform crusts. Color various shades of green.
**Physical properties** Semi-hard (4.5), heavy, no cleavage but uneven fracture. Streak green. Translucent with vitreous to resinous luster. Gives off water when heated in a closed tube. Fuses easily and is soluble in nitric or hydrochloric acid.
**Environment** A secondary mineral occurring in the oxidation zone of copper deposits associated with other copper arsenates, including olivenite, libethinite and clinoclase.
**Occurrence** Several localities produce fine specimens. In the USA the finest specimens are from the Tintic district, Utah, and Bisbee and Globe, Arizona. Several mines in Nevada have also produced notable specimens. Conichalcite from Mapimi (Durango, Mexico) and Tsemub (Namibia) are also well represented in mineral collections.
**Uses** Of interest to mineralogists and collectors.

## 138 DESCLOIZITE

PHOSPHATES, ARSENATES, VANADATES
PbZn(VO₄) (OH) (Hydrous lead zinc vanadate)

**System** Orthorhombic.
**Appearance** Crusts of small, frequently plumose groupings of crystals. Forms a solid solution series with mottramite, the copper-rich end member. Fibrous and botryoidal masses.
**Physical properties** Semi-hard (3.5), very heavy, fragile, with poor cleavage. Transparent or translucent with resinous to greasy luster. Orange to reddish-brown streak. Soluble in acids and easily fusible.
**Environment** In the oxidized alteration zone of lead, zinc and copper deposits, associated with vanadinite, cerussite, wulfenite and pyromorphite.
**Occurrence** Large masses at Otavi (Namibia). Beautiful crystals at Bisbee, Arizona, Lake Valley, New Mexico (USA), and Cordoba (Argentina). In the Alps has been found mainly in Austria (Obir, Carinthina). In Italy known at Bena de Padru (Sardinia) and in crusts in the lavas of Vesuvius, where it was described as a new mineral, "vesbine," which later turned out to be mottramite.
**Uses** An important but rare ore of vanadium.

▲ Green crust of conichalcite (ca. ×2.6). Pottamellera, Spain.

▲ Descloizite crystals (ca. ×1). Namibia.

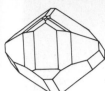

## 139 BRAZILIANITE

PHOSPHATES, ARSENATES, VANADATES
$NaAl_3(PO_4)_2(OH)_4$ (Hydrous sodium aluminum phosphate)

**System** Monoclinic.
**Appearance** Elongated or stubby, yellow or greenish-yellow, often very large, prismatic crystals.
**Physical properties** Hard (5.5), heavy, fragile, with perfect cleavage. Transparent with vitreous luster. Colorless streak. Dissolves with difficulty in strong acids and only fuses in small fragments. When heated quickly loses its color.
**Environment** Occurs in cavities in pegmatites associated with clay, lazulite and blue apatite.
**Occurrence** The finest crystals, up to 15 cm (6 in), are found in Minas Gerais, Brazil (Conselheiro Pena and Mendes Pimental). Smaller crystals have been found at the Palermo mine, North Groton, New Hampshire (USA).
**Uses** A rare and unusual gemstone. Of interest to mineralogists and collectors.

2.98

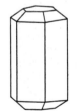

## 140 APATITE GROUP

PHOSPHATES, ARSENATES, VANADATES
$A_5(XO_4)_3(F,Cl,OH)$
A=Ca,Sr,Pb,Na,K X=P,As,V,Si

**System** Hexagonal.
**Appearance** Hexagonal, prismatic crystals, elongated or rather stubby, often terminated by dipyramidal faces. Granular masses and colloform and botryoidal crusts. Varies considerably in color from colorless to yellow, green, brown and occasionally red or blue.
**Physical properties** Hard (5), heavy, very fragile, with poor cleavage parallel to base. Transparent to opaque with vitreous to subresinous luster. Streak always white. Some apatites lose their color when heated, others have strong yellow fluorescence in ultraviolet light. The different members of the apatite group are classified on the basis of the predominant anion present, i.e.: hydroxylapatite (OH), chloroapatite (Cl) and fluorapatite (F).
**Environment** A very common mineral, stable in many environments. Occurs as an accessory mineral, mainly in the form of fluorapatite, in all types of eruptive rocks, in some hydrothermal veins and in iron-rich igneous deposits. Also common in marine sedimentary rocks formed by chemical deposition, in

3.16
3.22

▲ Brazilianite crystal (ca. ×1.5). Minas Gerais, Brazil.

▲ Apatite crystals (ca. ×1.5). Ehrenfriedersdorf, Germany.

organogenetic deposits (vertebrate fossils and phosphorites) and in metamorphic rocks of all kinds.

**Occurrence**  Tabular, blue crystals at Knappenwand (Austria) and Auburn, Maine (USA), in pegmatites. Enormous green crystals at Wilberforce, Ontario (Canada), in hydrothermal veins with purple fluorite. Well-formed, transparent, yellow crystals in the magnetite deposit at Cerro de Mercado, Durango (Mexico), and famous violet crystals from Ehrenfriedersdorf (Saxony, DDR). Large masses of apatite occur in nepheline-syenites in the Kola Peninsula (USSR) and in some carbonatites, for example at Alnö (Sweden) and Palabora (South Africa). Small, clear crystals are found in the effusions of Monte Somma (Italy) and Laacher See (Germany). The largest deposits of sedimentary phosphorites are in Morocco, Algeria, Tunisia, Togo, Nauru and Egypt. Nodular phosphatic rocks occur in Florida, Tennessee and other states in the USA. Semi-amorphous apatite is found in guano deposits in Chile. In Italy crystals from Ossola (Piedmont) and Alto Adige, opaque white crystals from Valle Aurina (Bolzano) and gray-green specimens from Olgiasca (Como) are popular with collectors.

**Uses**  In the manufacture of phosphate fertilizers and in the chemical industry for salts of phosphoric acid and phosphorus. Clear and beautifully colored apatites are used as gemstones, though apatite is rather soft.

---

## 141 PYROMORPHITE
PHOSPHATES, ARSENATES, VANADATES
$Pb_5(PO_4)_3Cl$ (Lead phosphate chloride)

**System**  Hexagonal.

**Appearance**  Small, stubby, barrel-shaped, prismatic crystals, often hollow. Green, brown or colorless. Sometimes parallel aggregates, reniform masses and crusts.

**Physical properties**  Semi-hard (3.5–4), very heavy, fragile, no cleavage and irregular fracture. Translucent with resinous to adamantine luster. White streak. Soluble in acids and fuses easily, forming a globule that shows crystalline faces when cool.

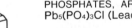

**Environment**  A typical secondary mineral found in the oxidation zone of lead deposits.

**Occurrence**  Beautiful, clear crystals found in Germany and DDR, green crystals in Bohemia (Czechoslovakia) and Cornwall (England), reddish at Broken Hill (Australia). Other fine specimens come from Mapimi (Mexico), the Coeur d'Alene, Idaho, area, and Phoenixville, Pennsylvania (USA), and Berezovskoy (Urals, USSR). Found in several places in Sardinia, in Valsugana (Trento) and at Monte Falo (Mottarone group, Novara) (Italy).

**Uses**  A minor ore of lead, sought after by mineral collectors and museums.

▲ Apatite crystals (ca. ×1.5). Durango, Mexico.

▲ Pyromorphite crystals (ca. ×2). Ural Mountains, USSR.

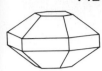

## 142   MIMETITE
PHOSPHATES, ARSENATES, VANADATES
Pb$_5$(AsO$_4$)$_3$Cl (Lead arsenate chloride)

7.1

**System**   Monoclinic.
**Appearance**   Small, hexagonal, prismatic crystals, sometimes with curved faces, often in groups or aggregates. Colorless to yellow-brown or orange. Reniform crusts.
**Physical properties**   Semi-hard (3.5), very heavy, fragile, no cleavage. Translucent with resinous to adamantine luster. White streak. Soluble in acids and fuses easily, giving off fumes with a strong smell of garlic.
**Environment**   Occurs as an alteration product in the oxidation zone of lead and zinc deposits with arsenopyrite or other arsenic minerals.
**Occurrence**   Fine colorless crystals at Johanngeorgstadt (near Freiberg, DDR) and at Pribram (Czechoslovakia). Superb specimens from Tsumeb (Namibia) and the Cornish mines (England). Brown barrel-shaped aggregates (campylite variety) are typical of Cumberland (England). In the USA found at Phoenixville, Pennsylvania. In Italy has been found in Valsugana (Trento), Valtrompia (Brescia), and various places in Sardinia (Malfidano, Iglesias and Bena de Padru, Ozieri).
**Uses**   A secondary ore of lead, of interest to mineralogists and collectors.

---

## 143   VANADINITE
PHOSPHATES, ARSENATES, VANADATES
Pb$_5$(VO$_4$)$_3$Cl (Lead vanadate chloride)

6.8
7.1

**System**   Hexagonal.
**Appearance**   Small, hexagonal prisms, usually barrel-shaped, often cavernous. Bright reddish-orange to yellow or brown. Sometimes also in fibrous, radiate masses or crusts.
**Physical properties**   Soft (2.75–3), very heavy, fragile, no cleavage. Transparent or translucent with resinous to adamantine luster. Yellowish-white streak. Fuses easily and soluble in acids. When dissolved in nitric acid will leave a deep red residue if allowed to evaporate on a glass slide, whereas other minerals in the group leave white residues.
**Environment**   A secondary mineral found in the oxidation zone of lead deposits.
**Occurrence**   Very large crystals 5 cm (2 in) long are found in South Africa (Grootfontein) and Morocco (Mibladen). Smaller but nonetheless fine crystals come from the Old Yuma mine (Arizona, USA), Argentina (Cordoba), and Mexico (Los Lamentos). Beautiful crystals found in the Alps at Obir (Carinthia, Austria). The yellow arsenic-rich variety (endlichite) is found in New Mexico (USA).
**Uses**   An ore of vanadium, used in metal alloys and in dyes and mordants.

▲ Mimetite crystals (ca. ×2). Cornwall, England.

▲ Vanadinite crystals (ca. ×2). Mibladen, Morocco.

## 144 VOLBORTHITE
PHOSPHATES, ARSENATES, VANADATES
$Cu_3(VO_4)_2 \cdot 3H_2O$ (Hydrated copper vanadate)

**System** Monoclinic.
**Appearance** Small, olive-green, pseudo-hexagonal plates, often arranged in globular form.
**Physical properties** Semi-hard (3), heavy, perfect basal cleavage. Translucent or transparent, with vitreous to pearly luster. Soluble in acids and fuses easily, turning the flame green (copper).
**Environment** Alteration crusts, probably formed by rising thermal waters, in sandstones and conglomerates or lavas.
**Occurrence** The main localities are Ferghana (in sands) and the Urals (USSR), but it is also common in sandstones in Utah and Arizona (USA). Found in exceptional conditions in a layer between two lava flows near Vancouver (Canada).
**Uses** Of interest to scientists and collectors.

3.5
3.8

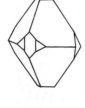

## 145 SCORODITE
PHOSPHATES, ARSENATES, VANADATES
$Fe^{+3}AsO_4 \cdot 2H_2O$ (Hydrated iron arsenate)

**System** Orthorhombic.
**Appearance** Crystals pyramidal, tabular or prismatic. Large crystals rare. Crusts of microcrystals more common. Color variable ranging from shades of green through colorless, blue, violet and yellow.
**Physical properties** Semi-hard (3.5–4), heavy, with indistinct cleavage and subconchoidal fracture. Streak white. Loses water when heated in closed tube. Soluble in most acids.
**Environment** A secondary mineral formed by the oxidation of arsenopyrite and other arsenic-bearing minerals. Occurs in the oxidation zone of copper-zinc deposits.
**Occurrence** The finest scorodite crystals in the world occur at Tsumeb (Namibia), where sharp, lustrous cobalt-blue crystals up to 5 cm (2 in) on edge have been discovered. Very fine crystals have also been found at Zacatecas (Mexico) and Ouro Preto, Minas Gerais (Brazil). In the USA small crystals (usually microcrystals) have been found at Tiger, Arizona, and Marajuba Hill, Nevada.
**Uses** Of interest to mineralogists and collectors.

3.3

▲ Volborthite (ca. ×2). Utah.

▲ Botryoidal crystals of scorodite (ca. ×1.5). Cornwall, England.

## 146 PHOSPHOPHYLLITE
PHOSPHATES, ARSENATES, VANADATES
$Zn_2(Fe^{+2},Mn)(PO_4)_2 \cdot 4H_2O$ (Hydrated zinc iron manganese phosphate)

**System** Monoclinic.
**Appearance** Long prismatic or thick tabular crystals, rare. Colorless to deep bluish-green. Polysynthetic twinning common.
**Physical properties** Semi-hard (3–3.5), light, with excellent prismatic cleavage. Translucent to transparent with vitreous luster. Decrepitates, turns gray and loses water when heated in a closed tube. Fuses easily and is soluble in most acids.
**Environment** A secondary mineral derived from the alteration of sphalerite and iron-manganese phosphates.
**Occurrence** The finest phosphophyllite crystals in the world, up to 10 cm (4 in) long, transparent and twinned, occur in Potosi, Bolivia. Other localities producing crystallized specimens include Hagendorf, Bavaria (Germany), and North Groton, New Hampshire (USA).
**Uses** Sometimes faceted into gems.

3.0
3.1

## 147 VARISCITE
PHOSPHATES, ARSENATES, VANADATES
$AlPO_4 \cdot 2H_2O$ (Hydrated aluminum phosphate)

**System** Orthorhombic.
**Appearance** Pseudo-octahedral crystals extremely rare. Pale green nodules and microcrystalline masses. Nodules usually veined with yellow crandallite and contain "eyes" of gray millisite, wardite and gordonite.
**Physical properties** Semi-hard (4), light, no cleavage but breaks, giving conchoidal fracture and very smooth surfaces. Translucent with vitreous to waxy luster. Infusible, but discolors when heated. Soluble only if heated beforehand.
**Environment** Formed by the infiltration of phosphatic waters into aluminous-rich rock (usually feldspar-rich igneous rocks).
**Occurrence** Masses over 1 m (3.3 ft) in diameter at Fairfield, Utah (USA) and nodules of different sizes in Bolivia, Austria and Arkansas and Nevada (USA).
**Uses** As an ornamental material. Smoothed and polished, it is sometimes sold as turquoise.

2.52

▲ Massive phosphophyllite (ca. ×3). Cerro Rico, Bolivia.

▲ Crystalline variscite (ca. ×1). Arkansas.

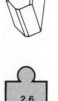

## 148 VIVIANITE

PHOSPHATES, ARSENATES, VANADATES
$Fe_3^{+2}(PO_4)_2 \cdot 8H_2O$ (Hydrated iron phosphate)

**System** Monoclinic.
**Appearance** Prismatic or tabular, deep blue or green crystals. Radiate, sometimes earthy, blue or blackish aggregates, films or massive.
**Physical properties** Soft (1.5–2), light, fragile, perfect cleavage into flexible plates. Transparent if fresh or translucent with vitreous to pearly luster. White streak. The mineral itself is transparent when just extracted and darkens when exposed to light. Easily fusible and soluble in strong acids.
**Environment** In the oxidation zone of sulfide deposits and as an alteration product of some ferromanganese-bearing phosphates in pegmatites. Also found in sedimentary rocks associated with fossil remains of bones, shells, etc.
**Occurrence** Splendid individual crystals up to 1 m (3.3 ft) long or star-shaped groups found in the Cameroons. Common at Richmond, Virginia, Bingham, Utah, Leadville, Colorado (USA), and Llallagua (Bolivia); in tin deposits in Cornwall (England) and in pegmatite at Hagendorf (Germany).
**Uses** When available in large quantites vivianite is mined for use as a cheap coloring agent. Usually of interest to mineralogists and mineral collectors.

2.6
2.7

---

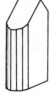

## 149 ERYTHRITE

PHOSPHATES, ARSENATES, VANADATES
$Co_3(AsO_4)_2 \cdot 8H_2O$ (Hydrated cobalt arsenate)

**System** Monoclinic.
**Appearance** Acicular, small crystals, in radiating aggregates. Crimson red, altered on the surface to pearly gray. Earthy masses and crusts.
**Physical properties** Soft (1.5–2.5), heavy, perfect cleavage into small, flexible plates. Translucent to transparent with vitreous luster or pearly on cleavage surfaces. Red streak which when heated turns lavender-blue and then fuses easily. Dissolves in acids, giving a red solution.
**Environment** An alteration product of cobalt minerals. Because of its bright color it is an excellent guide to scattered cobalt-bearing mineral deposits.
**Occurrence** Beautiful specimens from Schneeberg (Saxony, DDR), Wittichen (Germany), Allemont (France), Cobalt (Canada) and Idaho (USA). The largest crystals, though not necessarily beautiful ones, are from Talmessi (Iran) and Bou Azzer (Morocco). Found at Capo Calamità on Elba and at Valsugana (Trento, Italy).
**Uses** An important guide to cobalt mineral deposits.

3.07

Vivianite cleavage fragment (ca. ×1.5). Cameroons.

Erythrite crystals (ca. ×1). Bou Azzer, Morocco.

## 150 ANNABERGITE

PHOSPHATES, ARSENATES, VANADATES
$Ni_3(AsO_4)_2 \cdot 8H_2O$ (Hydrated nickel arsenate)

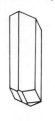

**System**  Monoclinic.

**Appearance**  Rarely in flattened, acicular, finely striated, green crystals. Usually in microcrystalline crusts or earthy masses.

**Physical properties**  Soft (2.5–3), heavy, very fragile. Translucent or earthy with vitreous to opaque luster. Soluble in acids. Easily fusible, changing color and turning black before fusing.

**Environment**  An alteration product which forms crusts on nickel minerals, especially chloanthite. Occurs in the upper portion of some deposits, as crusts and masses, sometimes replacing the original minerals completely.

**Occurrence**  The mineral is named after Annaberg (Saxony, DDR), but fine specimens have also been found at Cobalt (Canada), Allemont (France), Gonnosfanàdiga (Sardinia) and Valsassina (Como, Italy). A magnesium-bearing variety (cabrerite) is found at Laurium (Greece) and in Spain.

**Uses**  Rarely found in economic quantities. Of interest only to mineralogists and collectors.

## 151 WAVELLITE

PHOSPHATES, ARSENATES, VANADATES
$Al_3(PO_4)_2(OH)_3 \cdot 5H_2O$ (Hydrated aluminum phosphate)

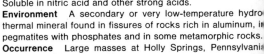

**System**  Orthorhombic.

**Appearance**  Spherulitic, radiate aggregates made up of small, fibrous, white, yellowish or pale-green crystals.

**Physical properties**  Semi-hard (3.3–4), light, perfect cleavage. Translucent with vitreous to silky luster. Infusible but slowly flakes and turns white as the temperature increases. Soluble in nitric acid and other strong acids.

**Environment**  A secondary or very low-temperature hydrothermal mineral found in fissures of rocks rich in aluminum, in pegmatites with phosphates and in some metamorphic rocks.

**Occurrence**  Large masses at Holly Springs, Pennsylvania (USA), fine spherulites in schists at Ouro Preto (Brazil) and in tin veins at Llallagua, Bolivia, and Cornwall (England). In Montgomery County, Arkansas (USA), there is a very distinctive phosphoritic breccia cemented with bluish-green spherulites of wavellite.

**Uses**  When available in economic amounts it is used in the making of phosphate fertilizers.

▲ Annabergite crystals (ca. ×2). Laurium, Greece.

▲ Radiating wavellite balls (ca. ×1.5). Vogtland, Germany.

## 152 EOSPHORITE

PHOSPHATES, ARSENATES, VANADATES

$MnAl(PO_4)(OH_2) \cdot H_2O$ (Hydrated manganese aluminum phosphate)

**System** Monoclinic.

**Appearance** Short-to-long prismatic crystals. Groupings of microcrystals or botyroidal masses. Color pink to pinkish-brown.

**Physical properties** Forms a solid solution series with childrenite $FeAl(PO_4)(OH)_2 \cdot H_2O$. Hard (5), light, with poor cleavage and uneven to subconchoidal fracture. Streak white. Transparent to translucent with vitreous to resinous luster. Gives off water when heated in a closed tube. Fuses readily into a black metallic bead and is soluble in acids.

**Environment** Occurs in granite pegmatites associated with apatite, manganese phosphates and rose quartz.

**Occurrence** Splendid prismatic crystals up to 1.2 cm (0.5 in) long perched on rose quartz crystals were found in Minas Gerais (Brazil). Small microcrystals are found at Branchville, Connecticut, and at Newry, Buckfield, Mt. Mica, Black Mountain and Hebron, Maine (USA).

**Uses** Of interest to mineralogists and collectors.

---

## 153 TURQUOISE

PHOSPHATES, ARSENATES, VANADATES

$CuAl_6(PO_4)_4(OH)_8 \cdot 5H_2O$ (Hydrated copper aluminum phosphate)

**System** Triclinic.

**Appearance** Prismatic crystals extremely rare. Usually occurs in light-blue or green microcrystalline masses, nodules and veins, sometimes filling cavities in various rocks.

**Physical properties** Hard (5–6), light, very fragile with good prismatic cleavage. Also has conchoidal fracture. Translucent only in thin sections, with a waxy or porcelaneous luster. Streak white or pale green. Infusible. Soluble in hydrochloric acid only when heated.

**Environment** A secondary mineral caused by alteration (in arid regions) of aluminum-bearing rocks rich in apatite and chalcopyrite, together with chalcedony and "limonite."

**Occurrence** Splendid sky-blue specimens, sometimes with white markings, from Iran (Nishapur), Egypt (Sinai) and Turkestan (Samarkand). Turquoise from Los Cerillos, New Mexico, and Nevada (USA) is also highly regarded, though less popular because of its greenish hue.

**Uses** A very valuable ornamental stone. Inferior stones are often colored artifically or "improved" with oil, wax or plastics, which are sometimes hard to recognize.

▲ Eosphorite crystals on quartz (ca. ×0.5). Minas Gerais, Brazil.

▲ Massive turquoise (ca. ×1.5). Nevada.

## 154 TORBERNITE

PHOSPHATES, ARSENATES, VANADATES

$Cu(UO_2)_2(PO_4)_2 \cdot 8\text{-}12H_2O$ (Hydrated copper uranium phosphate)

**System** Tetragonal.

**Appearance** Micaceous lamellae with square outline, emerald-green in color. Lamellar aggregates, sometimes in the form of crusts.

**Physical properties** Soft (2–2.5), heavy, perfect cleavage with pearly luster. Radioactive. Not fluorescent in ultraviolet light. Fuses easily and dissolves in strong acids. In the open air becomes partly dehydrated and turns into metatorbernite, which has only 8 molecules of water.

**Environment** Secondary mineral formed by alteration of pitchblende.

**Occurrence** Beautiful specimens occur at Shinkolobwe (Zaire), Mount Painter (Australia), Jachimov (Czechoslovakia), in Cornwall (England), the DDR and North Carolina (USA). Found at Bric Colmè in Val di Preit and at Peveragno (Cuneo) and San Leone (Cagliari) (Italy).

**Uses** A fairly important source of uranium.

## 155 AUTUNITE

PHOSPHATES, ARSENATES, VANADATES

$Ca(UO_2)_2(PO_4)_2 \cdot 10\text{-}12H_2O$ (Hydrated copper uranium phosphate)

**System** Tetragonal.

**Appearance** Micaceous lamellae with square outline, bright or greenish yellow. Intergrown fan-shaped aggregates.

**Physical properties** Soft (2–2.5), heavy, fragile, with perfect cleavage. Translucent with vitreous to pearly luster. Strongly radioactive and fluoresces yellowish-green in ultraviolet light. Fuses fairly easily and when heated turns into meta-autunite, which is tetragonal with 2–6 molecules of water.

**Environment** An alteration product of primary uranium minerals in pegmatites and hydrothermal veins. Also frequently in the alteration zone of granites.

**Occurrence** Large incrustations occur at Autun (France), Shinkolobwe (Zaire), Sabugal (Portugal), Lurisia (Cuneo, Italy) and above all at Mt. Spokane, Washington (USA). Crusts are common in Colorado (USA) and Australia.

**Uses** An important ore of uranium, the most widely used source of the element during World War II.

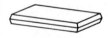

▲ Torbernite crystals (ca. ×1.5). Cornwall, England.

▲ Autunite crystals (ca. ×2). Lurisia, Italy.

## 156 CARNOTITE

PHOSPHATES, ARSENATES, VANADATES
$K_2(UO_2)_2(VO_4)_2 \cdot 3H_2O$ (Hydrated potassium uranium vanadate)

**System** Monoclinic.

**Appearance** Microcrystalline, earthy or powdery, canary-yellow aggregates. Often is disseminated in rocks. Flat, pseudo-hexagonal crystals rare.

**Physical properties** Very soft (indeterminate hardness), heavy, fragile, with perfect cleavage. Semi-opaque with earthy luster. Strongly radioactive. Not fluorescent in ultraviolet light. Infusible, dissolves slightly in acids.

4.7
5.0

**Environment** A secondary mineral, probably caused by deposition from meteoric waters, found as an impregnation in sands, in sandstones and around petrified trees.

**Occurrence** Very common in the desert regions of the Colorado Plateau (Paradox Valley), Utah, Arizona and New Mexico (USA) and in the Ferghana desert (USSR).

**Uses** One of the main ores of uranium and vanadium.

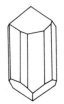

## 157 PHENAKITE

SILICATES (Nesosilicates)
$Be_2SiO_4$ (Beryllium silicate)

**System** Hexagonal.

**Appearance** White or colorless rhombohedral crystals or stubby prisms terminated by rhombohedral faces. Twinning common.

**Physical properties** Very hard (7.5–8), fairly light, with imperfect cleavage. Transparent with very bright vitreous luster. White streak. Infusible and insoluble. Looks very much like quartz, but is much harder and has different twinned forms.

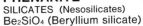

2.96
3.0

**Environment** In high-temperature pegmatite veins and in mica schists associated with quartz, chrysoberyl, beryl, apatite and topaz.

**Occurrence** The best crystals are found in emerald-bearing mica schists in the Urals (USSR), in the pegmatites of Minas Gerais (Brazil) and at Topaz Butte in Colorado (USA). Also found in Val Vigezzo (Domodossola, Italy).

**Uses** Transparent crystals are sometimes faceted as gems.

▲ Massive carnotite (ca. ×1.5). Colorado.

▲ Phenakite crystals (ca. ×1.5). Minas Gerais, Brazil.

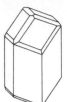

## 158 WILLEMITE

SILICATES (Nesosilicates)
$Zn_2SiO_4$ (Zinc silicate)

**System** Hexagonal.
**Appearance** Well-formed crystals rare. Generally found in compact, granular or massive aggregates; colorless, brown, yellow-green, flesh-red, or black.
**Physical properties** Hard (5.5), heavy, with good cleavage. Translucent with resinous to vitreous luster. Infusible. Decomposes in hydrochloric acid only as a powder, leaving a silica residue. Some varieties fluoresce bright green in ultraviolet light.
**Environment** Found in marbles, possibly the product of metamorphic recrystallization of smithsonite. Commonly associated with calcite, franklinite and zincite.
**Occurrence** Very important economic masses occur at Franklin and Sterling Hill, Ogdensburg, New Jersey (USA). Also common in many metamorphic lead and zinc deposits in the Congo, Zambia, South Africa and Greenland. Very rare idiomorphic crystals occur at Altenberg (Belgium).
**Uses** An ore of zinc.

## 159 "OLIVINE"

SILICATES (Nesosilicates)
$(Mg,Fe)_2SiO_4$ (Magnesium iron silicate)

**System** Orthorhombic.
**Appearance** Stubby crystals of prismatic habit, olive-green to yellowish, or brown through alteration. Granular masses.
**Physical properties** Hard (6.5–7), medium heavy, fragile, with conchoidal fracture. Transparent to translucent with vitreous luster. The term olivine refers to forsterite, $Mg_2SiO_4$, and fayalite, $Fe_2SiO_4$, which form a continuous series. Almost all olivine commonly found is forsterite (including gem-variety peridot). As the amount of iron in olivine increases its specific gravity, refractive indices and solubility increase and the fusing point is lowered. In hydrochloric acid forsterite dissolves very slowly, forming a silica gel; fayalite decomposes into a gelatinous reddish-brown suspension.
**Environment** Forsterites are typical of ultramafic and mafic intrusive or volcanic igneous rocks. They are the dominant constituents of dunites (over 90 percent olivine), peridotites, lherzolites and very abundant in gabbros and basalts. In metamorphic rocks they occur in high-temperature regional or

▲ Massive willemite (ca. ×1). Franklin, New Jersey.

▲ Crystalline "olivine" (ca. ×1). Dankalia, Ethiopia.

contact metamorphosed dolomitic limestones. Fayalites are rare and occur as a minor constituent of granites and in cavities of rhyolites and pegmatites. Forsterites rich in nickel are found in meteorites. Many minerals are formed by alteration of olivines (pseudomorphs). The commonest are serpentine, which also occurs in large masses as a result of the hydration of lherzolites under metamorphic conditions, and iddingsite, a reddish-brown association of serpentine, saponite and hematite, common in altered lavas.

**Occurrence** Clear crystals of a beautiful warm green color are found in lava on St. John's Island in the Red Sea (Egypt), in Norway, Eifel (Germany), San Carlos, Arizona (USA), Burma (Mogok) and Italy (Mt. Vesuvius). Certain varieties occur in the Urals (USSR) and Baden.

**Uses** Rocks with a high olivine but low iron content are used in the manufacture of basic refractories or as a magnesium ore. The transparent variety of forsterite is often faceted as a gem.

---

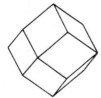

**160 PYROPE**

SILICATES (Nesosilicates, Garnet Group)
$Mg_3Al_2(SiO_4)_3$ (Magnesium aluminum silicate)

**System** Isometric.
**Appearance** Dodecahedral or trapezohedral crystals, dark red, usually very well formed. Rounded grains.
**Physical properties** Very hard (6.5–7.5), heavy, no cleavage but breaks into splinters (subconchoidal fracture). Often transparent with vitreous luster. Fuses fairly easily and is practically insoluble in acids.
**Environment** Typical of peridotites and serpentinized peridotites. Also occurs associated with diamond in kimberlite deposits and concentrated in placers.
**Occurrence** Splendid blood-red crystals are found in Bohemia (Czechoslovakia) and South Africa. Lighter-colored specimens come from Canton Ticino, Switzerland (Gorduno), Arizona and New Mexico (USA) and Scotland ("Ely ruby").
**Uses** Clear, uniformly colored crystals are used as gemstones.

3.58

▲ Fayalite (ca. ×1). Cuasso al Monte, Italy.

▲ Pyrope in peridotite (ca. ×1). Gorduno, Switzerland.

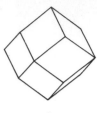

## 161 ALMANDINE

SILICATES (Nesosilicates, Garnet Group)
Fe₃Al₂(SiO₄)₃ (Iron aluminum silicate)

Fe$_3$Al$_2$(SiO$_4$)$_3$ (Iron aluminum silicate)

**System** Isometric.

**Appearance** Dodecahedral or trapezohedral well-formed crystals. Bright or dark red depending on chemical composition, sometimes with violet or brown tints. Rounded grains, altered on the surface.

**Physical properties** Very hard (6.5–7.5), heavy, fragile, no cleavage, breaks into splinters (subconchoidal fracture). Usually opaque but sometimes transparent with adamantine luster. White streak. Fuses fairly easily (3 on the Kobell scale), insoluble in acids.

**Environment** A common mineral in medium-grade metamorphic environments, stable in certain conditions (even in granulites). Less frequent in granites and pegmatites and contact metamorphic rocks. Because of its hardness and chemical resistance commonly a detrital mineral in sedimentary deposits of heavy minerals derived from the breakup of mica schists or gneisses.

4.32

**Occurrence** Clear, orange-red crystals found in sands in Sri Lanka, India (sometimes displaying asterism) and Brazil (Minas Gerais and Minas Novas). Outstanding crystals occur in Norway, in Alaska (Sticken River), the Adirondacks (New York), California, Idaho, South Dakota, Colorado, Michigan, Pennsylvania and Connecticut (USA), Greenland, Sweden, the Zillertal (Austria), Malagasy Republic and southern Australia. In Italy large opaque crystals are found in pegmatite at Olgiasca (Lake Como) and in the mica-schists of the Rombo pass (Bolzano). Perfect but fractured specimens occur in permatite at Alpe Siviglia in Val Codera (Sondrio).

**Uses** A medium-hard abrasive for garnet papers and cloth. Used to some extent in jewelry as an inexpensive gemstone. Of some interest to scientists and collectors.

▲ Trapezohedral almandine crystal (ca. × 1.5). Val Codera, Italy.

▲ Dodecahedral almandine crystal (ca. ×1). Tyrol, Austria.

## 162 GROSSULAR

SILICATES (Nesosilicates, Garnet Group)
$Ca_3Al_2(SiO_4)_3$ (Calcium aluminum silicate)

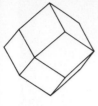

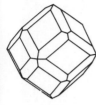

**System** Isometric.

**Appearance** Dodecahedral or trapezohedral crystals of various colors; colorless, pale green or milky when pure, cinnamon-brown to orange when containing iron (hessonite variety) and magnificent emerald-green with chromium (known commercially as tsavorite). Pink, red and black varieties are also found.

**Physical properties** Very hard (6.5–7.5), heavy, fragile, no cleavage, but subconchoidal fracture. Transparent with vitreous luster. Infusible and insoluble. Fluorescent in ultraviolet light, and when exposed to x-rays shows faint greenish-yellow fluorescence.

**Environment** A mineral typical of regional and contact metamorphism of calcareous rocks, associated with calcite, wollastonite, vesuvianite, diopside and scapolite. Less frequently found in metamorphosed basaltic lavas and serpentinites. Hessonite is also found lining fissures in metasomatic calcium-rich rocks (rodingites).

**Occurrence** Superb, clear hessonite crystals found in the gem-bearing sands of Sri Lanka, in the rodingite bodies of California (Ramona), the western Alps (for example in Val d'Ala, Val d'Aosta and Valle del Sangone, in Val della Gava, Liguria [Italy]), at Asbestos (Canada) and Maharitra (Malagasy Republic). In the USA beautiful crystals are found at Minot, Maine, Warren, New Hampshire, and Eden Mills, Vermont, and the mineral is common in several places in California and Colorado. Colorless crystals are found at Jordansmühl (Poland), Telemarken (Norway) and Tiriolo (Catanzaro, Italy). Pink grossular comes from Morelos and Chihuahua (Mexico) and the green variety from Tanzania and Kenya, also found at Rezbanya (Hungary) and Val di Fassa (Trento, Italy).

**Uses** Clear, finely colored crystals are cut and sold as gemstones of excellent quality, though still relatively unpromoted in the trade. A mineral much in demand by museums and collectors.

▲ Grossular (var. hessonite) crystals (ca. ×1.5). Val d'Ala, Italy.

▲ Grossular crystals (ca. ×1.5). Asbestos, Québec, Canada.

## 163 UVAROVITE

SILICATES (Nesosilicates, Garnet Group)
$Ca_3Cr_2(SiO_4)_3$ (Calcium chromium silicate)

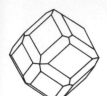

3.52

**System** Isometric.
**Appearance** Small, emerald-green, isometric crystals, usually with very complex forms.
**Physical properties** Very hard (6.5–7.5), heavy, fragile, no cleavage. Transparent with subadamantine luster. Almost infusible and insoluble in acids.
**Environment** In serpentine rocks rich in chromite.
**Occurrence** Clear crystals from the Urals (USSR), Finland (Outukumpu) and Turkey. In Italy has been found in Val Malenco (Sondrio) and at St. Marcel in Val d'Aosta.
**Uses** Used as a gemstone, but its commercial use is limited because it is so rare.

## 164 ANDRADITE

SILICATES (Nesosilicates, Garnet Group)
$Ca_3Fe_2^{+3}(SiO_4)_3$ (Calcium iron silicate)

3.86

**System** Isometric.
**Appearance** Dodecahedral crystals with a wide variety of colors: brown, red or black (melanite), yellow (topazolite) and emerald green (demantoid).
**Physical properties** Very hard (6.5–7.5), heavy, fragile, no cleavage, subconchoidal fracture. Transparent with adamantine luster. White streak. Insoluble in acids but fusible.
**Environment** In contact metamorphic limestones and oxidized skarns. Melanite is found in lavas and nepheline-syenites, topazolite and demantoid in serpentine rock bodies; demantoid generally associated with tremolite (var. ''mountain leather'').
**Occurrence** Blackish-brown crystals found on Elba and at Campiglia Marittima, Livorno (Italy), at Arendal (Norway), Henderson, North Carolina (USA), and reddish-black crystals at Franklin, New Jersey (USA). Melanite is common in the tufas of Lazio and in the volcanic effusions of Vesuvius (Italy), at Arendal (Norway), Kaisersthul (Germany) and in several places in the USA. Topazolite is found in Valle della Gave, Genoa, and Val d'Ala, Turin (Italy). Magnificent specimens of demantoid come from Val Malenco, Sondrio, and yellowish crystals from Val D'Aosta (Italy). Abundant in gold-bearing sands in the Urals (Bobrovka garnet or Urals emerald).
**Uses** A mineral of scientific interest. Demantoid variety is highly regarded as a gemstone.

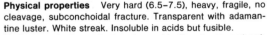

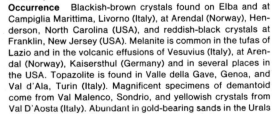

▲ Uvarovite crystals (ca. ×1). Ural Mountains, USSR.

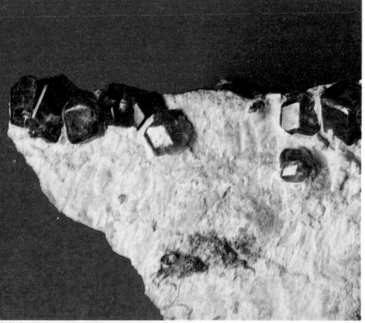

▲ Andradite (var. demantoid) crystals (ca. ×1). Val Malenco, Italy.

## 165  ZIRCON

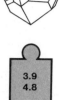

SILICATES (Nesosilicates)
ZrSiO₄ (Zirconium silicate)

**System**  Tetragonal.
**Appearance**  Stubby, prismatic, sometimes dipyramidal crystals, colorless to yellow, red, brown, gray or green. Irregular granules.
**Physical properties**  Very hard (7.5), heavy, with indistinct cleavage, conchoidal fracture. Sometimes perfectly transparent with adamantine luster and strong birefringence, but may also be opaque, dull and almost isotropic when metamict (structure destroyed by radioactive thorium and uranium, which substitute for up to 4 percent of the zirconium). Also contains up to 20 percent hafnium (then the mineral hafnon). Insoluble and infusible.

3.9
4.8

**Environment**  A typical accessory mineral of acidic igneous rocks and their metamorphic derivatives. Also concentrated in alluvial deposits in the form of small grains.
**Occurrence**  In beach sands in Australia, Brazil and Florida (USA). Splendid crystals are found in Alluvial deposits in Matura (Sri Lanka) and in gold gravels in the Urals (USSR). Large crystals at Renfrew (Canada), Litchfield, Maine (USA), in Norway and Sweden.

**Uses**  An important ore for zirconium, hafnium and thorium. Some varieties are used as gemstones.

---

## 166  SILLIMANITE

SILICATES (Nesosilicates)
Al₂SiO₅ (Aluminum silicate)

**System**  Orthorhombic.
**Appearance**  Long, slender crystals without distinct terminations. Gray, brown or pale green. Often in silky, fibrous aggregates (fibrolite). Also frequently occurs in acicular crystals as inclusions in other minerals, like quartz and feldspars.
**Physical properties**  Trimorphic with andalusite and kyanite. Hard (6–7), heavy, perfect cleavage parallel to fibers, with uneven fracture. Transparent to translucent with vitreous to pearly luster. Insoluble and infusible.
**Environment**  Widespread in high-temperature regional metamorphic rocks, though not very conspicuous. Also occurs in contact metamorphic rocks and occasionally in pegmatites.

3.2

**Occurrence**  Fine groups of greenish crystals found in Bohemia (Czechoslovakia). Yellowish crystals found in Val di Fassa (Trento, Italy), Bavaria and Saxony (DDR). Specimens of chatoyant fibrolite found in Brazil and South Carolina (USA). Blue crystals associated with rubies found in sands in Burma and Sri Lanka. Large, remarkably pure masses (85 percent sillimanite) have been mined in India.
**Uses**  An industrial mineral used in the manufacture of refractories and high-temperature crucibles. Chatoyant, green and blue varieties are used in jewelry. Sillimanite is a sensitive indicator of the temperature and pressure at which the host rock formed.

▲ Zircon crystals (ca. ×2). USSR.

▲ Fibrous sillimanite (ca. ×1). Unknown locality.

## 167 ANDALUSITE

SILICATES (Nesosilicates)
$Al_2SiO_5$ (Aluminum silicate)

3.16
3.20

**System** Orthorhombic.

**Appearance** Stubby, prismatic crystals, square in section, sometimes in rodlike aggregates. Flesh-red, brown, red or dark olive green, often with surface alteration into white mica. Chiastolite variety contains fine, dark-colored carbonaceous or clayey inclusions absorbed during the growth of the crystal and arranged in a regular pattern (cruciform in cross section). Alters frequently to kyanite and sillimanite, mica and quartz.

**Physical properties** Trimorphic with kyanite and sillimanite. Very hard (7.5) if unaltered, heavy, perfect cleavage parallel to the prism face. Translucent to opaque, very rarely transparent with vitreous luster, sometimes very pleochroic. White streak. Insoluble and infusible.

**Environment** Typical of low-pressure metamorphic rocks rich in aluminum and poor in calcium, potassium and sodium. Often associated with cordierite. Rarely found in pegmatites and hydrothermal veins.

**Occurrence** Large crystals found in Andalusia (Spain) and the Lisenz Alp (Austria), though these are completely altered into sericite and quartz. Transparent crystals found in sands in the state of Minas Gerais (Brazil). Sizable concentrations at Fresno, California, and in Pennsylvania, Massachussetts and Maine (USA), at Bimbowrie (Australia) and Senus-Bugu (USSR). The finest chiastolite comes from Westford, Massachusetts, Leiperville, Pennsylvania (USA), and Compostela (Spain). Found in Val Malenco (Sondrio), and minute pink crystals have been discovered at Musso (Como), Sonico (Brescia) and in tourmaline veins on Elba (Italy).

**Uses** When available in large quantities, used in industry in the manufacture of refractories, high-temperature electrical insulators and acid-resistent ceramic products. Transparent crystals are sometimes used as gems and are admired for their iridescence, caused by strong pink-green pleochroism. Chiastolite is used sometimes as an amulet. Scientifically, it is an important index of the type and metamorphic grade of the rocks in which it occurs.

▲ Andalusite crystals (ca. ×1). Alpe Lisenz (Tyrol), Austria.

▲ Andalusite (var. chiastolite) (ca. ×1.5). Australia.

## 168 KYANITE

SILICATES (Nesosilicates)
$Al_2SiO_5$ (Aluminum silicate)

3.67

**System** Triclinic.

**Appearance** Elongated, tabular crystals, rarely terminated, often in groups of light-blue crystals, darker toward the center. Less frequently white, gray and green with spots or stripes of color. Gray radiating aggregates. Sometimes intergrown with staurolite.

**Physical properties** Trimorphic with sillimanite and andalusite. Hard across cleavage planes (6–7), semi-hard (4–5) along cleavage planes. Heavy, fragile, perfect prismatic cleavage. Transparent or translucent, with vitreous to pearly luster on cleavage planes. White streak. Infusible, insoluble in acids.

**Environment** Found almost exclusively in pelitic rocks rich in aluminum and metamorphosed under high pressure (gneiss, mica schists, amphibolites and eclogites), associated with garnet, staurolite and micas. Occasionally found in pegmatitic veins running through these rocks. Also common in emery deposits and concentrated in sands produced by the breakup of crystalline schists.

**Occurrence** Splendid blue crystals with staurolite and paragonite found at Pizzo Forno (Candon Ticino, Switzerland). Gray crystals with radiating structure found in Val di Vizze and Val Passiria (Bolzano, Italy), the Tyrol (Austria), Morbihan (France) and more rarely in rocks round Musso (Como, Italy). Enormous opaque blue crystals found in Minas Gerais (Brazil) and green crystals at Machakos (Kenya), sometimes as long as 30 cm (12 in). In the USA there are large deposits at Chesterfield, Massachusetts, Litchfield, Connecticut, and Gaston, Lincoln and Yancey counties, North Carolina. There are also important economic alluvial deposits in India, Kenya and Australia.

**Uses** A raw material for the manufacture of high-temperature porcelain products, perfect electrical insulators and acid-resistant products, including hydrofluoric acid. Occasionally faceted as a gem. Very important as a means of identifying the grade and type of metamorphism of the host rock.

▲ Kyanite crystals in mica schist (ca. ×1). Pizzo Forno, Switzerland.

▲ Kyanite crystals (ca. ×1). Minas Gerais, Brazil.

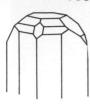

## 169 TOPAZ

SILICATES (Nesosilicates)
$Al_2SiO_4(F,OH)_2$ (Hydrous aluminum silicate)

3.49

**System** Orthorhombic.

**Appearance** Prismatic crystals, sometimes enormous (270 kg; 600 lb), often with vertical striations on prism faces. Color highly variable (colorless to yellow, blue, green, violet or reddish-yellow).

**Physical properties** Very hard (8), heavy, fragile, perfect basal cleavage. Transparent or translucent with vitreous luster. White streak. Insoluble and infusible. When some yellow topaz is heated to between 300° and 450°C (572°–842°F) it turns a characteristic and very attractive pinkish-brown color ("burnt topaz"), though at a higher temperature it often becomes cloudy.

**Environment** Typical of pegmatitic-pneumatolitic conditions found in greisen, granites and miarolitic cavities in rhyolites and some high-temperature hydrothermal veins. Because of its chemical resistance and hardness it is also concentrated in placers, especially in the form of round translucent pebbles.

**Occurrence** The largest and finest crystals come from pegmatites in Minas Gerais (Brazil) and from Alabaschka (Siberia, USSR), where natural "burnt topaz" is also found. Deep-yellow varieties, highly regarded as gems, occur in Burma and Sri Lanka, and light-blue varieties at Mursinka (Urals, USSR). Small, colorless or pink crystals are found in pegmatites on Elba (Livorno, Italy) and in the USA (especially at Pikes Peak, Colorado). Occasionally topaz has been found in quartz porphyry at Cuasso al Monte (Varese, Italy) and at Perda Majori (Sardinia). Compact masses associated with cassiterite and ferberite occur in greisen in the Erzgebirge in Bohemia and Saxony (DDR and Czechoslovakia), Cornwall (England) and New Brunswick (Canada).

**Uses** Clear or finely colored varieties have always been used as gems. The deep-golden-yellow variety is the most highly regarded, though blue or green stones are also very popular with collectors. Unfortunately, in the jewelry trade many faceted yellow stones are known as topaz (i.e., citrine quartz). Let the buyer beware.

▲▶ Topaz crystals (ca. ×1). Ouro Preto, Minas Gerais, Brazil.

▼ Topaz crystals (ca. ×1.5) Saxony, Germany.

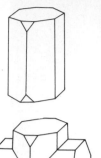

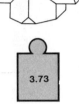

3.73

## 170 STAUROLITE

SILICATES (Nesosilicates)
$(Fe,Mg,Zn)_2Al_9Si_4O_{23}(OH)$ (Hydrous iron magnesium aluminum silicate)

**System** Orthorhombic.
**Appearance** Prismatic, sometimes stubby crystals, reddish-brown to black. Surfaces often rough or covered with an earthy coating caused by alteration. Frequently occurs in characteristic cruciform twins, in the form either of a Greek cross (90° angle between the arms) or of a St. Andrew's cross (60° angle). Shapeless granules in rocks. Very rarely massive. Often occurs as parallel intergrowths with kyanite.
**Physical properties** Very hard (7–7.5), semi-opaque, rarely transparent or translucent, with vitreous to resinous luster. Colorless streak. Infusible and virtually insoluble, though slightly affected by sulfuric acid.
**Environment** A metamorphic mineral typical of medium-temperature conditions (characteristic of the upper part of the amphibolite facies), associated with garnet and kyanite. Occasionally found in some pegmatites and contact metamorphic rocks. Because of its hardness and insolubility also often found in alluvial sands.
**Occurrence** Famous prismatic reddish-brown translucent crystals associated with blue kyanite and gray paragonite occur in mica schists at Pizzo Forno (Ticino, Switzerland). Large crystals also found in Goldenstein (Moravia), Bavaria (DDR), Scotland and various places in the USA. Cruciform twins found in Fannin County, Georgia, at Pilar, New Mexico (USA) and Morbihan (France). In Italy large crystals are common in schists at Monte Legnone (Como) and in Valtellina.
**Uses** Cruciform twins are sometimes worn as religious jewelry. Clear crystals are used to a small extent as cut stones. Important in petrology as a means of defining the metamorphic type and grade of the host rock.

Staurolite: single and twinned crystals (ca. ×1). Minas Gerais, Brazil.

Staurolite crystals in mica schist (ca. ×1). Ticino, Switzerland.

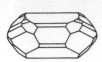

## 171 CHONDRODITE

SILICATES (Nesosilicates, Humite Group)
$(Mg,Fe)_3(SiO_4)(OH,F)_2$ (Hydrous magnesium iron silicate)

**System** Monoclinic.
**Appearance** Rare in small complex crystals. Usually in round grains and massive aggregates, yellow to dark reddish-brown.
**Physical properties** Hard (6–6.5), heavy, poor cleavage. Translucent with vitreous or resinous luster. White streak. Soluble in hot hydrochloric acid, producing a gelatinous precipitate as the solution cools. Infusible. Belongs to the humite group, which also includes norbergite, humite, clinohumite, alleghenyite, leucophoenicite and sonolite.
**Environment** A common metamorphic mineral in dolomitic marbles and skarns formed by regional and contact metamorphism. Other humites also occur in Alpine serpentinites.
**Occurrence** Fine crystals are found in Sweden (Kafveltorp and Åker) and the USA (Franklin, New Jersey, Brewster, New York). Common in the dolomitic effusions of Monte Somma (Vesuvius) and in the contact aureole at Predazzo (Trento) (Italy). Titanium-rich clinohumite is characteristic of serpentinites in Val Malenco (Sondrio), Val d'Ala (Turin) and Liguria (Voltri) (Italy).
**Uses** Of interest only to scientists and collectors.

3.14

## 172 TITANITE

SILICATES (Nesosilicates)
$CaTiSiO_5$ (Calcium titanium silicate)

**System** Monoclinic.
**Appearance** Prismatic, stubby, wedge-shaped and flattened crystals, or tabular and platy. Contact or penetration twins and granular aggregates common. Color varies from white to yellow-green, dark brown or black. In rare cases colorless or pink.
**Physical properties** Hard (5–5.5), heavy, distinct cleavage. Transparent or translucent with adamantine to resinous luster. Some deeply colored varieties display marked trichroism. White streak. Dissolves partially in hydrochloric acid and completely in sulfuric acid. If hydrogen peroxide is added, the solution turns yellow (titanium). Fuses fairly easily, giving dark-yellow glass.
**Environment** A common accessory in many acid and intermediate igneous rocks, both intrusive and extrusive, and in gneisses, mica schists and amphibolites. Also occurs in con-

3.45

▲ Chondrodite crystals (ca. ×1.5). Brewster, New York.

▲ Titanite crystals (ca. ×1.5). Minas Gerais, Brazil.

tact metamorphic limestones and fairly low-temperature hydrothermal veins. The largest crystals are found in rodingites and nepheline syenite pegmatites. Sometimes found in alluvial sands.

**Occurrence** Large economic masses in the Kola Peninsula (USSR). Fine crystals found in metamorphic dolomites in the Binnental (Switzerland) and at Renfrew, Ontario (Canada), in granite and gneiss lithoclases in the Gotthard region (Switzerland) and the Zillertal (Tyrol, Austria) and in rodingites in Val d'Ala (Turin) and the Gruppo di Voltri (Genoa, Italy). Occurs in the USA at Diana, New York, Franklin, New Jersey, and in exceptionally fine crystals at Brewster, New York. In Italy also found at Alpe Devero (Domodossola), in Val Malenco (Sondrio), at Prali in Val Germanasca (Turin) and in effusions of Vesuvius and the volcanoes of Lazio. Pink titanite occurs at Saint-Marcel in Val d'Aosta (Italy), associated with other manganese minerals (diopside [var. violane], piedmontite, etc.)

**Uses** Large masses are worked as an ore of titanium. Clear and attractively colored varieties are cut as cabochons or faceted into very handsome and valuable gems. Of interest to scientists and collectors.

---

## 173 CHLORITOID

SILICATES (Nesosilicates)
$(Fe,Mn)_2Al_4Si_2O_{10}(OH)_4$ (Hydrous iron magnesium aluminum silicate)

**System** Monoclinic and triclinic.
**Appearance** Lamellar pseudo-hexagonal crystals, sometimes polysynthetically twinned. Often in compact, scaly aggregates, yellow-green to black in color.
**Physical properties** Hard (6.5), heavy, fragile, with perfect cleavage into flexible, inelastic plates. Translucent with vitreous luster, pearly on cleavage surfaces. Pale-green streak. Insoluble in hydrochloric acid, soluble in concentrated sulfuric acid. Fuses with difficulty, giving a slightly magnetic black glass. There is a magnesium-bearing variety known as sismondine and a manganese-bearing variety ottrelite.

3.51
3.80

**Environment** Found only in low-grade regional metamorphic rocks, rich in aluminum, iron and possibly manganese, and poor in magnesium, calcium, sodium and potassium. Also in contact marbles with corundum and quartz.
**Occurrence** Chloritoid is found at Kossoibrod (Urals, USSR) in emery deposits and at Svalbard (Norway) in glaucophane schists. Occurs in the USA at Natick, Rhode Island, and in Marquette County, Michigan. Common in metamorphic rocks in the western Italian Alps. Sismondine is found at Zermatt (Switzerland), Biella (Vercelli, Italy) and Pregatten (Austria). Ottrelite comes from Ottré (Belgium).

**Uses** Of interest to scientists and collectors.

▲ Titanite crystals with adularia (ca. ×1.5). Berne, Switzerland.

▲ Chloritoid (var. ottrelite) (ca. ×1.5). Ottré, Belgium.

## 174 DATOLITE

SILICATES (Nesosilicates)
CaBSiO₄(OH) (Hydrous calcium borosilicate)

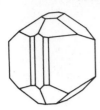

**System** Monoclinic.
**Appearance** Stubby, often large complex crystals, colorless, white or green. Granular aggregates and porcelaneous crusts.
**Physical properties** Hard (5–5.5), light, with conchoidal fracture. Transparent to translucent with vitreous luster. Fuses easily into a transparent glass, turning the flame green. Dissolves easily in acids, forming a silica gel.
**Environment** A secondary mineral in cavities in basalt and serpentinites and hydrothermal deposits, associated with zeolites, prehnite and calcite.
**Occurrence** Splendid crystals found at Andreasberg (Germany) and Arendal (Norway). Occurs at Alpe di Siusi and Riso (Bolzano), Serra dei Zanchetti (Bologna), and in chalcedony amygdales at Tizza (Alto Adige) (Italy). Also found with copper in basalt flows in the Lake Superior region, at Westfield, Massachusetts, and Paterson, New Jersey (USA).
**Uses** When available in large enough amounts, used as an ore of boron. Occasionally cut as a cabochon or faceted.

---

## 175 GADOLINITE

SILICATES (Nesosilicates)
Be₂FeY₂Si₂O₁₀ (Beryllium iron yttrium silicate)

**System** Monoclinic.
**Appearance** Rare prismatic crystals, often intergrown, usually poorly formed with deeply striated faces. Green or brown. Microgranular, earthy masses.
**Physical properties** Hard (6.5–7), heavy, no cleavage, fracture conchoidal or splintery. Transparent with vitreous luster and gray-green streak. Often metamict with some of the yttrium replaced by thorium, uranium and rare-earth elements. Then becomes blackish and opaque with a pitchlike or resinous luster. Radioactive. Soluble in acids, leaving a gelatinous residue. When heated, flakes and turns into a brown mass without fusing.
**Environment** Typical of granite or syenite pegmatites and also found in lithoclases of Alpine metamorphic rocks.
**Occurrence** The largest crystals (up to 500 kg; 1100 lb) come from Norway (Hitterö, Iveland and Hundholmen). Other localities are in Sweden (Finbo and Ytterby) and the USA (Llano County, Texas). Beautiful small crystals found in granite at Baveno and Montorfano (Novara, Italy), Val Nalps (Switzerland) and in Austria. Very rare at Cuasso al Monte (Varese, Italy) in quartz porphyry.
**Uses** An ore of yttrium, thorium and rare-earth elements; also of interest to scientists and collectors.

▲ Datolite crystals (ca. ×1). Ciano Val d'Enza, Italy.

▲ Gadolinite crystals in granite (ca. ×1.5). Baveno, Italy.

## 176 DUMORTIERITE

SILICATES (Nesosilicates)
$Al_7(BO_3)(SiO_4)_3O_3$ (Aluminum borosilicate)

**System** Orthorhombic.
**Appearance** Prismatic blue or violet crystals rare. Usually in columnar or fibrous, radiating aggregates, sometimes reddish-brown.
**Physical properties** Very hard (7), heavy, poor cleavage. Translucent with vitreous luster. White streak. Insoluble and infusible.
**Environment** In metamorphic rocks rich in aluminum. In some pegmatites and in contact metamorphic rocks.
**Occurrence** In the USA, commercial deposits at Oreana and Rochester mining districts (Nevada), Dehesa (California), and in Arizona; and also in Lyons (France), Malagasy Republic and Brazil. Found in pegmatites at Sondalo (Sondrio, Italy).
**Uses** For the manufacture of aluminous refractories.

3.45

---

## 177 ÅKERMANITE-GEHLENITE GROUP

SILICATES (Sorosilicates)
$Ca_2MgSi_2O_7 - Ca_2Al_2SiO_7$ (Calcium magnesium aluminum silicates)

**System** Tetragonal.
**Appearance** Åkermanite and gehlenite form a solid solution series. Melilite is a field term for åkermanite-gehlenite and is usually of intermediate composition but may contain some sodium. Small, prismatic, stubby or tabular crystals, white, gray, yellow and brownish-red or gray-green through alteration.
**Physical properties** Hard (5), heavy, good cleavage. Transparent or translucent with vitreous to greasy luster. Soluble in strong acids, leaving a gelatinous silica skeleton. Åkermanite and melilite fuse with difficulty, and gehlenite is practically infusible.
**Environment** Åkermanite has been found in contact metamorphosed calcareous blocks from Vesuvius and is also a common product in metallurgical slag from blast furnaces. Melilite usually occurs in undersaturated volcanic rocks and in the matrix of kimberlites and carbonatites. Gehlenite is typical of contact calcareous rocks.
**Occurrence** Åkermanite comes from the volcanic effusions of Monte Somma, Vesuvius (Italy). Melilite is found in lavas at Capo di Bove, Rome (Italy) and in other rocks in Germany, Malagasy Republic, Colorado (USA) and the Kola Peninsula (USSR). Gehlenite is found in Trento (Italy) and in Romania.
**Uses** Of interest only to scientists and collectors.

2.94

▲ Dumortierite (ca. ×1.5). California.

▲ Gehlenite crystals (ca. ×1.5). Val di Fassa, Italy.

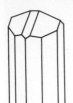

## 178 ILVAITE

SILICATES (Sorosilicates)
$CaFe_2^{+2}Fe^{+3}(SiO_4)_2(OH)$ (Hydrous calcium iron silicate)

**System** Orthorhombic and monoclinic.
**Appearance** Elongated prismatic crystals, with striations on the prism faces. Aggregates of acicular crystals with fibrous, radiating structure. Very dark brown, almost black.
**Physical properties** Semi-hard (5.5–6), heavy, distinct cleavage and fracture. Translucent with resinous to vitreous luster on fresh fracture surfaces, semi-opaque where the fracture is not recent. Fuses easily, giving a black magnetic globule. Dissolves in hydrochloric acid, forming a silica gel.
**Environment** In contact metamorphic rocks and metasomatic skarns. Less often in nepheline syenites.
**Occurrence** Fine crystals and fibrous, radiating aggregates found with hedenbergite, magnetite, andradite, and pyrite at Campiglia Marittima (Livorno) and the island of Elba (Capo Calamità) (Italy). Also found at Perda Niedda (Sardinia), Laxey, Idaho (USA), Trepča (Yugoslavia), Seriphos (Greece), Julianahaab (Greenland) and in the Urals (USSR).
**Uses** Of interest to scientists and collectors.

## 179 HEMIMORPHITE

SILICATES (Sorosilicates)
$Zn_4Si_2O_7(OH)_2 \cdot H_2O$ (Hydrated zinc silicate)

**System** Orthorhombic.
**Appearance** Rarely in distinct crystals, usually small and clearly hemimorphic (i.e., with different terminations at either end of the vertical polar C-axis). Fibrous, radiating crusts. Mammillary, stalactitic, compact granular and earthy masses. Colorless in crystals, usually blue-green, white, yellow or brown.
**Physical properties** Semi-hard (4.5–5), heavy, fragile, with perfect cleavage. Transparent in crystals, usually translucent with vitreous luster. When heated becomes strongly pyroelectric and when pressure is applied becomes piezoelectric (see Introduction to Minerals). Soluble in strong acids, forming a silica gel. When heated to 500°C (932°F) loses water without clouding (showing that the phenomenon occurs without much

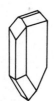

▲ Ilvaite crystal (ca. ×1.5). Seriphos, Greece.

▲ Hemimorphite crystals (ca. ×1). Mapimi, Durango, Mexico.

alteration in the crystal structure), then fuses with difficulty.

**Environment** In the oxidation zone of zinc and lead sulfide deposits, associated with smithsonite, cerussite, anglesite, sphalerite and galena.

**Occurrence** Fine specimens found at Chihuahua (Mexico), Moresnet (Belgium), in Carinthia (Austria), Cumberland and Derbyshire (England), Algeria, and at Granby, Missouri (USA). Common in all lead, zinc and silver deposits, for example at Sterling Hill, Ogdensburg, New Jersy, Friedensville, Pennsylvania, and Leadville, Colorado (USA); at Olkus (Poland) and in several localities in Kazakhstan (USSR). Fine concretions occur at Raibl (Udine), in the Bergamo mines of Val del Riso, Gorno, Oneta, Dossena and San Pietro d'Orzio (Italy) and in the Iglesiente mines (Sardinia).

**Uses** An ore of zinc.

---

## 180 CLINOZOISITE

SILICATES (Sorosilicates, Epidote Group)
$Ca_2Al_3Si_3O_{12}(OH)$ (Hydrous calcium aluminum silicate)

**System** Monoclinic.

**Appearance** Elongated prismatic crystals, often striated and usually poorly terminated. Roundish grains and rodlike aggregates common. Gray, pale green or pink.

**Physical properties** Hard (6.5), heavy, fragile, with perfect cleavage. Transparent or translucent, with vitreous luster. Insoluble in acids but fuses fairly easily into a bumpy white mass.

**Environment** In regional metamorphic rocks transitional between green schist and amphibolite facies. Also frequent in contact metamorphic rocks and plutonic rocks as a secondary hydrothermal alteration product of calcic plagioclases.

**Occurrence** Beautiful pink crystals found at Chiampernotto in Val d'Ala (Turin, Italy) and in Malagasy Republic. Massive variety found in Baja California (Mexico).

**Uses** Of interest to scientists and collectors.

3.3
3.5

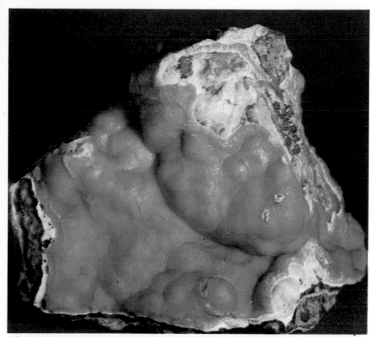

▲ Botryoidal hemimorphite (ca. ×1). Sa Duchessa, Sardinia, Italy.

▲ Clinozoisite crystals (ca. ×1). Pregatten, Austria.

## 181 EPIDOTE

SILICATES (Sorosilicates, Epidote Group)
Ca₂(Al,Fe)₃Si₃O₁₂(OH) (Hydrous calcium aluminum iron silicate)

3.3
3.5

**System** Monoclinic.
**Appearance** Prismatic, columnar crystals, often with shiny, finely striated faces. Some shades of green, sometimes with yellow or grayish tints. Granular masses or fibrous, radiating aggregates.
**Physical properties** Hard (6–7), heavy, fragile, distinct cleavage. Translucent with bright vitreous luster. Insoluble, fuses fairly easily.
**Environment** In regional and contact metamorphic rocks of mafic composition, especially in the green schist facies and in cracks in amphibolites and gabbros. May replace garnets, pyroxenes and amphiboles under conditions of retrograde metamorphism.
**Occurrence** A very common constituent of rocks. Splendid, transparent dark-green crystals occur in Untersulzbachtal (Austria). Fine crystals also found at Bourg d'Oisans (Isère, France), Arendal (Norway), the Naziamsky mountains (Urals, USSR) and in the USA at Riverside, California, Woburn, Massachusetts, and Haddam, Connecticut. In Italy found in Val d'Ala and at Traversella (Turin), in Val Codera and Val Malenco (Sondrio), in Val di Fassa (Trento) and on Elba.
**Uses** Occasionally used as a gemstone, but mainly of interest to scientists and collectors.

## 182 PIEMONTITE

SILICATES (Sorosilicates, Epidote Group)
Ca₂(Al,Mn⁺³,Fe⁺³)₃(SiO₄)₃(OH) (Hydrous calcium aluminum manganese silicate)

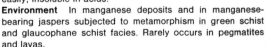

3.4

**System** Monoclinic.
**Appearance** Small, imperfect, rodlike crystals, reddish or purplish brown, often in granular aggregates or compact masses.
**Physical properties** Hard (6.5), heavy, distinct cleavage. Translucent with vitreous luster. Cherry-red streak. Fuses fairly easily, insoluble in acids.
**Environment** In manganese deposits and in manganese-bearing jaspers subjected to metamorphism in green schist and glaucophane schist facies. Rarely occurs in pegmatites and lavas.
**Occurrence** Common in the manganese-bearing mines of Saint Marcel (Val d'Aosta) and Ceres (Turin) (Italy). Also common in schists on the island of Groix (France), at Shikoku (Japan) and in the Jakobsberg deposit (Sweden). Found relatively often in small amounts in many rocks in the western Alps, Adams County, Pennsylvania (USA), and Japan.
**Uses** Of interest to scientists and collectors and has been used as a hard polished stone for inlay work.

▲ Epidote crystals (ca. ×1). Val Codera, Italy.

▲ Piemontite (ca. ×1). Saint-Marcel, Italy.

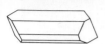

## 183 ALLANITE

SILICATES (Sorosilicates, Epidote Group)
$(Ce,Ca,Y)_2(Al,Fe)_3(SiO_4)_3(OH)$ (Rare-earth silicate)

**System** Monoclinic.
**Appearance** Rarely in stubby or elongated prismatic crystals. Usually in shapeless grains, compact masses or rodlike aggregates, brownish- or pitch-black in color.
**Physical properties** Hard (5.5–6), heavy, fragile, indistinct cleavage. Translucent to opaque with submetallic or resinous luster. Slightly radioactive as other rare-earth elements (yttrium, lanthanum) and thorium commonly substitute for cerium. When heated, expands and fuses easily, giving black magnetic glass. Soluble in hydrochloric acid, leaving a residue of colloidal silica.
**Environment** An accessory of felsic plutonic rocks in pegmatites and greisen, also found in crystalline schists and contact metamorphic calcareous rocks.
**Occurrence** Idiomorphic crystals found at Miask (Urals, USSR), Ytterby and Falun (Sweden), Malagasy Republic and Greenland. Also occurs in several places in the USA (Monroe and Edenville, New York, Franklin, New Jersey, Amelia Court House, Virginia, and Barringer Hill, Texas), Norway and Italy (Baveno, Novara and parts of Adamello).
**Uses** Of interest to scientists and collectors.

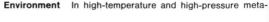

## 184 ZOISITE

SILICATES (Sorosilicates, Epidote Group)
$Ca_2Al_3(Si_3O_{12})(OH)$ (Hydrous calcium aluminum silicate)

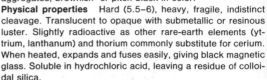

**System** Orthorhombic.
**Appearance** Elongated, prismatic crystals, finely striated on the prism faces and usually poorly terminated. Also frequent in formless grains, rodlike aggregates and granular masses. White, blue, pale green, sometimes pink (manganese-bearing thulite variety) and violet-blue (tanzanite variety, containing chromium and strontium). Lawsonite, a mineral typical of high-pressure metamorphic conditions, is similar in structure and composition but has an additional hydroxyl and also a water molecule.
**Physical properties** Hard (6–6.5), heavy, with distinct pinacoidal cleavage. Usually transparent with vitreous luster. Insoluble in acids. Fuses fairly easily into a blistered white glass. The violet-blue color typical of tanzanite can be obtained artificially by slowly heating green zoisite to 380°C (716°F).
**Environment** In high-temperature and high-pressure meta-

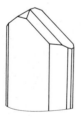

▲ Massive allanite (ca. ×1.5). Norway.

▲ Zoisite crystals in mica schist (ca. ×1.5). Eastern Alps.

morphic rocks (eclogites, granulites and some varieties of gneiss). Also in hydrothermal veins, associated with sulfides, and as an alteration product of calcic plagioclases in extrusive rocks (saussuritization). In very rare cases in metamorphic contact aureoles in impure calcareous rocks.

**Occurrence**  Zoisite comes from Sausalpe (Austria), Wyoming (USA), Kenya and Tanzania, and from the Yourma Mountains (Urals, USSR) in fine crystal aggregates. Sometimes present as an accessory mineral in rocks. All tanzanite comes from Tanga province (Tanzania), though large, clear crystals are becoming increasingly rare. Thulite is found in many manganese deposits, including those of Telemark (Norway), and Tennessee and South Carolina (USA). Found at Pra Isio (Sondrio) and Vipiteno (Bolzano) (Italy).

**Uses**  Zoisite is of interest mainly to scientists and collectors, but tanzanite, made famous by Tiffany and Co., who gave the variety its name, is an important gemstone in its own right, and is not considered simply a replacement for blue sapphire. Thulite is also used as an ornamental stone.

---

## 185  VESUVIANITE

SILICATES (Sorosilicates)
$Ca_{10}Mg_2Al_4(SiO_4)_5(Si_2O_7)_2(OH)_4$ (Hydrous calcium magnesium aluminum silicate)

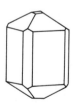

**System**  Tetragonal.
**Appearance**  Stubby, prismatic crystals, rarely with dipyramidal terminations. Aggregates, sometimes columnar, with finely striated faces. Compact granular masses, brown, olive green or more rarely yellow, red and blue.
**Physical properties**  Hard (6.5), heavy, fragile, with conchoidal fracture. Usually opaque, but translucent and transparent varieties have vitreous to resinous luster. Streak always white. Fuses fairly easily into greenish or brown blistered glass. Virtually insoluble in acids.

3.27
3.45

**Environment**  A mineral typical of contact metamorphic skarns (calcareous), associated with various kinds of garnets (grossular, andradite), wollastonite and diopside. Also occurs in metasomatic seams and lenses (rodingites) of serpentinized ultramafic rocks, associated with the same minerals and with chlorite.

**Occurrence**  Brown or yellow crystals found in the calcareous blocks ejected by Vesuvius and Monte Somma (Italy),

▲ Zoisite (var. thulite) (ca. ×1). Telemark, Norway.

▲ Vesuvianite crystals (ca. ×1.5). Canzoccoli, Italy.

where the mineral was first identified. Perfect crystals found in the contact aureole of Monzoni (Val di Fassa, Trento). Green or yellowish-green vesuvianite found in lithoclases of Alpine rodingites in Val d'Ala (Piedmont), Valle della Gava (Liguria), Val Malenco (Sondrio) (Italy) and at Zermatt (Switzerland); elsewhere, at Litchfield (Canada), Fresno, California (USA) and in various localities in the Urals. The pale-blue variety (cyprine) comes from Arendal (Norway), pale-green or whitish (wiluite) from Siberia (USSR) and yellow (xanthite) from Amity, New York (USA). A massive, microgranular variety, white speckled with green (californite or American jade), is found in Siskiyou, Fresno and Tulare counties, California, and in Oregon (USA). Fine vesuvianite also comes from Mexico (Morelos and Chiapas) and in Italy from tuffs in Lazio, from Sardinia and more rarely from contact metamorphic rocks in Adamello (Val Camonica) and Val d'Aosta (Bellecombe).

**Uses** Californite and attractively colored transparent varieties are used for jewelry. Also of interest to petrologists and collectors.

---

## 186 BENITOITE

SILICATES (Cyclosilicates)
$BaTiSi_3O_9$ (Barium titanium silicate)

**System** Hexagonal.
**Appearance** Stubby, prismatic, dipyramidal crystals, various shades of blue, often zoned.
**Physical properties** Hard (6.2–6.5), heavy. Translucent to transparent with vitreous to subadamantine luster. Exhibits pleochroism visible to the naked eye: a crystal looks blue if seen through the acute faces of the rhombohedron and colorless when viewed through the obtuse faces. Brightly fluorescent in ultraviolet light.
**Environment** Found in veins in the brecciated zone of a blue schist body associated with serpentinite. Associated minerals include neptunite, natrolite, crossite and albite.
**Occurrence** To date only found at Diablo Range in San Benito County, California (USA).
**Uses** Sapphire-blue specimens cut as gemstones are popular.

▲ Vesuvianite crystals (ca. ×1.5). Bellecombe, Italy.

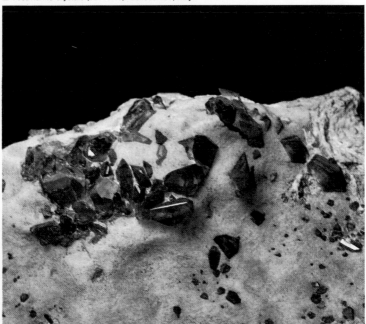

▲ Benitoite crystals (ca. ×1). San Benito County, California.

## 187 FERROAXINITE — MANGANAXINITE
SILICATES (Cyclosilicates)
$Ca_2FeAl_2BSi_4O_{15}(OH)$—$Ca_2MnAl_2BSi_4O_{15}(OH)$ (Hydrous calcium iron manganese borosilicates)

**System** Triclinic.
**Appearance** Crystals with very sharp edges, varied in habit. Reddish-brown to yellow, violet and gray. Sometimes green when incrusted with chlorite. Granular and platy masses.
**Physical properties** Very hard (6.5–7), heavy, fragile with perfect pinacoidal cleavage. Transparent or translucent with bright vitreous luster. Pyroelectric. Fuses easily, expanding to give a green glass that turns black in an oxidizing flame. Insoluble in acids.

3.25

**Environment** Typical of fissures and cavities in granitic rocks or contact zones around granitic masses.
**Occurrence** Splendid crystals found at Bourg d'Oisans (Isère, France), Scopi (Grisons, Switzerland), St. Just (Cornwall, England), Obira (Japan) and Luning and Pala, California (USA). Manganaxinite is found at Tinizon (Grisons, Switzerland), Franklin, New Jersey (USA), and at Cassagna and Gambaresa (Liguria, Italy). Ferroaxinite is also found in Italy at Baveno (Novara).
**Uses** Of interest to scientists and collectors. A rare and unusual gemstone.

## 188 BERYL
SILICATES (Cyclosilicates)
$Be_3Al_2Si_6O_{18}$ (Beryllium aluminum silicate)

**System** Hexagonal.
**Appearance** Individual crystals are sometimes enormous, up to 9 m (30 ft) long and weighing 25 tons. Hexagonal prisms, often without terminations or bounded only by the basal pinacoid or by the combination of basal pinacoids and hexagonal dipyramids. The prism faces are often finely striated. Color varies. Usually grayish-white, yellowish-white or pale blue-green, but some varieties are transparent yellow (heliodor), pink (morganite), blue (aquamarine) and green (emerald). Very rarely occurs in druses or compact rodlike masses.

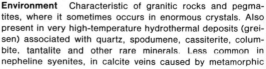
2.65
2.90

**Physical properties** Very hard (7.5–8), light, with imperfect basal cleavage. Transparent or translucent with vitreous luster. Often contains a great many inclusions, making it opaque. White streak. Insoluble in acids. Fuses with difficulty, and only rounds the edge of small chips. Heliodor often fluoresces blue in ultraviolet light, and morganite fluoresces lilac.
**Environment** Characteristic of granitic rocks and pegmatites, where it sometimes occurs in enormous crystals. Also present in very high-temperature hydrothermal deposits (greisen) associated with quartz, spodumene, cassiterite, columbite, tantalite and other rare minerals. Less common in nepheline syenites, in calcite veins caused by metamorphic

▲ Axinite (ca. ×1.5). Isère, France.

▲ Beryl (var. common) crystal (ca. ×1). Val Codera, Italy.

segregation and in biotitic schists in medium-to-high facies. Because of its hardness and great chemical resistance it remains unaltered in alluvial deposits derived from the rocks where it originally crystallized.

**Occurrence**  Opaque beryl is common in several places in Brazil (Minas Gerais), Malagasy Republic, the USA (Black Hills, South Dakota, Grafton, New Hampshire, Albany, Maine, Branchville, Connecticut, etc.), the DDR, Czechoslovakia, India and the USSR. Found in pegmatites in many places in Italy, including Craveggia (Val Vigezzo), Beura and Montescheno (Novara) and Piona (Lake Como). Emerald comes mainly from the mines at Muzo, Chivor and Somondoco (Colombia), where it is found in veins of white calcite running through blackish metamorphosed limestone. There are other famous deposits in Brazil, at Minas Gerais and Salininha (Bahia) in clays and limestones, in the USSR (the Urals and Siberia) and in Africa (the Transvaal and Rhodesia). Less important deposits are found in Australia, North Carolina (USA), Norway and Italy, where emerald has recently been found in Val Vigezzo (Domodossola) in albite and biotite rock. The best aquamarine for cut stones comes from Brazil, found there in alluvial deposits and

▲ Beryl (var. achroite) crystal (ca. ×2). Island of Elba, Italy.

▲ Beryl (var. aquamarine) crystal (ca. ×1.5). Val Codera, Italy.

▶ Beryl (var. emerald) (ca. ×1). Urals, USSR.

▼ Beryl (var. heliodor) crystal, (ca. ×1). Minas Gerais, Brazil.

pegmatites in Minas Gerais, the Urals (USSR), Malagasy Republic, India, Namibia and Ireland. In Italy clear aquamarines of a very light blue color with many fractures are found in Val Codera (Sondrio) and on Elba. Morganite, the only beryl variety found in stubby or tabular prisms, comes from Pala and Ramona, California (USA), where it is associated with exceptionally fine tourmaline, from Brazil and Malagasy Republic. Found in granite on Elba. Heliodor is found in Siberia (USSR), Namibia, Malagasy Republic, and Brazil. A very rare beryl variety rich in scandium, bazzite, was first identified in granites at Baveno (Novara, Italy), and later in Switzerland and the USSR (Kazakhstan).

**Uses**    The main industrial source of beryllium, used in the nuclear industry and in light, very strong alloys in the aircraft industry. The salts are used in fluorescent lamps, in x-ray tubes and as a deoxidizer in bronze metallurgy. Emerald is one of the most expensive and admired gemstones, and the other clear varieties are also used in jewelry. Heliodors and aquamarines have been cut into gigantic gems, weighing thousands of carats, whcih are now real "museum pieces." Synthetic emeralds are now manufactured with artificial inclusions similar to those found in natural emeralds.

## 189 CORDIERITE

SILICATES (Cyclosilicates)
$Mg_2Al_4Si_5O_{18}$ (Magnesium aluminum silicate)

**System**    Orthorhombic.
**Appearance**    Stubby, prismatic crystals, pseudo-hexagonal twins, usually gray (or more rarely blue) with a glassy appearance very much like quartz. Becomes opaque and greenish when altered into mica, chlorite or talc (pinite, gigantolite, prasiolite, etc.). There is also a black, iron-bearing variety (sekaninaite) that occurs in elongated, imperfectly developed crystals. Frequently microgranular and massive.
**Physical properties**    Very hard (7–7.5), light, poor cleavage and conchoidal fracture. Usually translucent with vitreous luster. Displays pleochroism visible to the naked eye, appearing blue or violet when viewed parallel to the prism base, colorless when viewed vertically. White streak. Can be distinguished

2.60
2.66

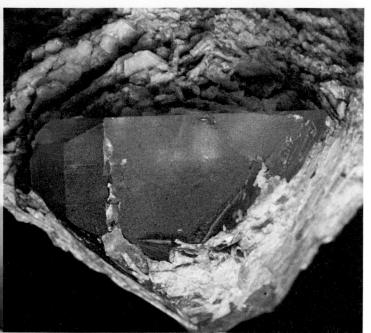

▲ Beryl (var. margarite) crystal (ca. ×1.5). Pala, California.

▲ Cordierite crystal (ca. ×1.5). Kragerö, Norway.

from quartz because it is fusible on thin edges and is twinned differently. Softer than corundum. Insoluble in acids.

**Environment** In contact metamorphic rocks and low-pressure regional metamorphic rocks. Less frequent in granitic or rhyolitic eruptive rocks. Sekaninaite has so far only been identified in pegmatites.

**Occurrence** The finest crystals come from Bodenmais (Bavaria), Orijarvi (Finland) and Norway. Small, round transparent pebbles, attractively colored, are found in sand and gravel in Sri Lanka and Malagasy Republic. Tuscany (Roccatederigi), and Valtellina (Sondalo). Other localities are the USA (Haddam, Connecticut), India (Madras), Canada (Great Slave Lake) and Brazil (Minas Novas). Sekaninaite has so far only been found in one place: Dolni Bory (Czechoslovakia).

**Uses** When transparent faceted as a gem known as "water sapphire," iolite or dichroite. Also of interest to scientists and collectors.

---

## 190 TOURMALINE GROUP

SILICATES (Cyclosilicates)
$(Na,Ca)$ $(Mg,Fe^{+2},Fe^{+3},Al,Mn,Li)_3Al_6(BO_3)_3(Si_6O_{18})(OH,F)_4$ (Complex borosilicates)

**System** Hexagonal.
**Appearance** Prismatic crystals, often very elongated with vertical striations and sometimes hemimorphic. Stubby prismatic crystals less frequent. Aggregates of parallel or radiating individuals. Compact masses rare. At present there are six members of the tourmaline group: elbaite (sodium lithium aluminum rich), schorl and buergerite (sodium iron rich), dravite (sodium magnesium rich), uvite (calcium magnesium rich) and liddicoatite (calcium lithium aluminum rich). A continuous solid solution series exists between elbaite and liddicoatite and also between uvite and dravite (calcium-sodium substitution in sodium- and magnesium-rich tourmalines respectively).

3.0
3.25

**Physical properties** Very hard (7), heavy, with conchoidal fracture. Transparent to translucent with vitreous luster, tending to resinous on fracture surfaces. Some varieties are pleochroic, others change color when moved from natural to artificial light or display attractive chatoyance caused by inclusions or long, parallel cavities. Strongly piezoelectric and pyroelectric. Insoluble in acids. Fusibility depends on the iron

▲ Cordierite (var. sekaninaite) (ca. ×1). Czechoslovakia.

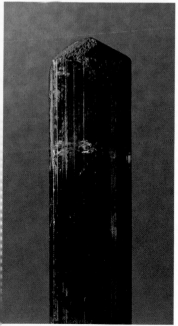

▲ Elbaite crystal (ca. ×1). Brazil.

▲ Elbaite (var. rubellite) crystal
(ca. ×1). Brazil.

and magnesium content. Lithium-bearing varieties are infusible.

**Environment**   A common accessory in igneous and metamorphic rocks and in sedimentary rocks as detrital or authigenic grains. Very common in pegmatites and greisen, in crystals of various colors and sizes, sometimes enormous. Sometimes found in marbles.

**Occurrence**   The finest elbaites come from Elba (Italy), Minas Gerais (Brazil), the Urals (USSR), Sri Lanka, Namibia, Mozambique and Newry, Maine, and Pala, California (USA). Famous pink elbaite (var. rubellite) crystals come from Pala and Ramona, California, and blue-green and red ("watermelon") crystals from Newry, Maine (USA). Uvite is found in well-formed crystals at Franklin and Hamburg, New Jersey, and Gouverneur, DeKalb and Pierrepont, New York (USA). Dravite crystals occur at Yinniethara (Australia) and Gouverneur, DeKalb and Pierrepont, New York (USA). Schorl is a common accessory mineral in granite pegmatites. Liddicoatite is found in Malagasy Republic.

**Uses**   High-pressure gauges, making use of the mineral's piezoelectric property. Some colored and transparent varieties of elbaite are widely sold as semiprecious stones, and synthetic stones, identifiable only by laboratory tests, are also available.

---

## 191  DIOPTASE

SILICATES (Cyclosilicates)
$CuSiO_2(OH)_2$ (Hydrous copper silicate)

**System**   Hexagonal.
**Appearance**   Stubby, prismatic, bright emerald-green crystals, with rhombohedral terminations.
**Physical properties**   Hard (5), heavy, fragile, perfect rhombohedral cleavage. Transparent to translucent with vitreous luster. Green streak. Infusible, expands and turns black when heated. Dissolves in hydrochloric and nitric acid, leaving a silica residue. Also soluble in ammonia.
**Environment**   In the oxidation zone of copper deposits, in cavities of massive copper minerals.
**Occurrence**   The first crystals were mined in the Altyn-Tubé deposits (central Kazakhstan, USSR) and were thought to be emerald. Crystals from Tsumeb (Namibia), Copiapo and Atacama (Chile), Mindouli (Zaire) and the Mammoth mine, Tiger, Arizona (USA), are now famous.
**Uses**   For jewelry, and very popular with collectors.

3.3

▲ Bi-colored elbaite crystals (ca. ×1.5), Island of Elba, Italy.

▲ Dioptase crystals (ca. ×1). Otavi, Namibia.

## 192 MILARITE

SILICATES (Cyclosilicates, Osumilite Group)
$K_2Ca_4Be_4Al_2Si_{24}O_{60} \cdot H_2O$ (Hydrated potassium beryllium aluminum silicate)

**System**  Hexagonal.
**Appearance**  Small, hexagonal, prismatic crystals, often with pointed terminations, colorless, whitish or greenish.
**Physical properties**  Hard (5.5–6), light, perfect basal cleavage. Transparent with greasy luster. White streak. Fuses easily, becoming cloudy and losing water when heated. Insoluble in acids.
**Environment**  Occurs as a hydrothermal mineral in Alpine veins.

**Occurrence**  Classic specimens from Val Giuv and Val Striem (Grisons, Switzerland), where crystals up to 3 cm (7.6 in) long have been found. Also found in other valleys in Austria and Bavaria. Clear crystals found at Swakopmund (Namibia) and the Kola Peninsula (USSR).
**Uses**  Has been used as a gemstone, but mainly of interest to mineralogists and collectors.

---

## 193 DIOPSIDE

SILICATES (Inosilicates, Pyroxene Group)
$CaMgSi_2O_6$ (Calcium magnesium silicate)

**System**  Monoclinic.
**Appearance**  Prismatic crystals, almost square or octagonal in cross section. More often in granular, rodlike or fibrous, radiating aggregates, pale green, blue, whitish, yellowish or brown. The purple manganese-bearing variety is known as violane. There are also dark-green varieties containing chromium (chromian diopside) and vanadium (lavrovite).
**Physical properties**  Hard (5–6), heavy, fragile, with perfect prismatic cleavage (angle between cleavage planes 87°). Transparent to translucent with vitreous luster. White streak. Fuses with difficulty (4 on the Kobell scale) to a green glass. Insoluble in acids. Diopside is the magnesium-bearing end member in the isomorphous, monoclinic diopside-hedenbergite series, a complete solid solution series in which two intermediate members, salite and ferrosalite, have been identified.
**Environment**  In contact metamorphic rocks, especially in dolomitic marbles associated with other calcium silicates. Also appears in metasomatized seams or lenses in serpentinized

▲ Milarite crystals (ca. ×1.5). Val Giuy (Grisons), Switzerland.

▲ Diopside crystals (ca. ×2). Val d'Ala, Italy.

rocks (rodingites). Chromium-bearing diopside is typical of kimberlites. Salite is typical of some hypabyssal rocks derived from alkaline basalt magmas. Ferrosalite has been discovered in gneiss and metasomatized contact rocks (skarns). Violane is typical of manganese-rich metamorphic rocks.

**Occurrence** Magnificent, clear, often complex crystals, found in Italy in Val di Fassa (Trento), in calcareous blocks ejected by Vesuvius, at Monte Cervandone (Val d'Ossola), Pian della Mussa in Val d'Ala (Turin) and in the Binnental (Switzerland), and at DeKalb and Gouverneur, New York (USA). There are other famous localities in the Zillertal (Austria), the Urals and Lake Baikal (USSR), the Tyrol (Austria) and Nordmarken (Sweden). Violane is typical of Saint-Marcel (Val d'Aosta, Italy), and green chromium-bearing diopside comes from kimberlites in South Africa and Siberia (USSR). Salite comes from Sala (Sweden) and other places in Finland, Scotland, Greenland and New Zealand. Ferrosalite has been found at Fabian Mine (Sweden) and in St. Lawrence County, New York (USA).

**Uses** Some transparent varieties are faceted into gems. The mineral is of interest to collectors and petrologists.

---

## 194 HEDENBERGITE

SILICATES (Inosilicates, Pyroxene Group)
$CaFeSi_2O_6$ (Calcium iron silicate)

3.58
3.60

**System** Monoclinic.
**Appearance** Fairly rare in stubby, prismatic crystals, with almost square cross section. More often in radiating aggregates.
**Physical properties** Hard (5–6), heavy, fragile, perfect cleavage parallel to the faces of the vertical prism. Almost opaque except when viewed on the edge of slivers. Light-brown streak with green tints. Insoluble. Fuses fairly easily to blackish magnetic glass. The iron-rich end member of the monoclinic diopside-hedenbergite solid solution series.
**Environment** In contact metamorphic rocks rich in iron minerals (skarns), associated with ilvaite, garnets, sulfides, epidote and calcite.
**Occurrence** Splendid radiating aggregates found at Capo Calamità (Elba) and Campiglia Marittima (Livorno) (Italy). Also found at Arendal (Norway) and Obira (Japan), in Nigeria, Australia and several places in the USSR (Kazakhstan, the Caucasus and Altai).
**Uses** Of interest to scientists and collectors.

▲ Diopside (var. violane) (ca. ×1). Saint-Marcel, Italy.

▲ Fibrous hedenbergite (ca. ×1). Campiglia Marittima, Italy.

## 195 JOHANNSENITE
SILICATES (Inosilicates, Pyroxene Group)
$CaMnSi_2O_6$ (Calcium manganese silicate)

**System** Monoclinic.
**Appearance** Elongated prismatic crystals, greenish-brown or gray, usually in radiating aggregates.
**Physical properties** Forms a solid solution series with diopside and hedenbergite. Hard (5–6), heavy, fragile, good cleavage. Translucent to almost opaque, with vitreous to greasy luster. When heated to about 830°C (1526°F) turns into a triclinic phase (bustamite) and then fuses to a transparent yellowish glass. Soluble in hot hydrochloric acid. The surface of the mineral is often marked by black manganese oxides and sometimes has been completely transformed into a triclinic manganese-rich inosilicate (rhodonite).
**Environment** In metasomatized limestones and in skarns, associated with manganese silicates. Has also been found in quartz veins running through rhyolite, probably as a late hydrothermal product.
**Occurrence** The most famous locality is Monte Civillina, near Recoaro (Vicenza, Italy), but johannsenite is also abundant at Broken Hill (Australia), Pueblo (Mexico), Vanadium, New Mexico, and Franklin, New Jersey (USA).
**Uses** Of interest only to scientists and collectors.

3.44
3.46

## 196 JADEITE
SILICATES (Inosilicates, Pyroxene Group)
$Na(Al,Fe^{+3})Si_2O_6$ (Sodium aluminum iron silicate)

**System** Monoclinic.
**Appearance** Crystals very rare. Usually in compact, felty, waxy masses, white or yellowish-white (sometimes with green markings), greenish or black (chloromelanite).
**Physical properties** Hard (6.5–7), heavy, tough, with difficult cleavage. Translucent with vitreous luster which becomes pearly on cleavage surfaces. Insoluble and fuses fairly easily (unlike nephrite jade, which is similar) into an almost transparent globule.
**Environment** Metasomatic concentrations in serpentinized ultramafic rocks, associated with nepheline. In schistose metamorphic rocks of the blue schist facies.
**Occurrence** Compact masses at Tawmaw (Burma), in Tibet, southwest China (Yunnan), Guatemala, California (USA) and Japan. In Italy has been found very rarely in the Ambin and Lanzo massif and at Bellino (Piedmont), and in dark pebbles between Brosso and Traversella (Turin).
**Uses** Jade is a term used in the gem industry to refer to two different minerals: jadeite and actinolite (variety nephrite). Precious jade is always jadeite.

3.3
3.5

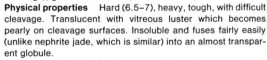

▲ Johannsenite (ca. ×1.5). M. Civilina, Italy

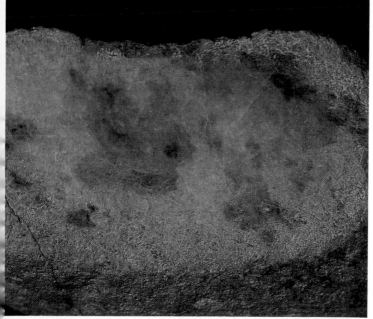

▲ Jadeite (ca. ×1.5). Burma.

## 197 ACMITE (AEGERINE)

SILICATES (Inosilicates, Pyroxene Group)
$NaFeSi_2O_6$ (Sodium iron silicate)

**System** Monoclinic.

**Appearance** Columnar, prismatic crystals, sometimes striated and with pointed terminations, green to brown. Fibrous, sometimes radiating aggregates. "Aegerine" and "aegerine-augite" are field terms used to describe pyroxenes or rocks bearing pyroxene with a large acmite component.

**Physical properties** Hard (6–6.5), heavy, imperfect cleavage parallel to the vertical prism. Translucent with vitreous luster. Pale-brown streak. Fuses easily, coloring the flame yellow (sodium), leaving a slightly magnetic spherule. Almost insoluble in acids. Solid solutions with other pyroxenes may form rare varieties typical of particular environmental conditions (aegirine-augite, urbanite, etc.).

**Environment** In intrusive and volcanic alkaline rocks, rich in sodium and oxidized iron and often undersaturated (nepheline syenites and phonolites). Rarely in regional metamorphic rocks.

**Occurrence** Large crystals common in nepheline syenites in the Kola Peninsula (USSR), at Mont Saint Hilaire, Québec (Canada), and in Greenland. Smaller but more beautiful crystals at Magnet Cove, Arkansas, and Libby, Montana (USA).

**Uses** Of interest to scientists and collectors.

3.55
3.60

---

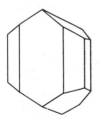

## 198 AUGITE

SILICATES (Inosilicates, Pyroxene Group)
$(Ca,Na)(Mg,Fe,Al,Ti)(Si,Al)_2O_6$ (Calcium magnesium iron aluminum silicate)

**System** Monoclinic.

**Appearance** Crystals of prismatic habit, often stubby, almost square or octagonal in cross section, terminated by small pinacoidal faces. Contact twins, frequently polysynthetic. Granular, black, greenish-black or dark-brown aggregates.

**Physical properties** Variable, since augite is an intermediate member of a solid solution series between pure end members like diopside and hedenbergite. Augite proper is hard (5–6) and heavy, with two directions of cleavage at nearly a right angle (87°) and conchoidal or splintery fracture. Opaque to translucent with vitreous to resinous luster. Gray-green streak. Insoluble in acids (except titanium-bearing varieties, which dissolve in heated hydrochloric acid) and fuses with great difficulty.

**Environment** A common mineral in plutonic rocks (gabbros, pyroxenites, peridotites), volcanic rocks (basalts, essexites), tuffs, high-temperature metamorphic rocks, contact or regional

3.2
3.5

▲ Acmite crystal (ca. ×1.5). Greenland.

▲ Augite crystal in tuff (ca. ×1.5). Ariccia, Italy.

(granulites). Easily altered by hydrothermal solutions into intergrowths of hornblende amphibole (known as uralite), chlorite and, less commonly, epidote and calcite.

**Occurrence** Augite is found in plutonic rocks in a number of localities. Typical specimens come from the Bushveld Complex (South Africa), Skaergaard (Greenland) and Stillwater Complex (USA). Titaniferous augite is characteristic of volcanic rocks and has a rather dark, slightly violet color. It also displays a zoned internal structure ("hourglass" structure). Fine crystals are found in the lavas and pyroclastic rocks of Vesuvius, the volcanoes of Lazio, Etna and Stromboli in Italy, in volcanic lavas at Kaiserstuhl (Germany) and in the Auvergne (France) and in essexites and shonkinites in Montana and Colorado (USA). Colorless augite is characteristic of Bathurst, Ontario (Canada). Augites poor in calcium are found in basalt and andesites in Japan and in other gabbro intrusions. Acmite-augite is found in the volcanic rocks of Oldojuo Langaj (Tanzania).

**Uses** Of considerable interest to scientists, petrologists, and collectors.

---

## 199 OMPHACITE

SILICATES (Inosilicates, Pyroxene Group)
$(Ca,Na)(Mg,Fe^{+2},Fe^{+3},Al)Si_2O_6$ (Calcium magnesium iron aluminum silicate)

**System** Monoclinic.

**Appearance** Rare prismatic crystals with almost square cross section. Usually in light-green grains.

**Physical properties** Hard (5–6), heavy, with poor cleavage. Translucent with vitreous luster. White streak, with a paler or darker greenish tinge, depending on the amount of ferric iron present. Insoluble and fuses with difficulty.

**Environment** With garnet, a typical mineral of eclogites. Also found in massive seams (omphacitites), where vacuoles lined with small crystals form. More rarely found in small crystals intergrown with quartz and mica in segregation lenses.

3.29
3.37

**Occurrence** Eclogites containing granular omphacite occur at Saualpe and Koralpe (Austria), Munchberg (Germany), in California (USA), and at Monviso (Val d'Aosta) and Biella (Vercelli) (Italy). Crystals are found in metasomatic omphacitite at Besshi (Japan). Fine crystals also found in quartz lenses near Ivrea (Turin, Italy).

**Uses** A mineral of interest to scientists and collectors.

▲ Augite crystals in basalt (ca. ×1). Buffaure, Italy.

▲ Omphacite crystals in quartz (ca. ×1.5). Ivrea, Italy.

## 200 SPODUMENE
SILICATES (Inosilicates, Pyroxene Group)
LiAlSi₂O₆ (Lithium aluminum silicate)

**System** Monoclinic.
**Appearance** Prismatic crystals, sometimes gigantic (16 m [53 ft] long and weighing several tons), with vertical striations. Color whitish, yellow, gray, pink (kunzite) or emerald-green (hiddenite). Rodlike aggregates of compact cryptocrystalline masses.
**Physical properties** Hard (6.5–7), heavy, perfect cleavage parallel to the vertical prism. Transparent or translucent, with vitreous luster. A trichroic mineral, changing color depending on the angle at which it is viewed. Insoluble. Fuses easily, coloring the flame crimson (lithium).
**Environment** In lithium-bearing pegmatites associated with quartz, feldspars, lepidolite, beryl and tourmaline. Often subject to later alteration, turning into mixtures of various minerals, including clays, albite, muscovite, eucryptite, etc.
**Occurrence** Mined intensively at Etta and Tin Mountain, South Dakota (USA), where enormous crystals are found, Bernic Lake, Manitoba (Canada) and in the Urals (USSR). Occurs in many other places in the USA, Brazil, Mexico, Scotland and Sweden. Kunzite is found at Pala, California (USA), and in Brazil, and hiddenite in North Carolina (USA), Malagasy Republic and Brazil.
**Uses** An important industrial source of lithium and its salts.

---

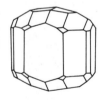

## 201 ENSTATITE
SILICATES (Inosilicates, Pyroxene Group)
Mg₂Si₂O₆ (Magnesium silicate)

**System** Orthorhombic.
**Appearance** Rare, stubby, prismatic crystals. More often fibrous or platy masses, yellowish, green or olive-colored.
**Physical properties** Hard (5.5), heavy, good cleavage. Translucent with vitreous luster, changing to pearly on cleavage surfaces. Insoluble and almost infusible. Forms a solid solution series with orthoferrosilite (FeMg)₂Si₂O₆, originally a synthetic material recently discovered in lunar rocks. Intermediate members include bronzite and hypersthene.
**Environment** In mafic and ultramafic plutonic and volcanic rocks. In high-grade metamorphic rocks (granulites) and meteorites.
**Occurrence** Large crystals found in the Sierra Nevada (California, USA), in Donegal (Northern Ireland), Greenland and at Cima di Gagnone (Switzerland). Common in many rocks in Scotland, Norway, Germany, South Africa, Japan, USA (Texas, Pennsylvania, Maryland, North Carolina, etc.) and the USSR (Caucasus, Urals, Siberia, etc.). Hypersthene is found in norites in the Cortland Complex (New York), in the Adirondacks, and in the Lake Baikal region (USSR).
**Uses** Of interest to scientists and collectors.

▲ Spodumene (var. kunzite) crystal (ca. ×1.5). Minas Gerais, Brazil.

▶ Enstatite crystal (ca. ×1.5).
Greenland.

## 202 "BRONZITE" (FERROAN ENSTATITE)

SILICATES (Inosilicates, Pyroxene Group)
$(Mg,Fe)_2Si_2O_6$ (Magnesium iron silicate)

**System** Orthorhombic.
**Appearance** Stubby, prismatic crystals. More often irregular grains or compact masses, greenish-brown or blackish.
**Physical properties** Hard (5), heavy, distinct cleavage. Translucent with bronzy, submetallic luster. Fuses more readily as the iron content increases. Insoluble. A member of the isomorphous series enstatite-hypersthene-orthoferrosilite containing 5–13 percent FeO. Frequently alters to green, lamellar or fibrous, serpentine-like minerals (bastites).
**Environment** In mafic and ultramafic plutonic and volcanic rocks. In some high-grade metamorphic rocks.
**Occurrence** Typical of gabbroic rocks in the Bushvelt Complex (Transvaal, South Africa), Stillwater Complex (Montana, USA) and Styria. Found at Sondalo (Sondrio) and in Valle d'Ultimo (Trento) (Italy).
**Uses** Of interest to scientists and collectors.

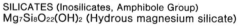

## 203 CUMMINGTONITE

SILICATES (Inosilicates, Amphibole Group)
$Mg_7Si_8O_{22}(OH)_2$ (Hydrous magnesium silicate)

**System** Monoclinic.
**Appearance** Very rarely in individual crystals. Usually rod-like or fibrous aggregates, light or dark brown depending on the percentage of iron.
**Physical properties** Hard (5–6), heavy, with good prismatic cleavage. Translucent with vitreous to silky luster in fibrous varieties. Insoluble in acids. Fuses with difficulty to a black glass. Forms a solid solution series with grunerite, $Fe_7Si_8O_{22}(OH)_2$. Dannemorite is a manganiferous cummingtonite.
**Environment** In medium- to high-grade metamorphic schists, in contact rocks and occasionally in plutonic rocks.
**Occurrence** Fine aggregates of crystals are found at Cummington, Massachusetts (USA), Large masses are typical of the iron deposits of Lake Superior, where grunerite is found—also common at Orejarvi (Finland), in Australia and at Rockport, Massachusetts (USA). Dannemorite is found mainly in Sweden. Cummingtonite has been found in rocks in Adamello, Sila and Val d'Ossola (Italy), and dannemorite in Valle Strona (Italy).
**Uses** Of interest to scientists and collectors.

▲ "Bronzite" (ca. ×1). Styria, Austria.

▲ Fibrous cummingtonite (ca. ×1.5). Proves, Italy.

## 204 TREMOLITE

SILICATES (Inosilicates, Amphibole Group)
$Ca_2Mg_5Si_8O_{22}(OH)_2$ (Hydrous calcium magnesium silicate)

**System** Monoclinic.
**Appearance** Elongated prismatic crystals, sometimes in columnar aggregates, colorless or white if pure, often pale grayish-green. The opaque fibrous variety is commonly known as asbestiform amphibole (asbestos). Forms a solid solution series with actinolite.
**Physical properties** Hard (5–6), light, fragile with perfect cleavage parallel to the vertical prism. Transparent or translucent with vitreous luster, silky in fibrous aggregates. Fuses with difficulty to a white glass. Insoluble in acids.
**Environment** In dolomitic marbles, serpentinites and talc schists associated with magnesite and calcite.
**Occurrence** Splendid crystals ranging in color from white to gray in saccharoidal dolomites at Campolongo and Val Tremola (Ticino, Switzerland), at Edwards and Balmat, New York (USA), and other places in the USA and the USSR. Very fine crystal groups also occur in Val Devero (Novara, Italy), Greiner (Austria) and several places in the Urals (USSR). Asbestos is common in several places in the USA and the USSR, where it is worked industrially.
**Uses** Of interest to scientists and collectors. The fibrous varieties, like asbestiform-serpentine (chrysotile), are used in fireproofing and are highly effective electrical insulators.

## 205 ACTINOLITE

SILICATES (Inosilicates, Amphibole Group)
$Ca_2(Mg,Fe)_5Si_8O_{22}(OH)_2$ (Hydrous calcium magnesium iron silicate)

**System** Monoclinic.
**Appearance** Elongated prismatic crystals, often in massive or felty aggregates, color dark to light green.
**Physical properties** Hard (5–6), heavy, perfect cleavage parallel to the vertical prism. Translucent with vitreous luster. Insoluble, small fragments fuse easily (4 on the Kobell scale). An intermediate member of the tremolite-ferroactinolite series, with magnesium predominating over iron. A tough, felted, greenish variety is known as nephrite.
**Environment** Common in mafic metamorphic rocks of the green schist facies and parts of the amphibolite facies. May also be produced by alteration of pyroxenes in gabbroic and diabasic rocks.
**Occurrence** Splendid crystals occur in the Zillertal (Austria), and in Val Malenco (Sondrio) and Val Germanasca (Turin) (Italy). Ornamental nephrite comes from central Asia, New Zealand and Siberia but is also found in the USA (Wyoming), Canada (British Columbia), Germany, Poland and Italy.
**Uses** Of interest to scientists and collectors. Nephrite is used as an ornamental stone in jewelry and is often confused with jadeite, which is much more valuable.

▲ Fibrous tremolite (ca. ×1). Elba, Italy.

▲ Actinolite crystals (ca. ×1). Val Malenco, Italy

## 206 RICHTERITE

SILICATES (Inosilicates, Amphibole Group)
$(Na,K,Ca)_3(Mg,Mn)_5Si_8O_{22}(OH)_2$ (Hydrous sodium magnesium silicate)

**System** Monoclinic.
**Appearance** Elongated, fibrous, prismatic crystals, often in feltlike aggregates, purplish-red to yellow.
**Physical properties** Hard (5–6), heavy, perfect prismatic cleavage. Transparent or translucent with vitreous luster. Pale-yellow streak. Almost insoluble in acids but fuses fairly easily.
**Environment** Contact metamorphic mineral found in metasomatic calcareous rocks. Also occurs as a hydrothermal product in extrusive alkaline rocks.
**Occurrence** Typically eruptive varieties at Libby, Montana and Leucite Hills, Wyoming (USA), and Mont Saint Hilaire (Québec, Canada). Found in manganese-bearing calcareous rocks at Långban (Sweden) and Saint-Marcel (Val d'Aosta, Italy).
**Uses** Of interest to scientists and collectors.

2.97
3.45

## 207 HORNBLENDE (VAR. PARGASITE)

SILICATES (Inosilicates, Amphibole Group)
$(Ca,Na)_{2-3}(Mg,Fe^{+2},Fe^{+3},Al)_5(Al,Si)_8O_{22}(OH)_2$ (Complex silicate)

**System** Monoclinic.
**Appearance** Prismatic, generally rather stubby crystals, light brown, greenish or gray. Compact, almost lamellar masses.
**Physical properties** Hard (5–6), heavy, good prismatic cleavage and uneven fracture. Transparent or translucent with vitreous to greasy luster. Yellow-white streak. Insoluble in hydrochloric acid and fuses with difficulty.
**Environment** Mainly contact metamorphic marbles and skarns. Also found in some intrusive mafic rocks (gabbros, diorite porphyries) and in eclogites.
**Occurrence** A characteristic mineral in marble at Pargas (Finland) and skarns at Mansjo and Långban (Sweden). A variety known as carinthine is common in eclogites (at Saualpe, Austria, Munchberg, Germany, and Nowa Wies, Poland), but recently there have been an increasing number of finds in gabbros (Burlington, Pennsylvania, USA), in peridotites (Tinaquillo, Venezuela) and in veined rocks (Arigna, Sondrio, Italy).
**Uses** Of interest to scientists and collectors.

3.07
3.18

▲ Richterite crystals (ca. ×1.5). Långban, Sweden.

▲ Hornblende (var. pargasite) crystals (ca. ×1.5). Pargas, Finland.

## 208 HORNBLENDE

SILICATES (Inosilicates, Amphibole Group)
$(Ca,Na)_{2-3}(Mg,Fe^{+2},Fe^{+3},Al)_5(Al,Si)_8O_{22}(OH)_2$ (Complex silicate)

**System** Monoclinic.
**Appearance** Prismatic, generally stubby crystals, sometimes acicular or fibrous in parallel or sheaflike aggregates, dark green or black.
**Physical properties** Hard (5–6), heavy, perfect prismatic cleavage, approx. 120°. Translucent with vitreous luster. Insoluble and fuses with some difficulty to a green glass.
**Environment** An important constituent of metamorphic rocks especially amphibolites, some granulites and eclogites. Also found in mafic and ultramafic igneous rocks, both plutonic (diorites, gabbros, hornblendites, etc.) and (less often) volcanic (basalts).
**Occurrence** Because hornblende forms a complex solid solution series with several other amphibole minerals, localities for other species in the series will be cited. Pale green edenite, poor in iron, is typical of contact calcareous rocks, for example at Edenville, New York (USA), Pargas (Finland) and Kotaki (Japan). Rare tschermakite, rich in calcium and aluminum, is found in amphibolites at Pernio (Finland), in eclogites at Hurry Inlet (Greenland) and in basalts in Germany. Hastingsite, richer in sodium and aluminum, is found in gabbros in Ontario (Canada), Ghana, the Urals, Norway and various places in the USA.
**Uses** Of interest to scientists and collectors.

## 209 GLAUCOPHANE

SILICATES (Inosilicates, Amphibole Group)
$Na_2(Mg,Fe)_3Al_2Si_8O_{22}(OH)_2$ (Hydrous sodium magnesium aluminum silicate)

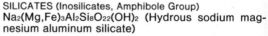
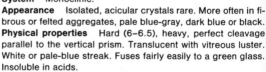
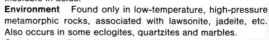

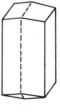

**System** Monoclinic.
**Appearance** Isolated, acicular crystals rare. More often in fibrous or felted aggregates, pale blue-gray, dark blue or black.
**Physical properties** Hard (6–6.5), heavy, perfect cleavage parallel to the vertical prism. Translucent with vitreous luster. White or pale-blue streak. Fuses fairly easily to a green glass. Insoluble in acids.
**Environment** Found only in low-temperature, high-pressure metamorphic rocks, associated with lawsonite, jadeite, etc. Also occurs in some eclogites, quartzites and marbles.
**Occurrence** Splendid, small, light-blue crystals (gastaldite variety) found in quartz nodules in schists in Piedmont (Italy). Fibrous and felted aggregates in California (USA), the island of Groix (France), at Svalbard (Norway), in eastern Kazakhstan and Azerbaidhan (USSR) and in many places in the western Alps.
**Uses** Of interest to petrologists as a means of defining the metamorphic conditions in which the surrounding rock formed.

▲ Hornblende crystals (ca. ×1). St. Gotthard, Switzerland.

▲ Glaucophane (ca. ×1.5). Piedmont, Italy.

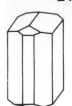

## 210 RIEBECKITE

SILICATES (Inosilicates, Amphibole Group)
$Na_2(Fe^{+2},Mg,Fe^{+3})_5Si_8O_{22}(OH)_2$ (Hydrous sodium iron magnesium silicate)

**System** Monoclinic.
**Appearance** Irregular, dark-blue or black, prismatic crystals. Also in rodlike or lamellar aggregates. Bluish, fibrous crystals in asbestiform aggregates (crocidolite), with characteristic chatoyancy.
**Physical properties** Semi-hard (4) with perfect prismatic cleavage (56° between the planes). Translucent with vitreous to silky luster. Fuses fairly easily, coloring the flame yellow (sodium). Insoluble in acids.
**Environment** In low-temperature and low-to-medium-pressure metamorphic rocks. Also present in felsic plutonic rocks rich in sodium (granites and pegmatites).
**Occurrence** Fine crystals in granite at Quincy, Massachusetts (USA) and in schists in Galicia (Spain). The fibrous variety (crocidolite) is found in South Africa, Australia and Brazil.
**Uses** Common riebeckite is of interest only to scientists and collectors, but crocidolite is an important industrial mineral, used like asbestiform serpentine (chrysotile) for fireproof products and heat insulation. Quartz pseudomorphous after crocidolite, with attractive chatoyancy, is used in jewelry under the name "tiger's-eye."

---

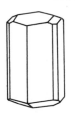

## 211 ARFVEDSONITE

SILICATES (Inosilicates, Amphibole Group)
$Na_{2-3}(Fe,Mg,Al)_5Si_8O_{22}(OH,F)_2$ (Hydrous sodium iron magnesium aluminum silicate)

**System** Monoclinic.
**Appearance** Prismatic, often tabular crystals, rarely well terminated, greenish-black to black. Rodlike or granular aggregates.
**Physical properties** Hard (6), heavy, perfect cleavage. Almost opaque, even in thin splinters, with vitreous luster. Dark blue-gray streak. Fuses easily to a magnetic black glass and insoluble in acids.
**Environment** Typical of alkaline plutonic rocks rich in iron (nepheline syenites and related pegmatites) and in rare metamorphic schistose rocks.
**Occurrence** Classic examples in Greenland (Julianehaab), Norway (near Oslo), Finland (Kiihtelysvaara) and the USSR (Mariupol, Ukraine, and the Kola Peninsula). Also found in the USA at Boulder, Colorado, and Red Hill, New Hampshire, and in Canada at Mont Saint Hilaire, Québec.
**Uses** Of interest to scientists and collectors.

▲ Fibrous riebeckite (var. crocidolite) (ca. ×1). South Africa.

▲ Arfvedsonite crystals (ca. ×1). Mont Saint Hilaire, Québec, Canada.

## 212 ANTHOPHYLLITE

SILICATES (Inosilicates, Amphibole Group)
$(Mg,Fe)_7Si_8O_{22}(OH)_2$ (Hydrous magnesium iron silicate)

**System** Orthorhombic.
**Appearance** Fibrous needles, lamellar, or felted aggregates, gray to light green or light brown.
**Physical properties** Hard (5.5–6), light to heavy, perfect prismatic cleavage. Translucent with vitreous luster. Insoluble and fusible with difficulty. Anthophyllite forms a solid solution series with gedrite $(Mg,Fe,Al)_7(Al,Si)_8O_{22}(OH)_2$. Cummingtonite $Mg_7Si_8O_{22}(OH)_2$ is the monoclinic polymorph of anthophyllite.

**Environment** Found in crystalline schists rich in magnesium, probably as a modification of olivine through hydration. Cummingtonite is also found in intrusive rocks.
**Occurrence** Splendid crystals found at Köngsberg (Norway). Also occurs at Fahlan (Sweden), in the Gothaab Fjord (Greenland), at Orijärvi (Finland), in Italy and on the island of Elba and in several places in the USA.
**Uses** Of interest to scientists and collectors.

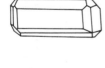

## 213 WOLLASTONITE

SILICATES (Inosilicates)
$CaSiO_3$ (Calcium silicate)

**System** Triclinic.
**Appearance** Rare tabular crystals. Usually in fibrous, acicular or radiating masses, white, grayish or rarely colorless.
**Physical properties** Hard (4.5–5), light to heavy, perfect cleavage. Translucent with vitreous or pearly luster on cleavage surfaces, silky when in fibers. Soluble in strong acids. Fuses fairly easily to a clear glass. Two other polymorphs exist: parawollastonite (very rare in nature and, like wollastonite, stable below 1126°C; 2059°F) and pseudowollastonite (rare in nature and stable above 1126°C).

**Environment** Typical of contact metamorphic calcareous and marl rocks, and also occurs in low-pressure regional metamorphic rocks.
**Occurrence** Sizable masses in the Black Forest (Germany), Brittany (France), Mexico and at Willsboro, New York (USA). Crystals found at Csiklowa (Romania), Chiapas (Mexico), Crestmore, California (USA), and Pargas (Finland). Occurs at Alpe Bazena (Adamello), in Valtellina, in the lavas of Vesuvius and at Capo di Bove (Rome) (Italy).
**Uses** Used in the manufacture of refractories.

▲ Fibrous anthophyllite (ca. ×1). Fahlan, Sweden.

▲ Wollastonite crystals (ca. ×1.5). Monte Somma, Italy.

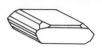

## 214 PECTOLITE

SILICATES (Inosilicates)
$NaCa_2Si_3O_8(OH)$ (Hydrous sodium calcium silicate)

**System** Triclinic.
**Appearance** Colorless, acicular crystals. Often in fibrous or fibrous-radiating spherical aggregates, white or gray. Compact masses.
**Physical properties** Hard (5), light, perfect cleavage. Transparent with vitreous to silky luster. White streak. Dissolves in hydrochloric acid, forming a silica gel. Fuses easily, coloring the flame yellow (sodium).
**Environment** A mineral formed by hydrothermal processes, filling cavities in basalts and associated with zeolites, prehnite, calcite, etc. Also found filling fractures in serpentinites and in contact metamorphosed limestones.
**Occurrence** Colorless and white crystals found at Franklin and Paterson, New Jersey (USA), and in mines at Thetford, Québec, and Ontario (Canada). Also occurs on the Kola Peninsula (USSR), in Bohemia (Czechoslovakia), and in Val di Fassa (Trento), and in cavities in basalt at Tierno (Trento) (Italy).
**Uses** Of interest to scientists and collectors.

2.89
2.90

## 215 RHODONITE

SILICATES (Inosilicates)
$(Mn,Fe,Mg)SiO_3$ (Manganese iron magnesium silicate)

**System** Triclinic.
**Appearance** Rare tabular crystals, sometimes with rounded edges and wrinkled faces, deep pink often tending to brown because of surface oxidation. Compact or granular masses with characteristic veins and black marks caused by manganese oxides.
**Physical properties** Hard (5.5–6.5), heavy, fragile, perfect prismatic cleavage (almost at right angles). Transparent to translucent, with vitreous luster, less frequently pearly on cleavage surfaces. However, often covered in places by opaque, blackish manganese oxides. Insoluble in acids (unlike similar, softer rhodochrosite). Fuses fairly easily to red or brown glass. Fowlerite is a zinc-bearing variety.

3.4
3.7

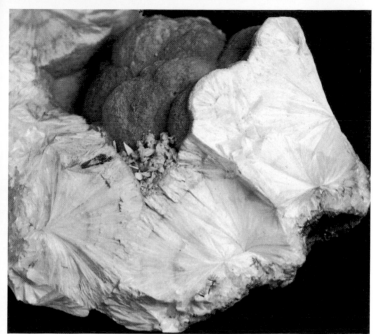

▲ Radiating pectolite balls (ca. ×1). New Jersey.

▲ Rhodonite crystals in calcite (ca. ×1.5). Vermland, Sweden.

**Environment** A mineral typical of metamorphism of impure limestones (manganese and silica-rich), which are often the result of contact metasomatic processes (skarns).

**Occurrence** Crystals and masses common at Långban and Pajsberg (Sweden) and in the Urals (Sverdlovsk district, USSR). Also found at Broken Hill (Australia), Chikla (India), Simsiö (Finland), in Arrow Valley (New Zealand), and in Brazil, South Africa and Japan. Found at Saint-Marcel (Val d'Aosta), Monte del Forno and in Valle di Scerscen (Sondrio) and at Libiola and Gambaresa (Genoa) (Italy). Fine fowlerite crystals are found in crystalline marbles at Franklin, New Jersey (USA).

**Uses** Faceted crystals are highly prized as collectors' pieces. Transparent or translucent varieties are cut as cabochons or as beads for necklaces and other types of jewelry. Ornaments made of massive rhodonite are more common and are particularly handsome when they are veined by manganese oxides.

## 216 BABINGTONITE

SILICATES (Inosilicates)
$Ca_2(Fe^{+2},Mn)Fe^{+3}Si_5O_{14}(OH)$ (Hydrous calcium iron manganese silicate)

**System** Triclinic.

**Appearance** Small, wedge-shaped crystals. Color black (blue or brown when altered).

**Physical properties** Hard (5.5–6), heavy, with good prismatic cleavage and uneven fracture. Opaque to translucent (splinters) with vitreous luster. Insoluble in hydrochloric acid. Fuses readily to a black magnetic globule.

**Environment** Occurs in cavities in basalt, gneiss and granite.

**Occurrence** Large, lustrous crystals on prehnite or calcite were found lining basalt cavities at Westfield, Massachusetts, and Paterson and Prospect Park, New Jersey (USA). Nice crystals on feldspar found in the granite pegmatites at Baveno (Italy).

**Uses** Of interest to mineralogists and collectors.

▲ Fibrous rhodonite (ca. ×1.5). Rocchetta Vära, Italy.

▲ Babingtonite crystals with calcite (ca. ×1). Westfield, Massachusetts.

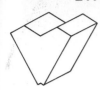

## 217 BERTRANDITE

SILICATES (Inosilicates—Reclassified Sorosilicates)
$Be_4Si_2O_7(OH)_2$ (Hydrous beryllium silicate)

**System** Orthorhombic.
**Appearance** Thin tabular or prismatic crystals usually pseudomorphous after beryl. Twinning common. Colorless to pale yellow.
**Physical properties** Hard (6–7), light, with excellent prismatic cleavage. Transparent to translucent with vitreous luster. Barely fusible. Insoluble in acids.
**Environment** Occurs in granite pegmatites, aplites and hydrothermal veins containing beryl. Forms through the breakdown of beryl, frequently crystallizing in cavities corroded out of beryl crystals by solutions.
**Occurrence** Fine bertrandite crystals are found in a number of pegmatites in the USA. Large crystals (up to 1 cm; 0.4 in) occur at Mount Antero, Colorado. Other localities include Portland, Connecticut, and Bedford, New York (USA); also the United Kingdom, Norway, Brazil, Germany, Czechoslovakia and the USSR.
**Uses** Of interest to mineralogists and collectors.

2.6

---

## 218 NEPTUNITE

SILICATES (Inosilicates)
$Na_2KLi(Fe,Mn)_2Ti_2Si_8O_{24}$ (Sodium potassium iron titanium silicate)

**System** Monoclinic.
**Appearance** Elongated, prismatic crystals with pointed terminations, black or very dark brown.
**Physical properties** Hard (5–6), heavy, perfect cleavage. Opaque or translucent with strong vitreous luster. Dark-red streak. Insoluble in hydrochloric acid and easily fusible.
**Environment** In nepheline syenites (foyaites) and in cavities in nepheline pegmatites. Associated with benitoite and natrolite in altered serpentinites. Though rarely identified, a common accessory in many undersaturated plutonic rocks.
**Occurrence** Splendid crystals at San Benito, California (USA). Abundant but less well crystallized in syenites in the Kola Peninsula (USSR) and in pegmatites in Greenland and Ireland.
**Uses** Of interest to scientists and collectors.

3.23

▲ Bertrandite crystals on microcline (ca. ×1). Brunswick, Maine.

▲ Neptunite crystals in natrolite (ca. ×1.5). San Benito County, California.

## 219 EUCLASE

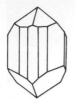

SILICATES (Phyllosilicates—Reclassified Nesosilicate)
BeAlSiO₄(OH) (Hydrous beryllium aluminum silicate)

$BeAlSiO_4(OH)$ (Hydrous beryllium aluminum silicate)

**System** Monoclinic.
**Appearance** Long or short prismatic crystals. Colorless, white, green or blue. Crystals often striated.
**Physical properties** Very hard (7.5), light, with perfect cleavage and conchoidal fracture. Transparent to translucent with vitreous luster. Insoluble. Fuses with difficulty.
**Environment** Occurs principally in granite pegmatites associated with topaz or in placer deposits.
**Occurrence** Currently the finest crystals of colorless, blue and green gem-quality euclase are mined near Ouro Préto, Minas Gerais (Brazil), where crystals up to 5 cm (2 in) or larger have been found. Other localities include Kenya, Tanzania, Sanarka River (USSR) and Park County, Colorado (USA).
**Uses** Occasionally faceted into gems.

## 220 PREHNITE

SILICATES (Phyllosilicates)
Ca₂Al₂Si₃O₁₀(OH)₂ (Hydrous calcium aluminum silicate)

$Ca_2Al_2Si_3O_{10}(OH)_2$ (Hydrous calcium aluminum silicate)

**System** Orthorhombic.
**Appearance** Rare, tabular, light-green crystals. More often in greenish-white reniform, stalactitic or mammillary aggregates.
**Physical properties** Hard (6–6.5), light, perfect basal cleavage. Translucent with vitreous luster. White streak. Fuses easily to a blistered yellowish-white glass and dissolves slowly in hydrochloric acid without leaving a residue of gelatinous silica.
**Environment** A hydrothermal mineral filling cavities in basaltic volcanic rocks. Also lines fractures in various intrusive and metamorphic rocks. Found in schistose rocks with zeolites.
**Occurrence** Splendid mammillated aggregates with barrel-shaped crystals found in the Dauphiné (France). In Italy has been found in Val Malenco (Sondrio), Val D'Ala (Turin), Val di Fassa (Trento), Alpe di Siusi (Bolzano), and the Modenese Apennines (Toggiano and Reggiano) (Ciano d'Enza). Botryoidal masses in basalts at Paterson, New Jersey, Westfield, Massachusetts, Fairfax, Virginia, and in copper deposits at Lake Superior (USA).
**Uses** Of interest to scientists and collectors.

▲ Euclase crystals in pegmatite (ca. ×2). Minas Gerais, Brazil.

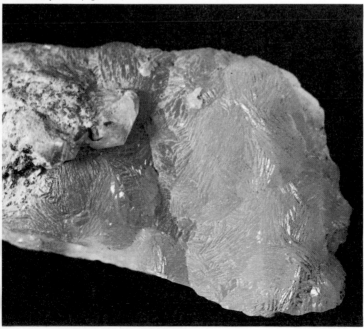

▲ Botryoidal prehnite (ca. ×1). Alpe di Siusi, Italy.

## 221 ASTROPHYLLITE

SILICATES (Phyllosilicates)
$(K,Na)_3(Fe,Mn)_7Ti_2Si_8O_{24}(O,OH)_7$ (Hydrous potassium iron titanium silicate)

**System**  Triclinic.
**Appearance**  Rare, small, tabular or platy crystals, often grouped in starlike aggregates, golden yellow or yellowish brown.
**Physical properties**  Semi-hard (3.5–4.5), heavy, fragile, with perfect cleavage into inflexible, inelastic plates. Transparent or translucent with vitreous luster tending to metallic, pearly on cleavage surfaces. Yellowish streak. Soluble in acids with some difficulty, fuses easily to a dark magnetic glass.
**Environment**  In cavities and fissures of nepheline syenites and other felsic alkali plutonic rocks, associated with acmite, feldspars, brown micas, zircon and titanite.
**Occurrence**  Found in Greenland (Narsarsuk and Kangerdluarsuk), the Kola Peninsula (Khibina, USSR), Norway (Brevig) and the USA (Pikes Peak, Colorado).
**Uses**  Of interest to scientists and collectors.

## 222 APOPHYLLITE

SILICATES (Phyllosilicates)
$KCa_4Si_8O_{20}(F,OH)\cdot8H_2O$ (Hydrated potassium calcium silicate)

**System**  Tetragonal.
**Appearance**  Tabular crystals of cubic or octahedral appearance, sometimes dipyramidal or platy. Colorless or white, or light shades of pink, green and yellow.
**Physical properties**  Semi-hard to hard (4.5–5), light, perfect basal cleavage. Translucent or transparent with pearly luster on basal faces, vitreous on prism faces. Fuses easily, coloring the flame violet (potassium) and giving a pale-colored glass. Dissolves in hydrochloric acid, forming gelatinous silica.
**Environment**  A hydrothermal mineral, filling cavities in basaltic and tufaceous rocks, associated with stilbite, scolecite, calcite, prehnite, analcime, etc.
**Occurrence**  Fine crystals found at Alpe di Siusi (Bolzano, Italy), Poona (India), Andreasberg (Germany), in Nova Scotia (Canada), Iceland, the Faroe Islands and Greenland and Rio Grande do Sul (Brazil). Also occurs at Traversella (Turin) and in basalts at San Pietro di Montecchio Maggiore (Vicenza) (Italy). Found in the copper deposits of Lake Superior and at Paterson, New Jersey (USA).
**Uses**  Of interest to scientists and collectors.

▲ Astrophyllite (ca. ×1.5). Kola Peninsula, USSR.

▲ Apophyllite crystals (ca. ×1.5). Poona, India.

## 223 PYROPHYLLITE

SILICATES (Phyllosilicates)
$Al_2Si_4O_{10}(OH)_2$ (Hydrous aluminium silicate)

**System** Monoclinic.
**Appearance** Never in distinct crystals. Usually in lamellar or radiating foliated aggregates, yellowish white, pale green or green-brown. Compact masses are used for carvings (variety agalmatolite or pagodite).
**Physical properties** Very soft (1–2), light, perfect cleavage. Small sections are flexible but not elastic. Pearly or greasy luster. Greasy feel. Almost insoluble. Infusible and flakes when heated. Very difficult to distinguish from talc except by chemical or x-ray analysis.
**Occurrence** Large masses found in Guilford and Orange counties, North Carolina, Arkansas and Georgia (USA), in South Africa and the USSR (southern Urals). Small amounts found in the Swiss Alps and also occurs in Sicily. Pyrophyllite rosettes come from Chesterfield, South Carolina, and California.
**Uses** As a dry lubricant and for heat and electrical insulation. Also used in paper-making and in the rubber, fabric and soap industries. Agalmatolite is used for ornamental purposes.

2.8

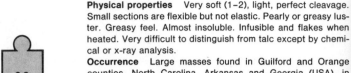

## 224 TALC

SILICATES (Phyllosilicates)
$Mg_3Si_4O_{10}(OH)_2$ (Hydrous magnesium silicate)

**System** Monoclinic.
**Appearance** Never occurs in distinct crystals. Sometimes in pseudo-hexagonal laminae, but usually in scaly, foliated aggregates, white, greenish-white, gray or brownish. Compact, felted light-gray masses (steatite or soapstone).
**Physical properties** Very soft (1), light, sectile, with perfect cleavage into laminae, sometimes thin, that are flexible but not elastic. Translucent with pearly luster in laminae, greasy when massive. White streak. Greasy feel. Poor conductor of heat. Insoluble and infusible. Steatite is easy to carve and can be worked with a lathe.
**Environment** An alteration product of magnesium silicates in ultramafic rocks, common in regionally metamorphosed rocks (schists). Also formed by metasomatism in impure dolomitic marbles.
**Occurrence** Large deposits in Styria (Austria), Canada, at Madras (India), in the Transvaal, Australia, the Pyrenees, Korea and various places in the USA and USSR. Quarried in Val Chisone and Val Germanasca (Piedmont), Val Malenco (Sondrio) and at Orani (Sardinia) (Italy).
**Uses** In powder form used as an ingredient in paper and rubber, in the textile, cosmetics and paint industries, etc.

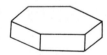

2.58
2.83

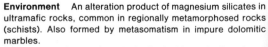

▲ Radiating pyrophyllite crystals (ca. ×1.5). California.

Massive talc (ca. ×1). USSR.

## 225 MUSCOVITE

SILICATES (Phyllosilicates, Mica Group)
$KAl_2(AlSi_3)O_{10}(OH)_2$ (Hydrous potassium aluminum silicate)

**System**  Monoclinic.

**Appearance**  Tabular crystals with pseudo-hexagonal or triangular outline, sometimes with deep striations on the prism faces. Foliated, scaly, lamellar masses. Sometimes microcrystalline (sericite). Laminae silvery-white, masses white or yellow, sometimes dark brown when small laminae of hematite or rutile are scattered over the cleavage planes.

**Physical properties**  Soft (2–2.25), light, perfect basal cleavage into laminae (sheets), sometimes very thin, which are flexible and elastic. Insoluble and fuses with difficulty. There are a number of distinct varieties containing chromium (fuchsite), iron (ferrianmuscovite) and manganese (alurgite). These display characteristic colors and other features which distinguish them from muscovite itself. Paragonite is a species similar to muscovite but containing sodium rather than potassium. Illite is a micalike mineral poor in potassium and typical of sedimentary rocks.

**Environment**  One of the most common minerals in rocks, especially plutonic igneous rocks rich in silica and aluminum (pegmatites, granites) and low- or medium- to high-grade metamorphic rocks (green schist and amphibolite facies). The varieties are less common and are characteristic of certain chemical environments. Muscovite is also a sedimentary mineral common in loose rocks (sands) and rocks affected by diagenesis (sandstones, marls, etc.).

**Occurrence**  Enormous crystals measuring 30–50 sq m (32–54 sq yds) have been found in pegmatites in Ontario (Canada), New Hampshire and South Dakota (USA), India and Brazil. Also frequent in many places in the Alps, especially in pegmatites. Pargonite is found at Pizzo Forno (Ticino, Switzerland), with kyanite and staurolite, and in several places in the Alps, especially the western Alps, as an essential component of calc-schist metamorphic rocks, and in compact nodules (cossaite variety).

**Uses**  Used for electrical and heat insulation, either as thin sheets or in the form of a manufactured paper made from powdered muscovite bound with cement, plastic, etc. In the manufacture of paper, rubber and fireproof paint. Also used in porcelain products and as a dry lubricant.

▲ Muscovite crystals (ca. ×1). Brazil.

◢ Muscovite crystals (ca. ×1.5). Piona, Italy.

### 226 MUSCOVITE (VAR. FUCHSITE)
SILICATES (Phyllosilicates, Mica Group)
$K(Al,Cr)_2(AlSi_3)O_{10}(OH)_2$ (Hydrous potassium aluminum silicate)

**System** Monoclinic.
**Appearance** Small laminae or scaly aggregates, color bright emerald green.
**Physical properties** As this is the chromium-bearing variety of muscovite, it too is soft (2–2.25), light, with perfect basal cleavage into translucent sheets, though these are never large or transparent. Fuses with difficulty and insoluble in acids.
**Environment** In medium-grade schists associated with biotite. In metamorphic dolomites or calc-schists in the western Alps, where its presence shows that there are periodotitic or serpentinite rocks nearby.
**Occurrence** Large aggregates at Pragatten and in the Zillertal (Austria), in the Vizze and Fundres valleys (Alto Adige, Bolzano, Italy). Also found in Italy in Val Lanterna (Val Malenco, Sondrio), in marble at Musso (Como), Ontario (Canada) and California (USA).
**Uses** Of interest to scientists and collectors.

2.88

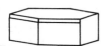

### 227 MUSCOVITE (VAR. ALURGITE)
SILICATES (Phyllosilicates, Mica Group)
$K(Al,Mn)_2(AlSi)_3O_{10}(OH)_2$ (Hydrous potassium aluminum silicate)

**System** Monoclinic.
**Appearance** Brick-red lamellae and foliated aggregates.
**Physical properties** Soft (2–2.25), heavy, perfect basal cleavage into thin, flexible, elastic lamellae. Transparent or translucent with vitreous luster. Pinkish-white streak. Insoluble in acids and infusible.
**Environment** In metamorphic rocks rich in manganese and partially oxidized iron.
**Occurrence** To date known only in the manganese-bearing deposit of Saint-Marcel (Val d'Aosta, Italy), where alurgite is associated with diopside (var. violane), braunite and quartz in metamorphosed jasper in glaucophane-schist facies.
**Uses** Of interest to scientists and collectors.

2.98
3.0

▲ Muscovite (var. fuchsite) crystals (ca. ×1.5). Zillertal, Austria.

▲ Muscovite (var. alurgite) (ca. ×1.5). Saint-Marcel, Italy.

## 228 GLAUCONITE

SILICATES (Phyllosilicates, Mica Group)
$(K,Na)(Al,Fe^{+3},Mg)_2(Al,Si)_4O_{10}(OH)_2$ (Hydrous potassium aluminum iron magnesium silicate)

**System** Monoclinic.
**Appearance** Very tiny crystals. Usually in spherulitic, earthy aggregates, sometimes platy, light blue-green.
**Physical properties** Soft (2–2.25), light, perfect basal cleavage. Translucent with rather greasy luster. Dissolves easily in hydrochloric acid, fuses with difficulty.
**Environment** A mineral of marine origin, found in sedimentary deposits of various kinds (sands, sandstones, silts, impure limestones), probably as a diagenetic mineral. Celadonite, a mineral of continental origin, has a similar composition and physical properties.
**Occurrence** Glauconite is common in Tertiary marine sands from the Baltic and in many other sedimentary rocks, recent and ancient. Celadonite is found in the Ukraine (USSR), and in cavities in basalts at Monte Baldo (Verona, Italy).
**Uses** Used in the textile, sugar and brewing industries, in the manufacture of fertilizers (because of its potassium content), and as a nontoxic coloring agent.

2.5
2.8

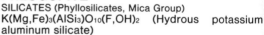

## 229 PHLOGOPITE

SILICATES (Phyllosilicates, Mica Group)
$K(Mg,Fe)_3(AlSi_3)O_{10}(F,OH)_2$ (Hydrous potassium aluminum silicate)

**System** Monoclinic.
**Appearance** Six-sided plates, sometimes of enormous size, up to 2 m (6.6 ft) in diameter. Light brown or yellowish in color. Often in groups or foliated aggregates.
**Physical properties** Soft (2.5–3), light, perfect basal cleavage into flexible, elastic, transparent plates with vitreous to pearly luster. Fuses with difficulty, decomposes in concentrated sulfuric acid (unlike muscovite). An excellent electrical insulator.
**Environment** In medium- to high-grade metamorphic rocks rich in magnesium (crystalline dolomites, altered peridotites, and serpentinites). Common as very large plates in kimberlites, in contact limestones and in pegmatites.
**Occurrence** Fine, small plates occur in saccharoidal dolomite in Campolongo (Switzerland), in marble at Pargas (Finland) and in Sweden. Enormous crystals in pegmatites in Ontario (Canada), Malagasy Republic and Sri Lanka. The Sri Lanka crystals are often asteriated due to tiny oriented inclusions of rutile. Found in crystalline dolomite at Crevola (Novara), in serpentinite in Val Malenco (Sondrio) and in various marbles (Italy).
**Uses** An industrial mineral, used in plates as an electrical insulator, and in powder form as a filler.

2.86

▲ Glauconite (ca. ×1). Grognardo, Italy.

▲ Phlogopite crystals (ca. ×1). Malagasy Republic.

## 230 BIOTITE

SILICATES (Phyllosilicates, Mica Group)
K(Mg,Fe)$_3$(Al,Fe)Si$_3$O$_{10}$(OH,F)$_2$ (Hydrous potassium aluminum silicate)

**System** Monoclinic.
**Appearance** Rarely in tabular pseudo-hexagonal crystals, black, brown or dark green. Commonly in small disseminated plates or platy aggregates.
**Physical properties** Soft (2.5–3), heavy, perfect basal cleavage into small, flexible, elastic, transparent or translucent, brownish sheets, with very pronounced vitreous luster. Fuses with difficulty, soluble only in heated, concentrated sulfuric acid.

**Environment** An important and common mineral of many intrusive igneous rocks, pegmatites, lamprophyres, some lavas and metamorphic rocks. Widespread as a sedimentary mineral in sands and sandstones, sometimes in small flakes or particles of a bright golden-yellow color.
**Occurrence** Very large crystals up to 70 sq. m (76.6 sq yds) found in pegmatites, for example in the Ilmen mountains (USSR), Greenland, Scandinavia and Brazil. Magnificent crystals occur in geodes in the lavas of Vesuvius and in Val di Fassa (Trento, Italy). Siderophyllite is found at Baveno (Novara, Italy) and in Alaska (USA).
**Uses** Of interest to petrologists and collectors.

## 231 LEPIDOLITE

SILICATES (Phyllosilicates, Mica Group)
K(Li,Al)$_3$(Si,Al)$_4$O$_{10}$(F,OH)$_2$ (Hydrous potassium lithium aluminum silicate)

**System** Monoclinic.
**Appearance** Six-sided, pink to lilac lamellae. More often in fine platy aggregates.
**Physical properties** Soft to semi-hard (2.5–4), light, perfect basal cleavage into flexible, elastic lamellae, translucent with pearly luster. Fuses easily, coloring the flame crimson (lithium), to an opaque white glass. Insoluble in acids.

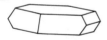

**Environment** In lithium-bearing pegmatites and in cavities in some greisen, associated with spodumene, amblygonite, tourmaline and other minerals.
**Occurrence** Small transparent laminae at Coolgardie (Australia). Fine aggregates in the USA at Pala, California, Dixon, New Mexico, Middletown, Connecticut, Maine and South Dakota. Also found in the Urals (USSR), the island of Elba (San Piero in Campo), Canada, Malagasy Republic, Rhodesia, in huge crystals at Virgem da Lape, Minas Gerais (Brazil), and Japan, at Rozna (Moravia) and Uto (Sweden).
**Uses** An ore of lithium.

▲ Biotite (ca. ×1). Ontario, Canada.

▲ Lepidolite (ca. ×1.5). Sweden.

## 232 LEPIDOLITE (VAR. ZINNWALDITE)

SILICATES (Phyllosilicates, Mica Group)
$KLiFe^{+3}Al(AlSi_3)O_{10}(F,OH)_2$ (Hydrous potassium aluminum silicate)

**System** Monoclinic.
**Appearance** Tabular crystals, often pseudo-hexagonal, color silvery-gray to brown.
**Physical properties** Soft (2.5–3), heavy, perfect basal cleavage into small, flexible and elastic laminae. Transparent with vitreous luster, pearly on cleavage planes. Fuses easily, coloring the flame crimson (lithium), to a slightly magnetic globule. Attacked by strong acids.
**Environment** In greisen, pegmatites and hydrothermal veins with quartz.
**Occurrence** Splendid aggregates found in Zinnwald (Saxony, DDR), the Erzgebirge (Bohemia, Czechoslovakia), and Cornwall (England). Found at Cuasso al Monte (Varese) in granophyre, and at Baveno (Novara) in granite (Italy). Common in many lithium-bearing igneous rocks in the USSR, Greenland, Canada, Malagasy Republic and the USA (Amelia Court House, Virginia, San Diego County, California).
**Uses** An ore of lithium.

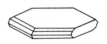

---

## 233 MARGARITE

SILICATES (Phyllosilicates, Mica Group)
$CaAl_2(Al_2Si_2)O_{10}(OH)_2$ (Hydrous calcium aluminum silicate)

**System** Monoclinic.
**Appearance** Individual lamellae with pseudo-hexagonal outline very rare. Usually in foliated or lamellar, pale-pink, white or gray aggregates.
**Physical properties** Semi-hard (3.5–4.5), light to heavy, perfect cleavage into inelastic lamellae ("brittle mica"). Transparent or translucent with pearly luster. White streak. Partially soluble in concentrated, hot hydrochloric acid. Fuses with difficulty around the edges.
**Environment** Medium-grade metamorphic mineral, common in emery deposits, but also found in chlorite schists and mica schists, associated with staurolite and tourmaline.
**Occurrence** Very common in emery deposits on Naxos (Greece), at Smyrna (Turkey) and in North Carolina (USA). In the Alps has been found in chlorite schists in the Zillertal (Austria) and recently in various schistose rocks in the St. Gotthard region (Switzerland), and in Val Formazza (Novara) and Val di Vizze (Bolzano) (Italy). Pink aggregates found at Chester, Massachusetts, and in Chester County, Pennsylvania (USA).
**Uses** Of interest to scientists and collectors.

▲ Lepidolite (var. zinnwaldite) crystals (ca. ×1.5). Zinnwald, Czechoslovakia.

▲ Margarite crystals (ca. ×1.5). Chester, Massachusetts.

## 234 MONTMORILLONITE

SILICATES (Phyllosilicates, Clay Group)
$(Na,Ca)_{0.33}(Al,Mg)_2Si_4O_{10}(OH)_2 \cdot nH_2O$ (Hydrated sodium calcium aluminum silicate)

**System** Monoclinic.

**Appearance** Never in crystals larger than 1 $\mu$ in diameter. Usually in earthy, dusty and scaly amorphous-looking masses. White or gray.

**Physical properties** Very soft (1), very light, disintegrates easily, greasy feel. Opaque because of its extremely fine particles. Two interesting characteristics of montmorillonite are its capacity to expand by absorbing water or other liquids and its potential for ion exchange when, under certain conditions, the cations sodium, potassium and calcium present between layers in the structure are replaced by other cations present in a solution.

**Environment** In a sedimentary environment in a tropical climate through alteration of feldspars in rocks poor in silica. In a hydrothermal environment at the expense of volcanic glasses and tuffs.

**Occurrence** A basic constituent of the rocks known as bentonites, found in large masses in France (Montmorillon), Germany, Japan, USA (Florida, Georgia, Alabama, California).

**Uses** An important industrial mineral, used as an absorbent to purify and decolor liquids, as a filler in paper and rubber, as a base for cosmetics and medicines and as a mud for drilling and quarrying.

---

## 235 "VERMICULITE"

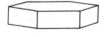

SILICATES (Phyllosilicates)
$(Mg,Fe,Al)_3(Al,Si)_4O_{10}(OH)_2 \cdot 4H_2O$ (Hydrated magnesium iron aluminum silicate)

**System** Monoclinic.

**Appearance** Platy, green or golden-yellow-to-brown crystals with pseudo-hexagonal outline. Scaly aggregates and frequently pseudomorphic after biotite.

**Physical properties** A generalized group of minerals. Soft (2–3), light, perfect basal cleavage into small, rather inelastic laminae. Translucent with vitreous to pearly luster. Pale-yellow streak. Slightly soluble in acids. When heated to about 300°C (572°F) loses water quickly, flakes and expands from 18 to 25 times its original volume. It develops into long, twisted, worm-like forms of a deep golden-yellow color, and the density decreases until it can float on water.

**Environment** A hydrothermal alteration product of biotite and phlogopite, especially at the contact between felsic intrusive rocks and ultramafic rocks.

**Occurrence** Large masses at Palabora (South Africa), Milbury, Massachusetts, Libby, Montana, and Macon, North Carolina (USA), in Argentina, western Australia and Canada.

**Uses** When partially heated becomes a very light thermal and acoustical insulating material, used in building and for technical insulating equipment. Also used in the paper, paint and plastics industries.

▲ Bentonite (containing montmorillonite) (ca. ×1). Daunia, Italy.

▲ Natural and expanded vermiculite (ca. ×1.5). Argentina.

## 236 CLINOCHLORE (VAR. PENNINITE)

SILICATES (Phyllosilicates, Chlorite Group)
$(Mg,Fe^{+2})_5Al(Si,Al)_4O_{10}(OH)_8$ (Hydrous magnesium iron aluminum silicate)

**System** Monoclinic.

**Appearance** Tabular crystals with hexagonal outline, dark green in color. Lamellar aggregates, scaly or radiate and massive.

**Physical properties** Soft (2–2.5), light, perfect cleavage into flexible, inelastic lamellae. Translucent with vitreous to pearly luster. Soluble in acids, especially hot, concentrated sulfuric acid. Does not fuse but flakes when heated. A pseudo-trigonal variety of clinochlore.

**Environment** A product of hydrothermal alteration or low-grade metamorphism of magnesium and iron silicates. The essential component of some massive metamorphic rocks (chlorite schists).

**Occurrence** Splendid crystals at Zermatt (Switzerland), in Val Malenco (Sondrio) and Val d'Ala (Turin) (Italy) and in the Zillertal (Austria). Penninite-rich rocks are common all over the world. It is quarried in Italy in Val Malenco, Val Chiavenna (Sondrio) and Val d'Ossola (Novara).

**Uses** Of interest to petrologists and collectors. It is easily carved and worked and is used for ornaments.

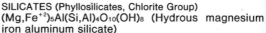

## 237 CLINOCHLORE

SILICATES (Phyllosilicates, Chlorite Group)
$(Mg,Fe^{+2})_5Al(Si,Al)_4O_{10}(OH)_8$ (Hydrous magnesium iron aluminum silicate)

**System** Monoclinic.

**Appearance** Lamellar crystals of pseudo-hexagonal habit. Rare, prismatic, barrel-shaped crystals, blue-green in color with yellow tints, or more rarely whitish (leuchtenbergite variety). Scaly or wormlike aggregates (helminthes variety).

**Physical properties** Soft (2–2.5), light, perfect basal cleavage into flexible, but inelastic lamellae. Translucent with pearly luster. Soluble in strong acids, leaving a silica skeleton. When heated flakes but does not fuse.

**Environment** An essential mineral in many chlorite schists and talc schists. Also occurs as isolated lamellae in fractures in rodingites and other metamorphic rocks.

**Occurrence** Splendid twisted aggregates of pseudo-hexagonal laminae found in crystalline schists in Val d'Ala (Turin, Italy). Also found in the western Alps, in Val Malenco (Sondrio), Val di Vizze (Bolzano) and in Liguria. There are other famous localities in Austria (the Zillertal), the USSR (Achmatowsk, in the Urals) and in the USA (Brewster, New York, and West Chester and Unionville, Pennsylvania).

**Uses** Of interest to scientists and collectors.

▲ Penninite (ca. ×1). Val Malenco, Italy.

▲ Clinochlore crystals (ca. ×1.5). Brewster, New York.

## 238 KAOLINITE
SILICATES (Phyllosilicates, Clay Group)
Al₂Si₂O₅(OH)₄ (Hydrous aluminum silicate)

$Al_2Si_2O_5(OH)_4$ (Hydrous aluminum silicate)

**System**   Monoclinic.

**Appearance**   Very rare, small, white, pseudo-hexagonal, lamellar crystals. Usually in rather friable, earthy or clayey aggregates (kaolin), varying in color from yellowish-gray to brown depending on the impurities present.

**Physical properties**   Soft (2–2.5), very light with perfect basal cleavage. Plates translucent with pearly luster, opaque when massive. When mixed with water becomes plastic and easy to mold. Soluble in hot, concentrated sulfuric acid. Infusible, loses water between 390° and 450°C (734°–842°F).

2.58
2.60

**Environment**   Formed by alteration, sometimes hydrothermal, of feldspars and other aluminum-bearing minerals. The largest deposits are beds of clay formed in lakes.

**Occurrence**   Large deposits in China (Janchu Fa) that were quarried in ancient times, in several places in Germany (Saxony, Bavaria), Czechoslovakia (Bohemia), Cornwall (England), France (near Limoges), the USSR (Urals, Ukraine) and Richmond, Virginia, and Georgia (USA). Also common in many places in Italy—Sardinia, at Torniella (Grosseto), Valle Pozzatello (Livorno), Oleggio (Novara) and S. Ulderico dei Tretti (Vicenza), though there it is very impure.

**Uses**   Essential in the china industry and also used as a filler for paper and rubber and in medicines and cosmetics.

---

## 239 SERPENTINE GROUP
SILICATES (Phyllosilicates)
(Mg,Fe)₃Si₂O₅(OH)₄ (Hydrous magnesium iron silicates)

$(Mg,Fe)_3Si_2O_5(OH)_4$ (Hydrous magnesium iron silicates)

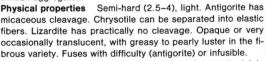

**System**   Monoclinic, orthorhombic and hexagonal.

**Appearance**   The serpentine group contains several polytypes, not always easily identifiable. Antigorite is lamellar and usually occurs in tough, compact, often pleated masses, some shade of dark green. Chrysotile (orthochrysotile and clinochrysotile) is almost always in fibrous yellowish-white or green aggregates (asbestos). Lizardite is in minute scales or compact whitish aggregates.

**Physical properties**   Semi-hard (2.5–4), light. Antigorite has micaceous cleavage. Chrysotile can be separated into elastic fibers. Lizardite has practically no cleavage. Opaque or very occasionally translucent, with greasy to pearly luster in the fibrous variety. Fuses with difficulty (antigorite) or infusible.

2.5
2.6

**Environment**   In low-grade metamorphic environments rich in water, derived from magnesium silicates like olivine, pyroxenes and amphiboles. Filling veins and pseudomorphic replacement of magnesium silicates in mafic and ultramafic rocks. The main minerals in serpentinites.

**Occurrence**   A common mineral. Occurs in the USA in numerous localities and at Val Malanco, Sondrio (Italy).

**Uses**   Serpentine-bearing rocks are used in building and as ballast material for railways. The fibrous variety, "chrysotile asbestos," is used for soundproofing and thermal insulation.

▲ Massive kaolinite (ca. ×1). Saxony, Germany.

▲ Serpentine (ca. ×1). Val Malenco, Italy.

## 240 "GARNIERITE"

SILICATES (Phyllosilicates, Serpentine Group)
$(Ni,Mg)_6Si_4O_{10}(OH)_8$ (Hydrous nickel magnesium silicate)

**System** Monoclinic.

**Appearance** A general term for hydrous nickel silicates. Lamellar or felted, microcrystalline, bright-green aggregates. Crusts or earthy masses.

**Physical properties** Soft to semi-hard (2–4), light, fragile with splintery fracture in massive varieties. Translucent with greasy luster, waxlike and earthy in microcrystalline varieties. Light-green streak. Soluble in hot hydrochloric acid and infusible.

**Environment** A product of hydrous alteration of ultramafic rocks rich in nickel sulfides (medium- to low-grade metamorphism).

**Occurrence** Large masses in serpentines in New Caledonia, the USSR and to a lesser extent in South Africa and Douglas County, Oregon (USA). Also occasionally found at Baldissero (Turin, Italy).

**Uses** One of the main nickel ores.

---

## 241 CHRYSOTILE

SILICATES (Phyllosilicates, Serpentine Group)
$Mg_3Si_2O_5(OH)_4$ (Hydrous magnesium silicate)

**System** Monoclinic (clinochrysotile) and orthorhombic (orthochrysotile).

**Appearance** Microcrystalline, fibrous aggregates, grayish-white, green, yellow or brown.

**Physical properties** The fibrous variety of serpentine (asbestos) has a characteristic structure with individual layers recurved, giving it an elongated tubelike or fibrous appearance. Semi-hard (2.5–4), light, no cleavage, but can be divided into elastic fibers flexible enough to be woven. Opaque with silky luster. Infusible. Dissolves in strong acids, with separation of silica. An insulator against heat and electricity.

**Environment** Produced by metamorphic alteration of ultramafic rocks in a low-grade environment rich in water. Associated with antigorite and lizardite in serpentinite, especially filling fractures with fibers more or less parallel or vertical to the fracture walls (long- or short-fiber asbestos).

**Occurrence** Large masses in Québec (Canada), Eden Mills, Vermont, and Brewster, New York (USA) and the Urals (USSR). Smaller deposits in Rhodesia, Cyprus and the Alps, for example in Italy at Balangero and the upper Val di Lanzo (Turin), in Val d'Aosta, Val Sesia (Vercelli), Val Malenco (Sondrio).

**Uses** An industrial mineral widely used for thermal or electrical insulation.

▲ Massive "garnierite" (ca. ×1). New Caledonia.

▲ Fibrous chrysotile asbestos (ca. ×1.5). Asbestos, Canada.

## 242 CHRYSOCOLLA

SILICATES (Phyllosilicates)
(Cu,Al)$_2$H$_2$Si$_2$O$_5$(OH)$_4$·nH$_2$O (Hydrous copper silicate)

**System** Monoclinic.
**Appearance** Stalactitic masses. Earthy, microcrystalline crusts, bright green or bluish.
**Physical properties** Soft to semi-hard (2.4), light, translucent with vitreous to greasy luster. When heated does not fuse but turns blackish and colors the flame green (copper). Decomposes in hydrochloric acid, producing a silica gel.

**Environment** In the oxidation zone of copper deposits, associated with azurite, malachite and cuprite. An important surface indicator pinpointing the presence of disseminated copper deposits (porphyry copper ore).
**Occurrence** Large masses at Chuquicamata (Chile) and in the desert mining regions of Arizona (Clifton-Morenci districts) and New Mexico (USA), Morocco and Rhodesia. Also found in the USSR and in Italy at Predarossa (Val Masino, Sondrio) and Monzoni in Val di Fassa (Trento).
**Uses** A useful ore of copper, though not of prime importance.

## 243 SEPIOLITE

SILICATES (Phyllosilicates)
Mg$_4$Si$_6$O$_{15}$(OH)$_2$·6H$_2$O (Hydrous magnesium silicate)

**System** Orthorhombic.
**Appearance** Compact, white or yellowish concretions or porcellanous masses, sometimes porous or earthy-looking.
**Physical properties** Soft (2-2.5), light, easily cut with a blade. So fine-grained that even very thin fragments are opaque, and the dry porous masses float on water (German *Meerschaum*, "sea foam"). Taste sets the teeth on edge. Becomes plastic when mixed with water. Almost infusible and insoluble in acids.
**Environment** A surface alteration product of magnesite and serpentine. Also occurs as an authigenic mineral in some lake sediments.
**Occurrence** Nodular masses at Esckischir (Turkey) and in various localities in Spain, Greece, Czechoslovakia, Morocco and the USA (Utah, California, Pennsylvania and New Mexico).
**Uses** Used in the manufacture of tobacco pipes.

▲ Massive chrysocolla (ca. ×1). Nevada.

▲ Massive sepiolite (ca. ×1). Turkey.

# QUARTZ

SILICATES (Tectosilicates, Silica Group)
α-SiO₂ (Silicon oxide)

**System**  Hexagonal.

**Appearance**  May occur either in well-formed crystals, sometimes enormous and weighing up to 130 kg (289 lb), or in compact and concretionary masses, microcrystalline (quartzites) to cryptocrystalline (agates, jaspers, carnelian, etc.). Colorless when pure (rock crystal). The presence of impurities may give a whole range of colors. Quartz crystals are usually hexagonal and prismatic, terminated by two (positive and negative) rhombohedra resembling hexagonal dipyramids. The faces of the prism may be horizontally striated, with the edges corroded randomly. "Scepter" crystals (oriented overgrowths whereby one quartz crystal caps another formed earlier) are rarer. Quartz is often twinned. Dauphiné twins are the most common followed by Brazil and Japan Law twins. Quartz is enantiomorphic or rotary polar. This means that a crystal is either right- or left-handed, which is caused by the spiral structure of linked silicon-oxygen tetrahedra in quartz. The handedness of a crystal is determined by the position of small (trigonal trapezohedral) faces adjacent to the large pyramidal and prismatic faces. Crystals often contain bubbles of gas or liquid clearly visible to the naked eye ("level quartz").

**Physical properties**  The common form is α quartz, which is stable up to 573°C (1063°F). Above this temperature hexagonal β quartz is stable and is preserved paramorphically, mainly in volcanic rocks. α quartz is very hard (7) and light. No cleavage but good conchoidal fracture. Transparent to translucent with vitreous luster. Strongly piezoelectric and pyroelectric (except in Brazil twins, where the charges of the two twins cancel one another since they are oriented in exactly opposite directions). Quartz is rotary polar; it has the ability to rotate the plane of polarization of light passing through a crystal. Infusible and insoluble in acids except for hydrofluoric acid, in which it decomposes giving off fumes of silicon tetrafluoride gas. Attacked by strong bases. Other polymorphs of silica have been identified. Three high-temperature forms are α and β cristobalite (tetragonal and isometric, respectively) and tridymite (hexagonal). Melanophlogite, a very rare and impure cubic form, is found in sulfur deposits in Sicily and rock fissures in Tuscany (Italy). Two high-pressure, high-temperature forms are coesite (monoclinic) and stishovite (tetragonal). Natural silica glass, derived from quartz by fusion of rock in meteoric craters, is called lechatelierite.

2.65

**Environment**  One of the commonest minerals of the earth's crust (12 percent by volume). It crystallizes directly from igneous magma, from the pegmatite-pneumatolytic to the low-temperature hydrothermal stage. Occurs in plutonic rocks (granites, granodiorites, tonalites) and in hypabyssal (granite porphyries, pegmatites, etc.) and volcanic rocks (quartz porphyry, rhyolites, etc.). Stable in sedimentary conditions either

▲ Quartz (var. rock crystal) crystals (ca. ×1). Dauphiné, France.

▲ Quartz (var. amethyst) crystals (ca. ×1). Brazil.

as a detrital mineral (alluvial, marine and desert sands) or as a cement in consolidated rocks (sandstones). It also crystallizes from hot solutions (geyserites) and cold solutions (hyalites) and occurs as a diagenetic mineral derived from the skeletons of certain organisms (diatomites, chert). Stable under both low- and high-grade metamorphic conditions (from phyllites and quartzites to granulites and eclogites), although in some cases easily mobilized (quartz segregation veins, secretion pegmatites). It is one of the first components to refuse in the process of anatexis and is also one of the first to recrystallize (migmatites, anatexis granites). Cristobalite forms in cavities of lava rocks where rapid cooling has occurred, as in volcanic glass (obsidiań), and through the action of high-temperature metamorphism of quartz-rich rocks. Tridymite is typical of cavities in acid volcanic rocks and devitrification of obsidian. Coesite is found in meteorite impact rocks and kimberlite xenoliths (also in diamonds), while stishovite has only been found in meteorite craters.

**Occurrence**  As quartz is so common it is impossible to list all the places where it is found, so a few localities for the most characteristic varieties are given.

*Milky quartz*  White. The commonest variety found in pegmatites and hydrothermal veins. The color is generally caused by numerous bubbles of gas and liquid in the crystal. Famous finds include a crystal weighing 13 metric tons from Siberia. Milky quartz is common in the Alps.

*Rock crystal*  Collectors of crystals used to believe that transparent, colorless rock crystal was petrified ice. It is very pure and highly prized by collectors. It occurs mainly in pegmatites, Alpine fissures and geodes in various rocks. Used for optical and piezoelectrical purposes in industry, though laboratory-grown quartz crystals are now generally used instead. Famous localities include Minas Gerais (Brazil), where crystals weighing up to 5 metric tons occur, Val Cristillina in the St. Gotthard region (Switzerland), Malagasy Republic, Italy (Dosso dei Cristalli, Val Malenco, Sondrio; Champoluc, Val d-Aosta; Valle di Vizze, Alto Adige), Stubachtal and the Zillertal (Tyrol, Austria), the Dauphiné (France) and Japan. Carrara marble from Italy contains unusually clear crystals. Small, perfect rock crystals are sometimes found in cavities in dolomitic rock (Selvino, Bergamo, Italy; Herkimer County, New York, USA).

*Smoky quartz*  Light or dark brown to black (black known as morion). Probably caused by exposure to natural radioactivity. When heated turns yellow then white. Has been found in crystals weighing up to 300 kg (667 lb) in high-temperature hydrothermal veins in Brazil, Malagasy Republic and Scotland (Cairngorm; hence its other name, "Cairngorm stone") and in Alpine fissures (St. Gotthard, Switzerland; Beura, Val d'Ossola, Italy).

*Blue quartz*  Caused by tiny rutile, tourmaline or zoizite inclu-

▲ Quartz (var. smoky) crystals (ca. ×1). St. Gotthard, Switzerland.

▶ Quartz (var. rutilated) crystals (ca. ×1). Grisons, Switzerland.

sions. Fairly common in metamorphic rocks.

**Citrine**  Yellow or brown because of inclusions of colloidal iron hydrates. Turns white if heated, and dark brown (morion) if exposed to x-rays. Widely used as an imitation of the more expensive gemstone topaz, under exotic names like "Brazilian topaz." Found in Brazil (Minas Gerais), France (Dauphiné) and the USSR.

**Amethyst**  Violet color caused by trace amounts of ferric iron. Turns white when heated to 300°C (572°F), then yellow (citrine) at 500°C (932°F), but becomes violet again if exposed to x-rays or bombarded with α particles. The most beautiful amethysts are from the state of Minas Gerais (Brazil). They are up to 30 cm (12 in) long and are found in geodes of volcanic rocks. Also found in Uruguay, Sri Lanka, Canada, the Urals (USSR), Zambia and Czechoslovakia.

**Rose quartz**  The color appears to be caused by traces of manganese or titanium. It occurs in massive form in many pegmatites, but well-formed crystals are very rare. Loses its color when heated and turns black if exposed to radiation. Found mainly at Pala, California, Newry, Maine (USA), and in the state of Minas Gerais (Brazil).

As already explained, some quartz varieties, including those listed below, contain inclusions of other minerals, creating strikingly beautiful effects.

**Tiger's eye**  Contains or contained fibers of crocidolite ("pseudocrocidolite") altered to a yellow color. Comes from South Africa. A less common variety called "falcon's eye" contains fibers of unaltered blue crocidolite

**Rutilated quartz**  Contains acicular yellow and red rutile crystals. Comes from Malagasy Republic, Minas Gerais (Brazil) and Switzerland.

**Aventurine quartz**  Contains scales of mica or goethite that give a spangled green or brownish-yellow look. Very small amounts of coesite and stishovite have been found in the Meteor Crater of Arizona (USA), and cristobalite is found at San Cristobal (Mexico), in the Rhineland (Germany), in Japan and in Yellowstone National Park (USA).

**Uses**  A very important mineral in industry because of its piezoelectric properties (special pressure gauges, oscillators, resonators and wave stabilizers), its ability to rotate the plane of polarization of light (polarimeters) and its transparency in ultraviolet rays (heat-ray lamps, prisms and spectrographic lenses, etc.). It is used in large quantities in the manufacture of glass, paints, abrasives, refractories and precision instruments (agate supports). It is also the raw material for silicon carbide, a very hard abrasive. Attractive varieties are used as semi-precious gems, ornamental stones and carvings.

▲ Quartz (var. chrysoprase) (ca. ×1). Brazil.

▲ Polished slab of quartz (var. tiger's eye) (ca. ×1.5). Brazil.

SILICATES (Tectosilicates, Silica Group)
$SiO_2$ (Silicon oxide)

**System** Hexagonal.

**Appearance** This compact, microcrystalline variety of quartz is usually banded. Bands with a fibrous structure alternate with microgranular bands or as separate fibrous and microgranular varieties. Also common in mammillary aggregates and nodules, often with more or less concentric bands of color. It has a number of different varieties which are listed below.

**Physical Properties** Very hard (7) and light. Smooth conchoidal fracture; transparent when colorless, otherwise translucent to opaque as colors get darker. Dull waxy to vitreous luster depending on granularity and the variety. Chalcedony is amorphous without the vectorial properties of quartz.

**Environment and Occurrence** Most chalcedonies develop as a microcrystalline precipitate from aqueous solutions or as dehydration products of opal, in low-temperature environments.

**Agate** A concentric, banded, fibrous variety formed by precipitation from watery solutions in rounded cavities in lava rocks (geodes), sometimes with beautiful clusters of rock crystal or amethyst at the center. Comes from Brazil (Rio Grande do Sul), Uruguay and formerly the Idar-Oberstein region (Germany), where a thriving and still famous gem-cutting and polishing industry developed. Also found in Val di Fassa (Trento, Italy).

**Jasper** Massive, fine-grained quartz with large amounts of admixed material, expecially iron oxides. The commonest forms are usually strong shades of red, but grayish-green (prase), yellow or black also occur. Also a very common sedimentary rock.

**Onyx** A variety of agate with alternating parallel layers of black and white; sardonyx when red and white. Often employed in cameo carving.

**Carnelian** Uniformly colored, light- to dark-brown chalcedony. The orange-red color is due to the presence of very fine particles of hematite or limonite. Traditionally used for seals. Comes from Brazil, Uruguay, India and California (USA).

**Chrysoprase** Translucent, greenish-yellow or apple-green because of traces of nickel. Found in serpentinite veins in Queensland (Australian jade), the Urals (USSR), California (USA) and Brazil.

**Heliotrope** Opaque green with red markings, like drops of blood (hence called "bloodstone"). The red spots are caused by iron oxides.

**Flint or Chert** Siliceous nodules frequently found in chalk and limestone. Flint has a compact microcrystalline granular texture and typically is dark gray to soot black in color. Fine flint is found in chalk from the cliffs of Dover, England.

**Uses** Used almost entirely as an attractive ornamental stone, relatively common and easy to work and polish. Most varieties of chalcedony can be colored artificially.

▲ Quartz (var. jasper) (ca. ×1). Sardinia, Italy.

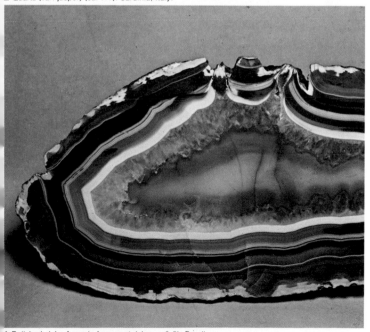

▲ Polished slab of quartz (var. agate) (ca. ×0.5). Brazil.

## 246 CRISTOBALITE
SILICATES (Tectosilicate, Silica Group)
SiO₂ (Silicon oxide)

**System** High-cristobalite (isometric); low-cristobalite (tetragonal).
**Appearance** Pseudo-isometric octahedral crystals rare. Usually massive submicrocrystalline aggregates. Colorless to white or gray.
**Physical properties** There are two polymorphs of cristobalite. High-cristobalite is stable from its melting point at 1728° down to 1470°C (3142°–2678°F). High-cristobalite can exist in a metastable form between 1470° and 268°C (2678°–514°F). Below 268°C low-cristobalite is the stable form. Very hard (7), light, no cleavage but conchoidal fracture. Opaque to translucent. Insoluble in all acids except hydrofluoric acid.
**Environment** Occurs in cavities in felsic igneous rocks such as andesites, rhyolites, trachytes and obsidians.
**Occurrence** Pseudo-octahedrons up to 4 mm (0.18 in) along an edge found in andesite at Cerro San Cristobal, Mexico. Found in spherulites in obsidian in Inyo County, California, Crater Lake, Oregon, and the San Juan region, Colorado (USA).
**Uses** Of interest to mineralogists and collectors.

---

## 247 TRIDYMITE
SILICATES (Tectosilicates, Silica Group)
β-SiO₂ (Silicon oxide)

**System** Monoclinic.
**Appearance** Small white blades (pseudo-hexagonal). Isolated, spherulitic, or forming rosettes in cavities in volcanic rocks.
**Physical properties** Very hard (7), light, prismatic cleavage. Transparent or translucent with vitreous or greasy luster. Insoluble and infusible. Pseudo-hexagonal β-tridymite is stable between 870° and 1470°C (1598°–2678°F) and turns into α-quartz at normal temperatures. α-tridymite, another low-temperature phase, is metastable, orthorhombic and never found in nature.
**Environment** A sublimation product in cavities in felsic volcanic rocks. Also occurs as a contact metamorphic mineral in sandstone fragments embedded in basaltic lava.
**Occurrence** Splendid crystals over 1 cm (0.5 in) long are found in trachytes in the Euganean Hills, Italy (Zovon, San Pietro in Montagna), but these are really α-quartz pseudomorphs after tridymite. Other localities where this mineral occurs are Pachuca (Mexico), Mont Doré (France), San Cristobal (Mexico), Obsidian Cliff, Yellowstone National Park, Wyoming (USA), and Siebengebirge (Germany).
**Uses** The natural mineral is of interest to scientists and collectors because of its rarity. Synthetic tridymite is used in making refractories and heat-resistant porcelain.

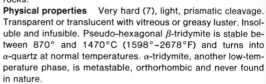

Cristobalite crystals in obsidian (ca. ×10). Pachuca, Mexico.

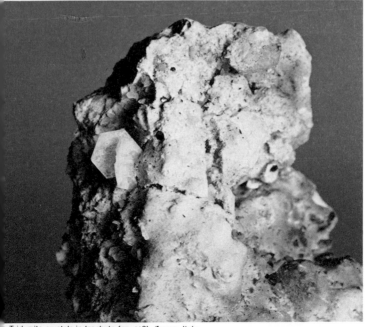

Tridymite crystals in trachyte (ca. ×2). Zovon, Italy.

**"OPAL"**
SILICATES (Tectosilicate)
$SiO_2 \cdot nH_2O$ (Hydrous silicon oxide)

**System** Consists of microspheroids of hydrous silica which may be cristobalite in some varieties. Previously thought to be amorphous.

**Appearance** Never found in crystals. Occurs as small veins, globules and crusts. Colorless, milky white, often hazy blue or black with splendid iridescence. Compact and earthy masses, often concretionary and stalactitic (diatomites and geyserites).

**Physical properties** Hard (5.5–6.5), light or very light, no cleavage. Very fragile, partly because when exposed to air it loses water and becomes filled with tiny conchoidal fractures (crazing). Transparent or translucent with greasy luster. Often fluorescent yellow or green in ultraviolet light. Insoluble and infusible, but decomposes when heated, altering to quartz. The following varieties are the best known. Precious opal is porcellanous and may have various colors, with a characteristic play of colors caused by the dispersion of light that varies according to the angle of incidence and the size of the ordered spherical particles of which the mineral is composed. Fire opal is a variety of precious opal, reddish in hue not necessarily iridescent. Hydrophane is a whitish opal which becomes transparent and sometimes iridescent if soaked in water. Hyalite is white or colorless and occurs mainly in globular and dendritic masses. Wood opal replaces the fibers in fossilized wood without destroying the texture or detail.

**Environment** Commonly precipitated from silica-rich solutions. It also replaces the skeletons of many marine organisms and produces accumulations of more or less coherent sedimentary rocks.

**Occurrence** Most black precious opal comes from Australia (Lightning Ridge, New South Wales), fire opal is from Mexico (Queretaro) and precious opal of various colors is found in Transylvania (Romania); Virgin Valley, Nevada, Idaho and Oregon (USA) and Caernowitza (Hungary). Geyserite is common in Iceland near large geysers. Stratified masses of diatomite are found in California (USA), Turkey and Italy (fiorite variety from Santa Fiora, Tuscany). Wood opal comes from the petrified forests of Yellowstone National Park (USA), Egypt and Lake Omodeo in Sardinia.

**Uses** Some black opals and fire opals are considered precious stones. Diatomites are used as a fine abrasive, a filtering powder, in insulators and in special porcelain.

▲ Polished slab of wood opal (ca. ×1). Egypt.

▶ Hyalite opal (ca. ×1). Santa Fiora, Italy.

▼ Precious opal (ca. ×1). Australia.

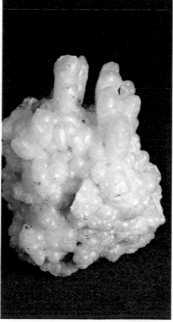

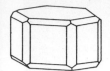

## 249 NEPHELINE

SILICATES (Tectosilicates, Feldspathoid Group)
(Na,K)AlSiO₄ (Sodium potassium aluminum silicate)

**System**  Hexagonal.
**Appearance**  Rare, white or yellowish, hexagonal, prismatic crystals with pinacoidal terminations. Generally in compact, granular aggregates, white, yellow, gray, green or even reddish (eleolite variety).
**Physical properties**  Hard (5.5–6), light, fragile, with poor cleavage. Translucent with greasy luster. Very occasionally transparent with vitreous luster. Soluble in hydrochloric acid, giving off a cloud of silica gel (hence the name from a Greek word meaning cloud). Fuses with difficulty, coloring the flame yellow (sodium).
**Environment**  In intrusive and volcanic rocks undersaturated with silica, and in their pegmatites. Very occasionally in mica schists and gneiss.
**Occurrence**  The largest masses are found in the Kola Peninsula and the Urals (USSR), in Norway, at Bancroft (Canada), Litchfield, Maine (USA), and in South Africa. Also found in the lavas of Vesuvius and the Alban Hills (Lazio, Italy).
**Uses**  In the glass, china, rubber, textile and oil industries.

2.56
2.66

---

## 250 PETALITE

SILICATES (Tectosilicates)
LiAlSi₄O₁₀ (Lithium aluminum silicate)

**System**  Monoclinic.
**Appearance**  Rare, colorless or white, tabular or columnar crystals. Usually in colorless, white or gray cleavage masses.
**Physical properties**  Hard (6–6.5), light, fragile, with perfect cleavage. Transparent to translucent with vitreous to pearly luster. Insoluble. When heated slowly displays blue phosphorescence. Fuses with difficulty, coloring the flame crimson (lithium).
**Environment**  In lithium-bearing pegmatites, associated with spodumene, lepidolite and tourmaline.
**Occurrence**  Fine transparent masses common at Varuträsk (Sweden), Bickita (Rhodesia), in South Africa, Finland and Australia. Fine terminated crystals to 20 cm (8 in) were found at Norwich, Massachusetts (USA). Also found in granite on the island of Elba and at Peru, Oxford County, Maine (USA).
**Uses**  An important ore of lithium.

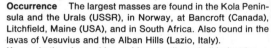

2.41

▲ Nepheline crystal (ca. ×1). Monte Somma, Italy.

▲ Petalite cleavage fragment (ca. ×1). Varuträsk, Sweden.

## 251 ANALCIME

SILICATES (Tectosilicates, Feldspathoid Group)
$NaAlSi_2O_6 \cdot 2H_2O$ (Hydrated sodium aluminum silicate)

**System** Isometric.
**Appearance** Trapezohedral crystals, or cubes modified by trapezohedrons. Colorless or white, or pink and yellow. Earthy or radiating granular aggregates.
**Physical properties** Hard (5–5.5), light, fragile, no cleavage. Transparent or translucent, with vitreous luster. Dissolves easily in strong acids and fuses into a transparent glass, coloring the flame yellow (sodium). When rubbed or heated becomes slightly charged with electricity.
**Environment** In cavities of intrusive and volcanic igneous rocks and in many undersaturated lavas where it replaces nepheline, associated with calcite and zeolites. Also authigenic in arenaceous sedimentary rocks.
**Occurrence** Magnificent clear crystals found in the Cyclopean Island (Aci Trezza, Catania, Sicily) and elsewhere in Italy in Val di Fassa (Trento), Alpe di Siusi (Bolzano), Montecchio Maggiore (Vicenza), Val di Cecina (Pisa) and in Campania. Also found in Australia, at Golden (Colorado), Bergen Hill and Paterson, New Jersey (USA), and in copper deposits in pillow lavas in Michigan (USA).
**Uses** Of interest to scientists and collectors.

---

## 252 LEUCITE

SILICATES (Tectosilicates, Feldspathoid Group)
$\alpha$-$KAlSi_2O_6$ (Potassium aluminum silicate)

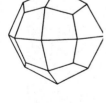

**System** Tetragonal.
**Appearance** Trapezohedral crystals, pseudomorphic after a high-temperature, cubic form (>605°C; 1121°F, $\beta$-leucite). Color white. Round grains.
**Physical properties** Hard (5.5–6), light, fragile, no cleavage, conchoidal fracture. Translucent with vitreous luster, often opaque. White streak. Dissolves easily in acids. Infusible, unlike analcime.
**Environment** Typical of undersaturated volcanic rocks rich in potassium (leucitic basalts, leucitic phonolites, leucitites, leucitophyres, etc.). Never found unaltered in old lavas or plutonic igneous rocks or in metamorphic rocks.
**Occurrence** Fine, large crystals found in lavas in the Alban Hills and Vesuvius (Italy). Also occurs at Leucite Hills (Wyoming) and in Montana (USA), in some volcanos of the African Rift Valley (Zaire, Uganda), in Germany (Kaiserstuhl) and Australia.
**Uses** In the past used as a fertilizer because it was so common and because of its potassium content, but now of interest mainly to scientists and collectors.

▲ Analcime crystals (ca. ×1). Alpe di Siusi, Italy.

▲ Leucite crystal (ca. ×1). Ariccia, Italy.

## 253　SANIDINE

SILICATES (Tectosilicates, Feldspar Group, K-Feldspars)
KAlSi₃O₈ (Potassium aluminum silicate)

2.53
2.56

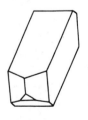

**System**　Monoclinic.

**Appearance**　Tabular, prismatic crystals, often twinned, colorless or whitish, and less often gray or yellow.

**Physical properties**　The formula $KAlSi_3O_8$ covers several polymorphous modifications grouped together as potassium feldspars (K-feldspars) in the alkali-feldspar series (sodium-potassium feldspars). They may be triclinic or monoclinic and are also characterized by their degree of structural order, which is related to such factors as temperature, chemical environment and speed of crystallization. Sanidine is a high-temperature, monoclinic polymorph with a disordered structure. Hard (6), light, distinct 90° cleavage. Infusible and insoluble in acids, except hydrofluoric acid, which decomposes it completely. Forms a partial isomorphous series with albite, $NaAlSi_3O_8$, constituting up to 20 percent of its composition. Crystals normally contain 10 percent or more albite component.

**Environment**　Typical of recent volcanic igneous rocks of the trachytic or high-potassium alkaline type. Also found in basalts and argillites.

**Occurrence**　In Italy classic examples found at Pantelleria (Sicily), Soriano (Viterbo) and Capo d'Enfola (Elba). Also found in Germany, Czechoslovakia and the Caucasus (USSR).

**Uses**　Of interest to scientists and collectors.

## 254　ORTHOCLASE

SILICATES (Tectosilicates, Feldspar Group, K-Feldspars)
KAlSi₃O₈ (Potassium aluminum silicate)

2.55
2.63

**System**　Monoclinic.

**Appearance**　Prismatic, columnar or tabular crystals, frequently twinned: penetration twins (Carlsbad habit) or contact twins (Manebach and Baveno). Compact, granular masses, sometimes large. Usually colorless or white, but sometimes pale yellow, pink, blue or gray.

**Physical properties**　The medium- to high-temperature, monoclinic polymorph of $KAlSi_3O_8$, with a partially ordered structure, may form by slow cooling (and ordering) of sanidines. Hard (6), light, fragile, with perfect 90° cleavage (hence the name), parallel to the base and to the side pinacoid. Transparent or translucent, with vitreous luster, tending to pearly. White streak. Insoluble in acids, except hydrofluoric acid. Fuses with difficulty, coloring the flame violet (potassium). Easily altered through the action of hot water rich in carbonic acid into secondary products like kaolinite, sericite, zeolites, epidotes, and quartz. If sodium is present in sufficient quantities, perthites will develop from the unmixing of a single feldspar into two feldspars (one Na rich, one K rich). These are lamellar intergrowths which may be invisible (cryptoperthite), microscopic (microperthite) or visible with the unaided eye (macroperthite). Celsian ($BaAl_2Si_2O_8$, monoclinic), is a rare feldspar which

▶ Twinned sanidine crystal (ca. ×1). Elba, Italy.

▼ Orthoclase crystals (ca. ×1). Baveno, Italy.

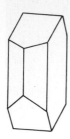

forms a solid solution series with orthoclase, with the intermediate member hyalophane. Celsian has high specific gravity, fuses with difficulty in a blowpipe flame, coloring the flame green (barium), and dissolves in concentrated hydrochloric acid, producing colloidal silica. The specific gravity of hyalophane (monoclinic) varies according to its composition.

**Environment** An essential component of many intrusive, plutonic rocks formed at a medium-to-high temperature and cooled slowly (granites, granodiorites, syenites, monzonites), and pegmatites. Also occurs in some crystalline schistose rocks (gneiss, migmatites) and, as a detrital mineral, in arkose or arenaceous sedimentary rocks. Celsian is typical of contact rocks, rich in manganese. Hyalophane has been found in gneissic rocks, manganese deposits and metamorphic dolomites.

**Occurrence** Clear, yellow orthoclase from the pegmatites of the Malagasy Republic is well known and is used as a gemstone. Fine pink crystals are found in cavities in granite at Baveno (Novara) and in granophyre at Cussao al Monte (Varese), and white crystals at San Piero in Campo (Elba) (Italy). Classic twinned forms come from Carlsbad (Czechoslovakia), and well-formed red crystals from Kirkpatrick (Scotland), Flims (Grisons, Switzerland), Marienburg (Germany), Striegau (Silesia) and various localities in the USSR. Blue crystals occur near Lake Baikal (USSR), gray at Villadreu (Spain). Remarkable crystals are found in lithoclases of crystalline-schistose rocks in the Zillertal (Austria) and Val di Vizze (Bolzano, Italy). Celsian has been identified in the Binnental (Switzerland), in Sweden, Japan, South Africa, Australia, at Candoglia (Val d'Ossola, Novara, Italy) and several places in the USA. Hyalophane, sometimes associated with celsian, is found in Japan, South Africa, Australia, Canada, Siberia (USSR) and in the metamorphic dolomite of Binna (Switzerland).

**Uses** An important industrial mineral. The mixture orthoclase, kaolin and quartz is easy to mold and fuses at a relatively low temperature (1100–1300°C; 2012°–2372°F) to a light, glassy, white, translucent, slightly porous mass known as porcelain. When very pure, it is used to make special porcelains (high-tension electrical insulators, ceramic glazes and dental products) and opalescent glass. For some years it has been used, finely ground and mixed with detergents, as a scouring powder. Transparent yellow or attractively colored varieties or varieties displaying schiller luster (a bronzy iridescence) are used as gemstones.

▲ Orthoclase crystal (ca. ×1). Baveno, Italy.

▲ Twinned orthoclase crystal (ca. ×1). Spain.

▲ Orthoclase crystals (ca. ×1). Elba, Italy.

## 255 MICROCLINE

SILICATES (Tectosilicates, Feldspar Group, K-Feldspars)
KAlSi₃O₈ (Potassium aluminum silicate)

2.55
2.63

**System** Triclinic.

**Appearance** Prismatic crystals, frequently twinned according to the Carlsbad law (penetration twins) or more rarely according to the Manebach and Baveno laws (contact twins). Polysynthetic twins also common, giving a characteristic "Tartan twinning" structure. White, pink, red, yellowish or blue-green (amazonite variety). Compact aggregates.

**Physical properties** A triclinic, highly ordered polymorph of KAlSi₃O₈. Hard (6–6.5), light, distinct, nearly right-angle cleavage (89°30′). Translucent or transparent with vitreous luster. White streak. Infusible and insoluble in acids, except hydrofluoric.

**Environment** In granitic pegmatites and metamorphic rocks (gneiss) formed at medium-to-low temperatures. Often intimately associated with quartz (myrmekite) and albite (perthite).

**Occurrence** Fine crystals in granite at Cala Francese (La Maddalena, Sardinia). Also found in various localities in the USA and the Urals (USSR). Amazonite comes from Pikes Peak, Colorado, and Amelia Court House, Virginia (USA), Brazil, India, Tanzania and Canada.

**Uses** Microcline is of particular interest to petrologists. Amazonite of a good blue-green color is cut as a cabochon or into beads, and polished.

---

## 256 ORTHOCLASE (VAR. ADULARIA)

SILICATES (Tectosilicates, Feldspar Group, K-feldspars)
KAlSi₃O₈ (Potassium aluminum silicate)

2.55
2.63

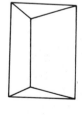

**System** Monoclinic.

**Appearance** Prismatic, pseudo-orthorhombic, colorless or whitish crystals, often striated, full of fluid inclusions often coated with green chlorite.

**Physical properties** A partially ordered low-temperature polymorph of KAlSi₃O₈, probably crystallized metastably and relatively quickly. Hard (6), light, distinct 90° cleavage. Transparent or translucent with vitreous luster. White streak. Infusible and insoluble in acids, except hydrofluoric acid. The opalescent variety, moonstone, displays a characteristic white or blue internal play of color caused by the microscopic exsolution lamellae of albite.

**Environment** Formed in low-temperature hydrothermal veins and Alpine lithoclases.

**Occurrence** There have been well-known finds in the Swiss and Austrian Alps, for example at Mount Fibia (Ticcino), the Maderanertal (Uri), Lucomagno (Grisons) Guttanen (Berne) and especially St. Gotthard (Switzerland) and in the Zillertal (Austria). In Italy found in Val di Vizze (Bolzano), and in small, imperfect crystals in aplites and granite on Elba. The best moonstone comes from Sri Lanka and Burma.

**Uses** Of interest to scientists and collectors. Moonstone is cut as a cabochon and used in jewelry.

Microcline (var. amazonite) crystals (ca. ×1). Pikes Peak, Colorado.

Orthoclase (var. adularia) crystals (ca. ×1). St. Gotthard, Switzerland.

SILICATES (Tectosilicates, Feldspar Group)
KAlSi₃O₈ and NaAlSi₃O₈ (Potassium aluminum silicate and Sodium aluminum silicate)

**System** Triclinic and monoclinic (orthoclase).

**Appearance** Not an individual mineral, but a regular association of undulating layers (exsolution lamellae) of albite (or oligoclase) in a microcline crystal host (or less often orthoclase). Antiperthites are associations in which the potassium phase forms the exsolution lamellae and the sodium phase is the crystal host. In mesoperthites the sodium and potassium phases are present in equal amounts so the lamellae are the same size. Superficially perthite looks like its feldspar host which may be pink or blue-green (perthite amazonite) with clearly visible white stripes of albite, but often perthites are only visible under the microscope (microperthites) or by means of x-ray diffraction (cryptoperthites).

**Physical properties** Very similar to microcline and orthoclase, though the physical discontinuity between plates makes perthites fragile. Other types display characteristics intermediate between the phases of which they are composed.

**Environment** Common in pegmatites, syenites, granites, porphyries and acid metamorphic rocks. Perthites are usually formed by exsolution of albite, during cooling, from microcline or orthoclase, starting from a single homogeneous crystal. At high temperatures the alkali feldspars (potassium and sodium bearing) form a complete solid solution series, but this capacity gradually decreases with cooling until there is almost none at room temperature. Sometimes perthites are formed by hydrothermal replacement of K-feldspar through the introduction of albite or oligoclase along latent planes of discontinuity (cleavages, microfractures).

**Occurrence** The most typical perthites, with white albite in a yellow or pink ground mass, come from Canada and Norway. The green variety comes mainly from Pikes Peak, Colorado and Amelia Court House, Virginia (USA), and also from Brazil, Malagasy Republic and Karelia (USSR).

**Uses** Of scientific interest to petrologists and to collectors. Sometimes used in jewelry, as it shows a fine play of color when properly cut.

"Perthite" (orthoclase and albite) (ca. ×1). Ontario, Canada.

"Perthite" (microcline and albite) (ca. ×1). Karelia, USSR.

## 258  ALBITE

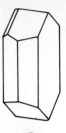

SILICATES (Tectosilicates, Feldspar Group, Plagioclase Series)

NaAlSi$_3$O$_8$ (Sodium aluminum silicate)

2.62

**System**  Triclinic.

**Appearance**  Crystals prismatic, usually bladed, often with polysynthetic twinning, recognizable by the presence of fine striations on basal cleavage. The lamellar variety is called cleavelandite and the elongated, prismatic variety, usually in polysynthetic twins, is called pericline. Colorless or white.

**Physical properties**  The end member of the isomorphous series of plagioclases, solid solutions varying in composition from albite (sodic) to anorthite (calcic) with the intermediate terms oligoclase, andesine, labradorite and bytownite. Albite occurs in both high-temperature disordered forms and low-temperature ordered forms. The disordered form may have up to 40 percent K-feldspar component and then becomes anorthoclase. Albite is hard (6–6.5), light, with perfect basal cleavage, less perfect parallel to the side pinacoid. Transparent or translucent with vitreous luster. White streak. Insoluble in acids, except hydrofluoric acid, in which it decomposes leaving a gelatinous silica residue. Fuses with some difficulty, coloring the flame yellow (sodium).

**Environment**  An essential constituent of many acid, plutonic igneous rocks (granites, syenites, pegmatites) in the ordered form. Also found in volcanic, acid or undersaturated rocks (rhyolites, trachytes) in the disordered form or as anorthoclase. Occurs in mica schists and gneiss and in segregation veins, as well as in many Alpine lithoclases. Derived from other feldspars by the process of albitization, which involves their modification by sodium-rich fluids.

**Occurrence**  Small but very pure crystals are found in veins in glaucophane schists in Cazadero, California (USA) and in the western Alps (Val Soana, Piedmont, Italy). Large cleavelandite crystals occur at Amelia Court House, Virginia (USA) and in pegmatitic veins at Rio Grande do Sul (Brazil). Well-formed pericline specimens occur mainly in Alpine lithoclases in Grisons (Switzerland) and in the Tyrol (Austria). There are fine crystals in granite at Baveno (Novara), on Elba, and in crystalline limestone in the Apuanian Alps (Massa Carrara) (Italy).

**Uses**  An important industrial mineral used in ceramics and refractories.

▲ Albite (var. clevelandite) crystals (ca. ×1). Rio Grande do Sul, Brazil.

▲ Albite crystals (ca. ×1). St. Gotthard, Switzerland.

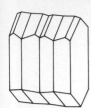

## 259 OLIGOCLASE, LABRADORITE, ANORTHITE

SILICATES (Tectosilicates, Feldspar Group, Plagioclase Series)

(Na,Ca)AlSi$_3$O$_8$ (Sodium calcium aluminum silicates)

**System** Triclinic.

**Appearance** Free-standing tabular crystals, rare. Often twinned. Masses, sometimes large, with parallel or criss-cross twinning striations. Colorless, white, yellowish, green or sometimes pink or reddish. Labradorite is sometimes strongly iridescent. Some varieties have an internal play of color (sunstone, peristerite) caused by inclusions or submicroscopic lamellae.

**Physical properties** Members of the isomorphous plagioclase series. The physical properties vary according to the calcium content. Density is particularly variable, though these minerals are always light. The cleavage angle between the basal pinacoid (perfect) and the side pinacoid (good) also varies but is close to 90°. Fusibility is more difficult toward the calcium-rich end of the series. Finally, solubility in hydrochloric acid varies: albite is insoluble, anorthite totally soluble in HCl.

**Environment** In igneous rocks, both plutonic and volcanic. Plagioclase becomes increasingly calcic in composition as the rock in which it occurs becomes more mafic. In metamorphic rocks the same thing occurs but anorthite is only found in calcareous rocks which have undergone contact metamorphism.

**Occurrence** Labradorite is typical of some eruptive rocks (anorthosites) and metamorphic rocks in Norway, Labrador (Canada) and the USSR. Anorthite is very rare. It is found in some Alpine metamorphic rocks (Val Schiesone, Sondrio, Italy), in basalts on the island of Hokkaido (Japan) and in the volcanic lavas of Vesuvius, in contact rocks at Monzoni (Val di Fassa, Trento) and on the Cyclopean Islands (Catania) (Italy). Plagioclases, especially the sodic varieties, are widely distributed rock-forming minerals. Sunstone is typical of some pegmatites in Norway, Canada and the USSR.

**Uses** Of interest to petrologists and value to collectors. Used in industry in ceramics, glazes and basic refractories and, in the case of labradorite, as freestone in building. Polished sunstone and labradorite are used in jewelry.

Oligoclase crystal (ca. ×1.5). Maine.

Anorthite crystal (ca. ×1). Japan.

Iridescent "labradorite" slab (ca. ×1). Ukraine, Russia.

## 260 DANBURITE

SILICATES (Tectosilicates)
$CaB_2Si_2O_8$ (Calcium borosilicate)

**System** Orthorhombic.

**Appearance** Clear, prismatic crystals, with wedge-shaped terminations. Occasionally pale yellow.

**Physical properties** Very hard (7–7.3), heavy, poor cleavage. Transparent with vitreous luster. Fuses fairly easily into a colorless glass, coloring the flame green (boron). Insoluble in acids.

**Environment** Found in fissures and lining Alpine lithoclases, especially as an incrustation on albite.

**Occurrence** Small, clear crystals with many faces found in Val Medel (Grisons, Switzerland). Found as a rarity at St. Barthélemy (Val d'Aosta) and in the Monti Cimini (Viterbo) (Italy). Larger but not so fine crystals found in Mexico (San Luis Potosí), the USSR, Japan (Obira), Malagasy Republic (Maharita), and in Danbury, Connecticut (USA), which gives the mineral its name.

**Uses** Of interest only to scientists and collectors.

---

## 261 CANCRINITE

SILICATES (Tectosilicates)
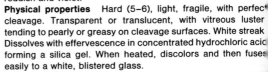
$(Na,K,Ca)_{6-8}(Al,Si)_{12}O_{24}(SO_4,CO_3,Cl)_{1-2} \cdot nH_2O$ (Complex silicate)

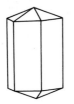

**System** Hexagonal.

**Appearance** Prismatic, stubby crystals very rare, generally massive or in disseminated grains. Colorless, yellow, white, reddish and violet.

**Physical properties** Hard (5–6), light, fragile, with perfect cleavage. Transparent or translucent, with vitreous luster tending to pearly or greasy on cleavage surfaces. White streak. Dissolves with effervescence in concentrated hydrochloric acid forming a silica gel. When heated, discolors and then fuses easily to a white, blistered glass.

**Environment** A primary constituent of plutonic igneous rocks undersaturated in silica. More often formed by metasomatic action of carbon dioxide at the expense of preexisting nepheline crystals.

**Occurrence** Associated with blue sodalite at Bancroft (Ontario, Canada) in nepheline syenite gneiss produced by metasomatism in an environment poor in silica. Also found in nepheline syenite pegmatites at Litchfield, Maine (USA), in the Kola Peninsula and the Ilmen Mountains (USSR) and in Greenland. Very rare in the lavas of Vesuvius (Italy).

**Uses** Of interest to scientists and collectors.

▲ Danburite crystal (ca. ×1.5). USSR.

▲ Cancrinite with sodalite (ca. ×1). Bancroft, Ontario, Canada.

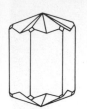

## 262 SODALITE
SILICATES (Tectosilicates, Sodalite Group)
Na₄Al₃(SiO₄)₃Cl (Sodium aluminum silicate chloride)

**System** Isometric.
**Appearance** Very rare dodecahedral crystals. Usually in compact masses, bright blue, white or gray with green tints.
**Physical properties** Hard (5.5–6), light, fragile, with poor cleavage. Translucent with vitreous luster. Soluble in hydrochloric and nitric acid, leaving a silica gel. Fuses fairly easily to a colorless glass, coloring the flame yellow (sodium).
**Environment** In undersaturated plutonic igneous rocks (nepheline syenites and phonolites), associated with leucite, nepheline and cancrinite. In desilicified, metasomatized limestones and in volcanic blocks.
**Occurrence** Beautifully colored masses at Bancroft (Canada), Litchfield, Maine, and Magnet Cove, Arkansas (USA), and in smaller amounts in Brazil, Bolivia, Greenland, Romania, Portugal, Rhodesia, Burma and the USSR. Magnificent, clear crystals found in the calcareous lavas of Vesuvius, in the Campi Flegrei (Naples) and in the Monti Cimini (Viterbo) (Italy).
**Uses** In jewelry, as polished slabs and for carved ornaments.

2.27
2.33

---

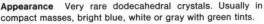

## 263 HAUYNE
SILICATES (Tectosilicates, Sodalite Group)
(Na,Ca)₄₋₈(Al₆Si₆)O₂₄(SO₄,S)₁₋₂ (Sodium calcium aluminum silicate sulfate)

**System** Isometric.
**Appearance** Rare octahedral or dodecahedral crystals. More often in round blue, white and occasionally green grains.
**Physical properties** Hard (5.5–6), light, perfect dodecahedral cleavage. Transparent or translucent with vitreous or greasy luster on cleavage surfaces. Fuses easily to a blue-green glass. Decomposes quickly in acids, leaving a silica gel.
**Environment** In undersaturated, igneous plutonic and volcanic rocks (nepheline syenites, phonolites). An essential component of hauynophyres.
**Occurrence** Found in the lavas of Vesuvius and the volcanoes of Lazio, associated with leucite and melilite in tuffs in the Alban hills and in hauynophyres at Monte Vulture (Basilicata) (Italy). Also found in volcanic bombs in the Rhineland, in lavas in Morocco and in andesites in France.
**Uses** Of interest only to scientists and collectors.

2.44
2.50

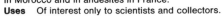
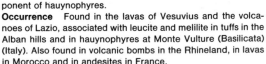

▲ Polished sodalite slab (ca. ×1.5). Brazil.

▲ Hauyne crystals (ca. ×1.5). Lazio, Italy.

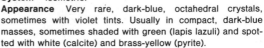

## 264 LAZURITE

SILICATES (Tectosilicates, Sodalite Group)
$(Na,Ca)_8(Al,Si)_{12}O_{24}(S,SO_4)$ (Sodium calcium aluminum silicate sulfate)

**System** Isometric.
**Appearance** Very rare, dark-blue, octahedral crystals, sometimes with violet tints. Usually in compact, dark-blue masses, sometimes shaded with green (lapis lazuli) and spotted with white (calcite) and brass-yellow (pyrite).
**Physical properties** Hard (5–5.3), light, fragile, imperfect cleavage and uneven fracture. Opaque in crystals, with vitreous luster. Light-blue, almost colorless streak. Soluble in hydrochloric acid, usually giving off hydrogen sulfide fumes caused by associated pyrite. Fuses to a whitish glass.
**Environment** In contact metamorphosed limestones characteristically associated with calcite and pyrite. Also in high-temperature granulites.
**Occurrence** The very important Sar-e-Sang deposits in Badakhshan in the Kokscha Valley (Afghanistan) were worked as long as six thousand years ago. Small masses found in the lavas of Vesuvius and in the Alban Hills (Lazio) (Italy), in Chile (Ovalle), Burma (Mogok), Siberia (USSR), Angola, Labrador (Canada), Pakistan and California and Colorado (USA).
**Uses** A highly prized gem, used in the past in large slabs as an ornamental stone, especially in Russia. It was also used as the pigment ultramarine blue.

---

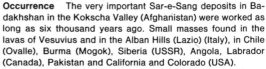

## 265 SCAPOLITE GROUP

SILICATES (Tectosilicates)
$(Na,Ca,K)_4Al_3(Al,Si)_3Si_6O_{24}(Cl,SO_4,CO_3)$ (Complex silicates)

**System** Tetragonal.
**Appearance** Prismatic crystals with vertical striations. Microgranular masses or fibrous aggregates, whitish, blue, gray or pink.
**Physical properties** Variable, since this is a complex isomorphous series with hypothetical end members marialite and meionite and intermediate member mizzonite. Hard (5–6.5), light, with specific gravity increasing from marialite to meionite. Poor cleavage. Translucent or transparent, with vitreous luster. Often strong orange-yellow fluorescence in ultraviolet light. Fuses easily to a blistered mass. Soluble in hydrochloric acid, leaving silica.
**Environment** In crystalline-schistose rocks of the mica-schist, gneiss and amphibolite types. In metamorphosed limestones and skarns. In some pegmatites and granulites.
**Occurrence** Crystals 50 cm (20 in) long are found at Rossie and Pierrepoint, New York (USA), and fine crystals at Renfrew (Ontario) and Grenville (Québec) (Canada), and at Lake Tremorgio (Switzerland). Transparent yellow crystals occur in pegmatites in Minas Gerais (Brazil) and Malagasy Republic. Found in Val Malenco (Sondrio), and other places in the central Alps, in blocks at Mt. Somma, Pianura (Naples) and Elba (Italy).
**Uses** Of interest to scientists and collectors.

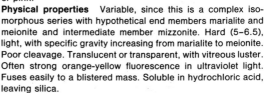
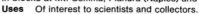

Massive lazurite (ca. ×1). Afghanistan.

Scapolite crystals (ca. ×1.5). Monte Somma, Italy.

**266**   **NATROLITE**

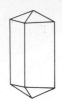

SILICATES (Tectosilicates, Zeolite Group)
$Na_2(Al_2Si_3)O_{10} \cdot 2H_2O$ (Hydrated sodium aluminum silicate)

**System**   Orthorhombic.
**Appearance**   Slender, prismatic crystals with vertical striations. Usually in globular aggregates of fibrous, radiating needles or more rarely felted and compact. Colorless, white, pink
or yellowish.

**Physical properties**   Hard (5–5.5), light, perfect cleavage.
Transparent with vitreous to pearly luster, silky in fibrous
varieties. White streak. Soluble in strong acids, leaving silica
gel. Fuses easily to a transparent glass, coloring the flame yellow (sodium). Loses water at about 300°C (572°F) and reab-
sorbs it quickly from atmospheric humidity. Sometimes
fluoresces orange in ultraviolet light. Natrolite is one of the
zeolite minerals occurring in acicular crystals. Mesolite and
gonnardite are similar species, intermediate in composition
between natrolite and scolecite $Ca(Al_2Si_3)O_{10} \cdot 3H_2O$.

**Environment**   Found lining cavities in basalts and other lavas,
associated with calcite and zeolites. Also as an alteration product of plagioclase in aplites and syenites and of nepheline and
sodalite in nepheline syenites and phonolites. Hydrothermal
veins of natrolite are found in serpentinite associated with
asbestos.
**Occurrence**   Enormous crystals 1 m x 15 cm (3 ft x 6 in)
found at Asbestos (Canada), and very fine crystals come from
Nova Scotia and British Columbia (Canada), Paterson and
Bound Brook, New Jersey, and San Benito County, California
(USA), Rio Grande do Sul (Brazil), Northern Ireland, Green-
land, Puy-de-Dôme (France), Aussig and Salesí (Bohemia) and
India. Found in basalts at Montecchio Maggiore, Solcedo and
Altavilla (Vicenza), in pink, radiate aggregates in Val di Fassa
(Trento) and at Tierno (Trento) (Italy). Mesolite is found in cavi-
ties in basalts in the Faroes, at Antrim (Northern Ireland) and on
Skye (Scotland), in Ireland, India and various localities in the
USA including Paterson, New Jersey. Gonnardite is found at
Puy-de-Dôme (France), in lavas at Aci Trezza and Aci Castello
(Catania, Sicily) and at Capo di Bove (Rome, Italy), in Germany
Styria and various places in the USA.
**Uses**   A mineral of interest to scientists and collectors.

▲ Acicular natrolite crystals in basalt (ca. ×1). Gambellara, Italy.

▲ Natrolite crystals (ca. ×1.5). India.

## 267 THOMSONITE

SILICATES (Tectosilicates, Zeolite Group)
$NaCa_2(Al_5Si_5)O_{20} \cdot 6H_2O$ (Hydrated sodium calcium aluminum silicate)

**System** Orthorhombic.

**Appearance** Distinct crystals rare. Usually forms radiating balls or globular aggregates. Color white (brown when impure).

**Physical properties** Hard (5–5.5), light, with prismatic cleavage and uneven-to-subconchoidal fracture. Streak colorless. Transparent to translucent with vitreous to pearly luster. Fuses with intumescence to a white enamel. Gelatinizes with hydrochloric acid.

**Environment** A secondary mineral deposited in cavities in basalts and other igneous rocks by hydrothermal fluids. Associated with calcite, prehnite and other zeolites.

**Occurrence** Occurs in cavities in basalt, notably at Paterson, New Jersey, where globular aggregates up to 5 cm (2 in) across, on prehnite, were found; at Table Mountain, Colorado, and several localities in Oregon (USA). Agate-like pebbles from Cook County, Minnesota, have been cut as cabochons. Nova Scotia (Canada), Germany and Scotland have produced specimens.

**Uses** Occasionally cut as cabochons for jewelry.

---

## 268 SCOLECITE

SILICATES (Tectosilicates, Zeolite Group)
$Ca(Al_2Si_3)O_{10} \cdot 3H_2O$ (Hydrated calcium aluminum silicate)

**System** Monoclinic.

**Appearance** Slender, striated, colorless or white, prismatic crystals, often grouped in fibrous, radiating masses.

**Physical properties** Hard (5–5.5), light, fragile, with perfect cleavage. Transparent with vitreous to silky luster. When heated, first curls into wormlike forms then fuses to a blistered glass. Soluble in hydrochloric acid, leaving silica gel.

**Environment** A mineral found lining cavities in lavas, especially basalts. Also found in cracks in schists and limestones affected by contact metamorphism.

**Occurrence** Beautiful crystallized forms found in basalts in Iceland; the Faroes; northern Scotland; Eritrea; Table Mountain, Colorado, and Great Notch, New Jersey (USA); India (Poona); and Rio Grande do Sul (Brazil). Found in blocks ejected by Vesuvius, in basalts in Val di Fassa (Trento) and in metamorphic rocks in several places in the Alps (Italy).

**Uses** Of interest to scientists and collectors.

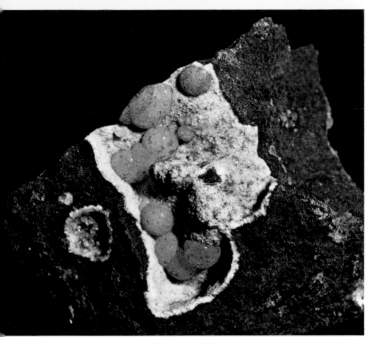

Radiating thomsonite crystals in basalt cavity (ca. ×3). Iceland.

Scolecite crystals (ca. ×1.5). Iceland.

## 269 LAUMONTITE

SILICATES (Tectosilicates, Zeolite Group)
$Ca(Al_2Si_4)O_{12} \cdot 4H_2O$ (Hydrated calcium aluminum silicate)

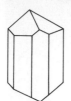

**System** Monoclinic.
**Appearance** Elongated, prismatic crystals with vertical striations, white or delicately colored yellow or pink. Columnar, fibrous and radiating aggregates.
**Physical properties** Semi-hard (3.5–4), light, fragile with perfect cleavage. Transparent with vitreous luster, tending to pearly. When exposed to dry air and light becomes opaque and powdery through partial dehydration, so in collections has to be kept in a sealed container or sprayed with Krylon. Fuses easily and dissolves in hydrochloric acid, forming silica gel.
**Environment** In veins and cavities in igneous rocks, both intrusive and volcanic, felsic or mafic. Found in veins in sedimentary and metamorphic rocks and in seams of metallic minerals.
**Occurrence** Outstanding 30 cm (12 in) crystals found at Bishop, California (USA). Very fine crystallized forms in the eastern Pyrenees (France); New Zealand; the Ardennes (France); northern Norway; Paterson, New Jersey (USA); and in Alpine gneiss (Ticino, Switzerland; Beura, Val d'Ossola, Italy). In Italy has also been found in cavities in granite at Baveno (Novara) and in the Montecatini mines in Val di Cecina (Pisa).
**Uses** Of interest to scientists and collectors.

---

## 270 HEULANDITE

SILICATES (Tectosilicates, Zeolite Group)
$(Na,Ca)_{4-6}Al_6(Al,Si)_4Si_{26}O_{72} \cdot 24H_2O$ (Hydrated sodium calcium aluminum silicate)

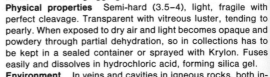

**System** Monoclinic.
**Appearance** Tabular crystals, often in parallel aggregates. Colorless, yellow, green or reddish-orange.
**Physical properties** Semi-hard (3.5–4), light, fragile with perfect basal cleavage. Transparent or translucent with vitreous or pearly luster. White streak. Dissolves easily in hydrochloric acid, leaving silica gel. Fuses easily and when heated intumesces and loses water.
**Environment** In cavities in volcanic rocks, especially basalts, associated with calcite and other zeolites. In veins in schists and gneiss. Disseminated in sedimentary rocks and some metal-bearing veins.
**Occurrence** Fine crystals in cavities in basalts at Paterson, New Jersey (USA), Iceland, Nova Scotia (Canada), Rio Grande do Sul (Brazil), India, Hawaii (USA), the Tyrol (Austria) and in silver-bearing seams at Andreasberg (Harz, Germany) and Köngsberg (Norway). Found in Val di Fassa (Trento), Agordino (Belluno), at Montecchio Maggiore (Vicenza), in augitic andesite at Montresta (Oristano, Sardinia) and more rarely in granites at Baveno (Novara) and on Elba (Italy).
**Uses** Of interest to scientists and collectors.

White laumontite crystals with pink chabazite crystals (ca. ×1). Paterson, New Jersey.

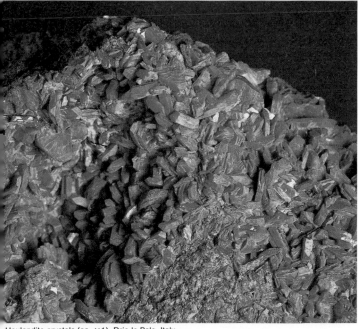

Heulandite crystals (ca. ×1). Drio le Pale, Italy.

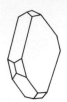

## 271 STILBITE

SILICATES (Tectosilicates, Zeolite Group)

NaCa$_2$(Al$_5$Si$_{13}$)O$_{36}$·14H$_2$O (Hydrated sodium calcium aluminum silicate)

**System** Monoclinic.

**Appearance** White, gray or reddish-brown prismatic crystals, usually in sheaflike aggregates. Pseudo-orthorhombic cruciform twins and fibrous, radiating masses.

**Physical properties** Semi-hard (3.5–4), light, with perfect cleavage. Transparent or translucent, with vitreous to pearly luster. White streak. Soluble in hydrochloric acid and fuses easily to an opaque, white glass.

**Environment** A late hydrothermal mineral in cavities in basaltic rocks, associated with calcite and other zeolites. More rarely in lithoclases in metamorphic rocks.

**Occurrence** Fine crystals at Paterson, New Jersey (USA), Kilpatrick and the Isle of Skye (Scotland), Nova Scotia (Canada), Rio Grande do Sul (Brazil), Teigarhorn (Iceland) and Poona (India). Found near Oristano (Sardinia) in tonalite in Adamello and in granites at Baveno (Novara) and Elba (Italy).

**Uses** Of interest to scientists and collectors.

## 272 PHILLIPSITE

SILICATES (Tectosilicates, Zeolite Group)

(K$_2$,Na$_2$Ca)(Al$_2$Si$_4$)O$_{12}$·4–5H$_2$O (Hydrated potassium sodium aluminum silicate)

**System** Monoclinic.

**Appearance** Small, rodlike, prismatic crystals. Usually complexly twinned (often cruciform with fine striations). Colorless or white, yellow or gray.

**Physical properties** Semi-hard (4–4.5), light, fragile, with distinct cleavage. Translucent or transparent with vitreous luster. Dissolves easily in acids. Fuses with some difficulty to a white glass, after intumescing.

**Environment** A hydrothermal mineral found lining cavities in basalt rocks, associated with chabazite. An alteration product of feldspars and volcanic ashes.

**Occurrence** Very fine crystals in lavas at Capo di Bove (Rome) and Aci Castello (Catania) (Italy). Common in the lava of Vesuvius and in some basalts in the region around Vicenza (Italy). Also found in Germany, Iceland, Ireland, the USA and in the Pacific Ocean as the nucleus of ferromanganese nodules.

**Uses** Of interest to scientists and collectors.

▲ Stilbite crystals (ca. ×1). India.

▲ Phillipsite balls (ca. ×1.5). Aci Castello, Sicily.

## 273 HARMOTOME

SILICATES (Tectosilicates, Zeolite Group)
$(Ba,K)(Al,Si)_2Si_6O_{16} \cdot 6H_2O$ (Hydrated barium potassium aluminum silicate)

**System** Monoclinic.
**Appearance** Cruciform penetration twins, either simple or double twins ("fourlings"). Color white, gray, yellow, red or brown.
**Physical properties** Semi-hard (4.5), light, with distinct cleavage and uneven-to-subconchoidal fracture. Streak white. Transparent to translucent with vitreous to pearly luster. Fuses into a translucent glass. Decomposed by hydrochloric acid.
**Environment** Occurs in volcanic rocks, especially basalt, but also phonolites and trachytes. Also in gneiss and low-temperature hydrothermal veins.
**Occurrence** Extremely fine crystal groups found in Andreasburg (Germany) and Argyll (Scotland). Small crystals are found in gneiss in Manhattan, New York City (USA).
**Uses** Of interest to mineralogists and collectors.

---

## 274 CHABAZITE

SILICATES (Tectosilicates, Zeolite Group)
$Ca(Al_2Si_4)O_{12} \cdot 6H_2O$ (Hydrated calcium aluminum silicate)

**System** Hexagonal.
**Appearance** Pseudo-cubic, rhombohedral crystals often as penetration twins. Colorless, white, greenish or reddish.
**Physical properties** Semi-hard (4-5), light, fragile, with fairly good rhombohedral cleavage. Transparent or translucent, with vitreous luster. White streak. Dissolves in hydrochloric acid, leaving a silica gel. Fuses easily, with intumescence, to a whitish, blistered glass.
**Environment** In cavities in volcanic and intrusive igneous rocks. In fractures of crystalline schistose rocks, associated with other zeolites. Also formed by chemical deposition from thermal waters.
**Occurrence** Famous specimens from the Faroe Islands, Scotland, Ireland, Aussig (Bohemia), Czechoslavakia, Oberstein (Germany), Nova Scotia (Canada) and Paterson, New Jersey (USA). Magnificent crystals have been found at Alpe di Siusi (Bolzano), the Cyclopean Islands (Aci Trezza, Catania), in basalts near Vicenza, in cavities in granites at Baveno (Novara) and on Elba, in several places in the volcanoes of Lazio and near Oristano in Sardinia (Italy).
**Uses** Of interest to scientists and collectors.

Harmotome crystals (ca. ×0.5). Andreasburg, Harz, Germany.

Chabazite crystals (ca. ×1). Nova Scotia, Canada.

## 275 WHEWELLITE
ORGANIC MINERALS
$CaC_2O_4 \cdot H_2O$ (Calcium oxalate)

**System** Monoclinic.
**Appearance** Small, white or colorless crystals, sometimes in heart-shaped twins.
**Physical properties** Soft (2.5), light, with distinct basal cleavage and conchoidal fracture. Transparent with pearly luster.
**Environment** An authigenic sedimentary mineral which crystallizes in the lowest parts of coal deposits or in the top (outer zone) of hydrocarbon deposits.
**Occurrence** Like all organic minerals has been identified with certainty in only a few places. Found in lignite at Zwickau (DDR). In oil deposits, found particularly at Maikop (northern Caucasus, USSR). Also found in Bohemia (Czechoslovakia) and Hungary. Found in Sulcis (Sardinia) and Tuscany, in lignites (Italy).
**Uses** Of interest to scientists and collectors. In the human body, one of the main crystalline components of kidney stones and urinary precipitates.

## 276 AMBER
ORGANIC MINERALS

**Appearance** Nodules or small masses of various shapes and sizes, sometimes rough or cracked on the surface. Orange to dark brown, occasionally green, violet or black. Often contain fossil remains of insects and plants which are clearly visible through transparent amber.
**Physical properties** Amorphous, soft, very light, fairly fragile, with splintery fracture. Easy to carve. Transparent to translucent with resinous to pitchy luster. Floats on water and burns giving off a strong smell of incense. In the trade there are many distinct varieties distinguished by their transparency (clear amber), by the presence of air bubbles (nebulous or bastard amber) or by origin (Baltic, Sicilian or sumetite, Romanian and Burmese).
**Environment** Fossilized resin, residual or derived from transportation in sedimentary deposits of the Tertiary period.
**Occurrence** The most famous deposits are in the Baltic region (in the Blaue Erde and sands derived from them), in Romania, Burma and the river Simeto (Sicily). Other localities are France, Spain, the USSR, Canada, the Dominican Republic and at Loiano (Bologna, Italy).
**Uses** A precious ornamental material. There are many synthetic, organic and inorganic imitations, including an aggregate made by hot-pressing the scraps left after working natural amber.

▲ Whewellite crystals (ca. ×1.5). Bohemia, Czechoslovakia.

▲ Amber (ca. ×1.5). Kaliningrad, USSR.

# INTRODUCTION

# TO ROCKS

Rocks are natural aggregates of one or more minerals, and sometimes noncrystalline substances, constituting masses that are geologically independent and can be represented on a map.

Their description and classification is the subject of the science of *petrography*; the interpretation of their genesis and evolution and the thermodynamic study of the processes that led to their present state are the study of *petrology*.

The study of rocks is based on methods drawn from mineralogy, geology, chemistry and physics, and the first step in their study requires the precise identification of their components, i.e., of the minerals present in them. Most rocks are *heterogeneous*, formed from several types of minerals; only a few are *homogeneous* or monomineralic. In addition to examining their composition, it is important to establish the proportions in which their various components are present, both on a microscopic and on a geological scale. The *texture* of a rock is the totality of the characteristics, given by the size, the shape and the mutual arrangement of its components, its *grains*. The *structure* of a rock is the totality of its characteristics on a geological scale, and mainly describes features that result from deformation experienced on or near the surface of the Earth. (For igneous [originally molten] rocks there is still a tendency to apply the nomenclature developed by earlier German petrologists

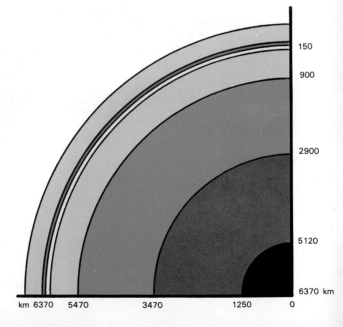

*Structure of the interior of the Earth: atmosphere (blue); crust (brown); upper mantle (light green); transition zone (medium green); lower mantle (bright green); outer core, liquid Ni-Fe (gray); inner core, solid Ni-Fe (black).*

150
900
2900
5120
6370 km

km 6370    5470    3470    1250    0

whereby the term "structure" describes the shape, dimensions and articulation of the components of a rock, and the term "texture" describes the spatial arrangement of the minerals as developed at the time of crystallization. In this text the better established international system described above is used.)

Crystalline rocks are formed mainly through three basic physical-chemical processes: crystallization from a hot molten state, precipitation from solution and recrystallization in the solid state. Each of these processes develops its own course and results in different types of end products (rocks) by means of variations in the sets of conditions or because the rocks may represent an incomplete ("congealed") stage of the entire development process and so are present as residual products. These types of rocks are very useful because they sometimes tell us more about the operative processes than the completed rocks. In studying the genesis of rocks it is also necessary to take a fourth, very complex, process into account: the decomposition of all rock types by forces acting on the surface of the Earth and their adjustment to temperature and pressure conditions.

## The igneous process

Igneous rocks are the final product of the consolidation of a *magma*, a hot molten mass of essen-

*Aplitic dikes in amphibolite in upper Val Malenco (Sondrio, Italy).*

tially silicate composition, rich in volatile element
and formed deep in the Earth by the fusion of pre
existing solid masses. The partial melting of th
layer of the Earth immediately below the crust o
which we live—the mantle—produces primar
magma, usually basaltic in composition, tha
comes to the surface of the Earth by means c
eruptions (*volcanic* or *extrusive* rocks) or by ir
jection into layers or cracks in the crust at variou
depths (*hypabyssal* rocks). Other magmas
derived from the basaltic melt by means of *di*
*ferentiation processes*, also come up to or near th
Earth's surface. In contrast, rock masses of su
face origin, or presently residing on the surface
may slowly sink to considerable depths (becaus
they are out of isostatic equilibrium) and ma
reach temperatures and pressures at which som
minerals with lower melting points are fused c
melted, and these magmatic masses or *anate*
*tic magmas* produce new igneous rocks. Thes

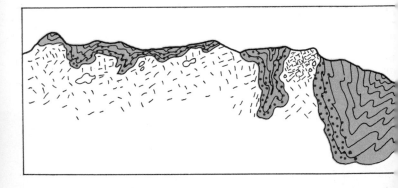

*Batholith (dashed pattern) in a
simplified geological cross
section with apophyses and
stock (closely spaced dashed
pattern), contact aureole (red
dots), surrounding rocks (light
brown). At various places in the
batholith inclusions of foreign
rock (or xenoliths) are visible.*

are often highly viscous melts, containing mar
components that are still solid (unmelted) held t
gether by a membrane of "melt." They have litt
mobility, or else move upward into fractures to a
pear only as *apophyses* and veins. Their compos
tion is not basaltic but distinctly granitic. Thes
magmas thus tend to recrystallize under conc
tions found at greater depth, forming the *plutor*
or *intrusive* rocks. Primary basaltic magma ar
granitic anatectic magma develop in differe
ways and show little affinity with one another, ;
though it can be shown experimentally that a ro
of granitic composition can be obtained from a b
saltic magma by fractional crystallization (thou)
the reverse process is not possible). Howeve
only a small amount of granite can be derived
this manner. Geologically, too, the two classes
rocks tend to occur apart from each other; basa
and rocks derived from them account for a lar)

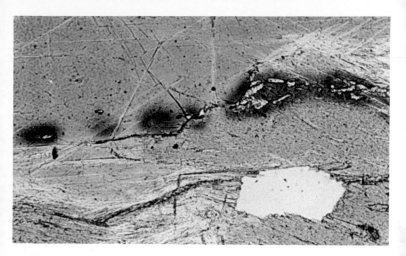

Photomicrograph of accessory minerals: zircon crystals (gray) included in biotite (light brown), surrounded by pleochroic halos (dark brown) caused by the radioactive elements (U, Th) in the zircon.

percentage of the volcanic rocks and are widespread on the ocean beds; granites and their derivatives make up the greater part of the plutonic rocks and are found mainly in continental masses in the form of large bodies, some deeper than others, called *batholiths*. Rocks of intermediate composition between basalts and granites, resulting sometimes from a mixture of unmelted matter with new magma (*hybrid* rocks) or from conditions in which equilibrium has not been reached, are not as common; they are found only in exceptional, geologically limited areas. Primary basalt, after its formation by partial melting of the ultramafic (very high in iron-magnesium silicates) rock constituting the mantle, often comes to the surface directly through long, deep fissures, typical for example of Iceland and the Deccan lava fields of India.

The first minerals to crystallize from basalt (at temperatures of 1100°–1200°C; 2012°–2192°F) are very small amounts of spinels (chromites) and sulfides, and, rarely, precious metals such as platinum, which constitute what are called the accessory minerals—not essential, that is, to establish the type of rock. At somewhat lower

Bowen's continuous and discontinuous reaction series.

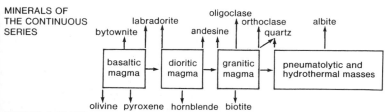

MINERALS OF THE CONTINUOUS SERIES

bytownite → labradorite → andesine → oligoclase → orthoclase → albite; quartz

basaltic magma → dioritic magma → granitic magma → pneumatolytic and hydrothermal masses

olivine → pyroxene → hornblende → biotite

MINERALS OF THE DISCONTINUOUS SERIES

temperatures the silicate rich in iron and magnesium (olivine) crystallizes, then, in order, those containing iron and magnesium as well as calcium and sodium (pyroxene) and, sometimes, those that are similar compositionally but also contain potassium and water (amphiboles and micas). This series of mineralogical reactions in basaltic rocks, first studied experimentally by N. L. Bowen, is called *Bowen's Reaction Series* and can be represented by two parallel branches. First, the "continuous" series, which concerns the plagioclase types of feldspar (felsic minerals), each of which after crystallization reacts with the melt again, adapting its composition to the lowering of the temperature and so passing in a continuous process from compositions rich in calcium (anorthite) to those rich in sodium (albite). The second branch is the "discontinuous" series, in which the iron-magnesium (mafic) minerals formed first react with the liquid and are completely resorbed by it, producing a new mineral, in such a reaction sequence as:

olivine → hypersthene → augite → hornblende → biotite.

The residue of both series, consisting essentially of silica, alkalis and water, crystallizes last, giving the *pegmatites* (rocks formed of quartz, albite, orthoclase and muscovite) at temperatures around 600°C (1112°F). The remaining fluids, essentially aqueous, produce hot springs, fumaroles and steam.

Differentiation may be interrupted at any stage by mechanical forces leading to intrusion or extrusion and hence to sudden changes in the conditions of temperature and pressure and a comparatively rapid cooling of the mass, which freezes in that particular developmental stage. This is expressed in the form of sills, laccoliths, dikes or directly as lava. The rocks in these masses sometimes acquire a *porphyritic texture*, with large crystals of the earlier-formed minerals in a groundmass of more rapidly crystallized fine-grained minerals or glass, which is the result of quenching the melt with no chance for crystallization of minerals.

Granitic magma is formed by the partial melting of preexisting rocks whose compositions are far more heterogeneous than those of mantle rocks, so that magmas can vary much more in composition. The final rock type which develops is determined by: (a) the composition of the original (or parent) rocks, and their possible heterogeneity, (b) the temperature at which melting takes place (the higher the temperature, the richer the rock will be in basaltic components), (c) the duration of the anatectic (or melting) phenomenon and (d) the availability, mobility and type of volatile components, whose presence promotes both the melting and homogenization of the masses. Only small portions of the melt, in which the gases are highly concentrated, are capable of intruding in the form of small plutons often of granitic or syenitic composition as veins, dikes or sills (called *pegmatites* if coarse-grained and *aplites* if fine-grained) or by actually coming to the surface, often explosively as thick layers of ignimbrite (or welded ash) and sometimes as lava beds (rhyolite). The greater part of the molten mass remains stationary in the form of enormous, deep structures tapering off into the rocks above or beside them, which become permeated by them (migmatites).

The crystallization order in the parts almost entirely melted, rich in volatiles and able to intrude, is somewhat similar to the later history in the intrusive rocks of basaltic origin: first, the accessory components (often inherited from the original material of the anatexis), then those poor in silica and rich in iron and magnesium (hornblende, biotite) and finally the alkali feldspars and quartz. Since the volatile elements are present in such abundance, however, the relationships between individual minerals are even more complex, sometimes even contradictory, including re-equilibration phenomena (called the *auto-metamorphic process*), comprising recrystallization of preexisting minerals resulting in new minerals, the result of low temperatures and abundance of water.

## The sedimentary process
Sedimentary rocks, which cover two-thirds of the

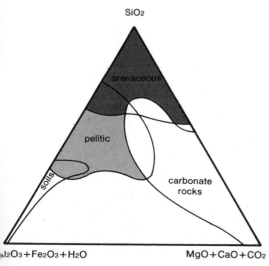

Triangular diagram of the variation in chemical composition of the sedimentary rocks.

**419**

Earth's surface, are produced by the transformation of preexisting rocks by gravity, atmospheric agencies and living organisms. They are the result of the consolidation of sediments, loose materials derived from the mechanical accumulation of fine and coarse fragments of rock (*clastic sediments*) or from precipitation from solutions, with or without intervention by living organisms (*organogenetic and chemical deposits*).

The clastic sedimentary process has four stages. First, it begins with *alteration* of the original material by the agencies named above. This leads to the formation of *soil* on top of the parent rock, its thickness governed by the particular alteration process, the duration of the process, the nature of the parent material and the possible washing away of the components before cohesion took place. The second stage, called *transportation*, normally takes place in water: first in streams, then in rivers and finally in the sea, by currents and wave motions. But it can also be due to the effects of wind, frost, gravity and also living organisms. Transportation normally separates the different components of the sediment according to size, specific gravity or chemical composition. In transport by water we must distinguish between materials simply rolled and those carried in suspension or solution. The third stage of the erosion cycle is *deposition*, the most important stage in some ways because it gives the sediment its special textural characteristics. Sediments are classified according to the environment in which they are deposited, such as *continental* and *marine.*

Continental deposits may be *subaerial* (on the Earth's surface), like the detritus (broken particles) from landslides, *aeolian* (wind-blown) sands of the desert and periglacial (end-stage glacial) loess, or *subaqueous* (under water). These include *fluviatile* (river bed) types consisting mainly of shingle (flattened pebbles often partially overlaying other pebbles) and rounded sands; *lacustrine* (lake) deposits with sands, muds and clays *lagoonal*, also consisting of muds and clays but sometimes having layers of *evaporites* (salts precipitated from supersaturated solutions) between layers of the deposit; and *deltaic* (the mouth of a large river), formed from a variety of materials thoroughly sorted by size.

Marine sediments consist of a mixture of detrital (clastic) material, fine or coarse-grained, often reconstituted from preexisting continental sediments, together with materials coming from the chemical or biochemical precipitation of salts contained in the sea water and the remains of organic activity present in those waters—skeletons shells, etc. They are classified according to the depth of the water in which they are deposited and their distance from the coast as follows: *pelagic*, mainly fine-grained and siliceous; *neritic*, coarsergrained with complex structures due to the movement of detritus by water and the activity of organisms; and *intertidal*, formed in deltas, lagoons and coral reefs, often chaotic, and generally mixed with organic material.

The chemical and biochemical sedimentary process consists of precipitation of inorganic salts or of substances that living organisms need for their survival—primarily calcium carbonate, secondarily calcium phosphate and hydroxides of iron and silica. The first of these is precipitated in continental areas and particularly in marine environments, often mixed with magnesium carbonate and very fine siliceous muds, in relatively shallow waters. Animal and vegetable organisms play a large role in the chemical processes; they use the inorganic compounds in the formation of their skeletons, shells and branches, and very extensive stratified areas are formed from these materials after their death. Below a certain depth of water calcium carbonate is redissolved, so that abyssal (deep) deposits are formed almost exclusively of silica, largely from the accumulation of remains of organisms and from precipitation from hot (hydrothermal) solutions of volcanic origin, which also contain abundant manganese and iron, this results, in part, in the formation of *cherts* (finegrained siliceous rock). Phosphatic and ferruginous deposits, on the other hand, are mainly continental in origin; the former are derived from the accumulation of skeletons of vertebrates and of excrement, the latter from bacterial fixation of iron in solution in swamp water. A rarer, but still important, type of chemical sediment consists of *evaporites*, derived from the evaporation of salt water (mainly sea water) in enclosed basins with the precipitation of salts (particularly chlorides and sulfates of alkaline elements) which would otherwise remain in solution.

The final stage of the sedimentary process is *lithification*: the transformation of the loose sediment into a hard, coherent rock by the elimination of the intergranular space. This takes place either by simple compaction or by the chemical precipitation of a *cement* (e.g., silica, calcite) which binds the alluvial grains together. This stage is completed by *diagenesis*, a partial recrystallization caused by the pressure of the sedimentary overburden and by the solution and selective chemical exchange carried out by the surrounding water, which often leads to the formation of rocks such as the *dolomitic limestones*.

## The metamorphic process

Metamorphism is the totality of the chemical and

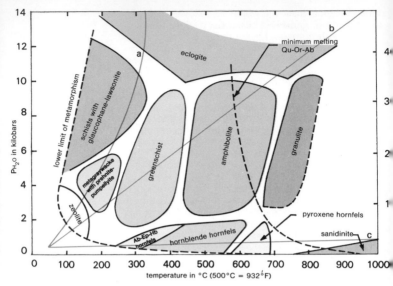

*Diagram of the conditions of metamorphism with various pressure and temperature gradients and areas of stability of the metamorphic facies. $P_{H_2O}$ indicates water pressure; gradient (a) is for load metamorphism; gradient (b) is for regional metamorphism; gradient (c) is for contact metamorphism.*

physical reactions and processes in the solid state by which a preexisting rock of any kind adapts to new conditions resulting in a new rock type. The change in condition is usually the result of a slow shift of the Earth's crust (diastrophism) as the result of plate tectonics. Every igneous or sedimentary rock is only in equilibrium (stable) within a limited field of temperatures and pressures. As soon as the environmental conditions change beyond a certain point, these rocks tend to be modified (metamorphosed) toward a new mineralogical association (usually with the aid of fluids) bringing them into equilibrium with the new conditions, especially new temperature and pressure conditions. Most commonly these rocks experience *recrystallization*. For instance, if a basaltic dike and clay are buried together under some 3000 m (10,000 ft) of sediment, which would produce a pressure equal to 1 kilobar (about 1000 times the atmospheric pressure at the Earth's surface) at a temperature of about 450°C (842°F), both of them would recrystallize. In the basalt the plagioclase and pyroxene, which had been formed magmatically at about 1000°C (1832°F), will be transformed into new, lower-temperature minerals. In the clay, which consists of minerals typical of lower (atmospheric) temperature, minerals of higher temperature will appear with the loss of the water which diffused into the basalt. From the latter a chlorite schist will be formed composed of albite and chlorite and having a schistose texture, and from the former a phyllite will be formed consisting of chlorite, muscovite and quartz and hav-

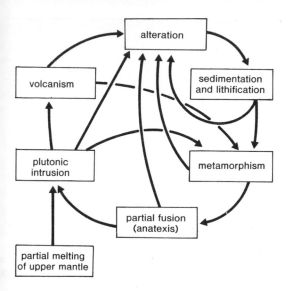

ing a phyllitic texture. The parent rocks will only be traceable by the presence of mineralogical and structural remains. For the clay, the metamorphism leading to the minerals of higher temperature and pressure will have acted in a *prograde* manner, for the basalt in a *retrograde* manner.

Metamorphism generally occurs during the sinking of surface or near-surface rock masses deeper into the crust, and as a result of the rise in temperature due to the geothermal gradient and the rise in pressure due to the weight of sediments deposited above them (static metamorphism). Often, however, tangential pressures come into play, the result of differential movement of geological masses in continental movements in the Earth's crust (*regional metamorphism*). The reactions that characterize these two types of metamorphism are of the solid-solid type; that is, they owe little to the aqueous solutions in the parent rocks, apart from the secondary phenomenon of dehydration. The regionally metamorphosed rocks have textures related to the presence of directional forces; they are characterized by a general flattening of the minerals or by a preferential increase of lamellar minerals; the prismatic minerals, in contrast, tend to be arranged according to the forces acting on them. When there are mainly flat or lamellar minerals schistose structures are formed, and when there are mainly linear minerals a linear structure is formed; both often having crenulations. Beyond a certain temperature, dehydration reaches the point at which it promotes the disappearance of hydrated schis-

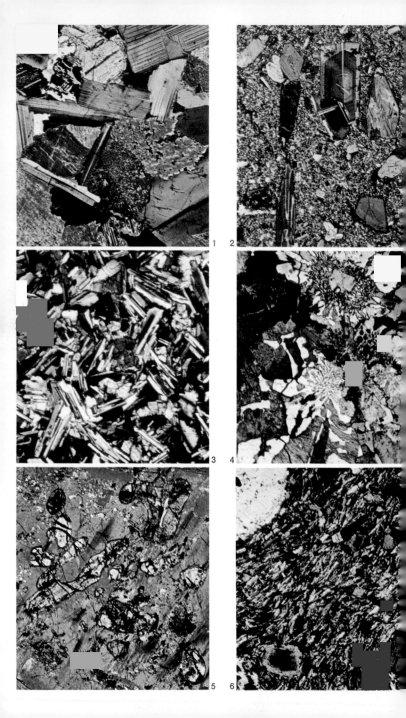

1    2

tose materials (muscovite, for instance) and the crystallization of new anhydrous minerals which are not lamellar (for example potassic feldspar). The result is that the schistosity is lost and more solid-looking rocks, which are tabular or "eyed" (*Augen*, from the German word for eyes) but are still divisible into sheets or layers (that is, of gneissic texture) are generated. In the presence of residual water the rise in temperature and pressure can lead to the remelting of potassic feldspar, quartz and albite, all at about the same point, producing an anatectic magma of granitic composition which impregnates the parent rock in the form of little veins, narrow "folded" layers and nodules (*neosomes*) scattered through unmelted parts of the preexisting country rock (*paleosome*); this results in a mixed rock, called a *migmatite*. As the

*Typical textures of the igneous and metamorphic rocks.*
*Opposite page: igneous rocks*
*1. idiomorphic (good crystalline forms)*
*2. porphyritic (large crystals, fine-grained groundmass)*
*3. ophitic (pyroxene and plagioclase intergrown)*
*4. micropegmatitic (quartz and feldspar intergrown)*
*5. poikilitic (earlier minerals included in later mineral)*
*6. fluidal (gray fine plagioclase crystals aligned by flow)*
*This page: metamorphic rocks*
*1. granoblastic (recrystallized, uniform size)*
*2. lepidoblastic (recrystallized, minerals aligned)*
*3. nematoblastic (radiating crystals)*
*4. porphyroblastic (recrystallized, some grains much larger than groundmass)*

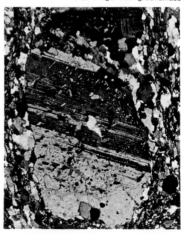

*Structures in igneous (left: layered structure in gabbro) and sedimentary rocks (right: stratification in jasper).*

process goes on the neosome component increases until it becomes predominant, giving the rock a new mobility and transforming it into a true plutonic magmatic, anatectic granite.

An entirely different form of metamorphism takes place when a magmatic mass at high temperature (it may be an apophysis of anatectic granite rising again) comes into contact with metamorphic or sedimentary rocks, or, less often, with preexisting igneous rocks. Conditions of high temperature and fairly low pressure are created which cause the expulsion of the volatiles from the surrounding rock and their infiltration for a greater or lesser distance from the point of contact——the *metamorphic aureole*. High-temperature minerals are then formed, often in large crystals because their growth is facilitated by the transport of material in the circulating gases. This is called *thermal metamorphism* or *contact metamorphism*. When induced by magmas already rich in unusual gases, it can result in magnificent crystals of less common minerals, and sometimes in economically useful mineral concentrations (*skarns*). Much commoner but much more localized are the conditions for *cataclastic metamorphism*, which occurs when two masses of rock move in relation to each other as a result of geological pressures; this leads to the formation of more or less minutely fractured areas and sometimes actually to remelting as a result of the heat generated by the friction (*mylonite, tachylite*). Another type is *impact metamorphism*, caused by meteorites striking the surface of the earth. These are not as rare as previously thought, but older impact events tend to be quickly destroyed by erosive forces and therefore only the most recent events are preserved. It is the major type of metamorphism on the Moon.

## The petrogenetic cycle

From the formation of primary basaltic magma by the partial melting of the mantle to the regeneration of anatectic granitic magma in the depth of the crust, we see the completion of the whole petrogenetic cycle to which all the rocks on the Earth are subjected, to a greater or lesser extent. The possible successive stages of the cycle comprise the intrusion and extrusion of magma or lava, which covers the whole of the igneous process; the decomposition of the rocks so formed and the redeposition of the products in the sedimentary phase; and metamorphic recrystallization or reconstitution in successive stages up to the formation of migmatites and anatectic plutonic rocks. The fact that from the original basalt we get a granite as the final product indicates which parts of the material generated in the mantle do not complete the cycle—the parts that make up the atmosphere and the hydrosphere, and which remain permanently on the surface in the form of residual sedimentary rocks. If this residual part of the cycle, the hydrosphere and atmosphere, had never been formed, we would not have continents, or life itself.

## RECOGNITION AND STUDY OF ROCKS

Rocks are aggregates of minerals. Therefore, in order to recognize rock types, we must be able to identify the minerals they contain. But that is not all we need. It is also necessary to determine the proportions in which the minerals are present and their spatial relationships to one another: their *texture*. In other words, to recognize a rock we need to: (1) identify the number and types of minerals present; (2) determine the size and shape of the crystals and their spatial relationship with each other, properties which establish the texture of a rock; (3) determine the quantitative proportions of the mineral constituents (the *mode*); and finally (4) determine the chemical composition of the rock. It may sometimes be necessary to combine these data with field observations of the *structure* of the rock, i.e., the large-scale characteristics it has acquired in the course of its development through geological processes (stratification, bedding, folds, faults, etc.).

The identification of the mineral constituting a rock can often be done directly in the field, by eye or with a hand lens. In the case of a fine- or very fine-grained rock, and of those that contain glass, it is necessary to make use of a petrographic microscope, x-ray or other laboratory methods. Observations of the texture and determination of the mode are also often possible in the field, especially for coarse-grained rocks and if one has had

some experience. But it is much easier under the microscope, particularly with the use of *thin sections*——slices of rock ground down to a thickness of about 30 $\mu$. This renders nearly all the minerals in them transparent, and it is possible then to undertake a series of optical observations as described in the Introduction to Minerals, making it possible to identify the species or at least the family to which each of the components belongs. The petrographic microscope is thus the essential instrument in the study of the petrography of rocks; nothing has yet been found to take its place. It can also be used, in combination with a *mechanical stage* and *point-counter*, for the precise determination of the mode, using one or more thin sec-

*Structures in metamorphic rocks: microfolds in mica schist (above) and faulted microfolds in polished slab of marble (right).*

428

tions from the same sample in such a way that the final count is sufficiently representative of the whole rock (generally 2000–3000 points, i.e., 2000–3000 grains identified and counted). Another useful accessory for use with the microscope is the *universal stage*, or Fedorov's table, with which it is possible to tilt the thin section at different angles to the optical path. Thus the observer may carry out highly accurate optical measurements and also measure the orientation of the individual grains in relation to the plane of the thin section, from a sample previously oriented in the field. Only occasionally, for rocks with very fine or very coarse grain size, will chemical analysis be necessary in order to compare their chemical composition with established patterns. This is particularly true of rocks that contain large amounts of glassy components. Similarly, chemical analysis is useful in classifying rocks formed from minerals of complex composition which cannot be identified accurately by their optical properties (amphiboles, pyroxenes, olivines, etc.). For coarse-grained rocks chemical analysis may turn out to be quicker and more efficient than modal analysis, which would have to be performed on an excessive number of thin sections to get a truly representative result. Some rocks, especially coarse-grained rocks or those made up of heterogeneous zones, can only be classified on the basis of structural observations in the field, taking a whole outcrop into consideration. This is the case with migmatites or with some polygenetic breccias, for which the usual samples, measuring 10 × 5 × 3 cm (4 × 2 × 1.2 in), are inadequate, and even rough or polished slabs of 1–2 sq m (3–6 sq ft) are insufficient. Other properties not usually measured in rocks, such as specific gravity, color, smell, compressibility, radioactivity, the way they lie as interpreted from geological maps, can all constitute useful diagnostic elements. (These properties often prove fundamental in the mechanical study of rocks; we can work out from them the possibility of using a given rock as building material. Some rocks, however compact and attractive in appearance, can only be used under quite special conditions. Evaporite rocks, for instance, like anhydrite, are compact, massive and durable in dry climates, but cannot be used on the exterior of a building in damp climates because their solubility makes them easily plastic and their surface deteriorates.)

## THE CLASSIFICATION OF ROCKS

Rocks are divided into three main groups according to the process by which they are formed: igne-

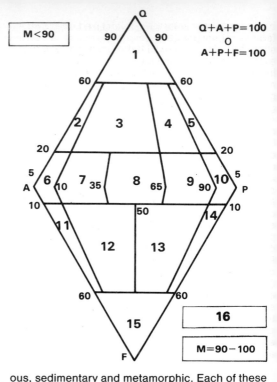

ous, sedimentary and metamorphic. Each of these groups is further subdivided according to several criteria, which have not been accepted universally, since rocks and scientists are both heterogeneous. A certain uniformity of approach has been reached for the igneous rocks, based on quantitative mineralogical methods, but there is still a long way to go with sedimentary rocks and, to a lesser extent, with metamorphic rocks, in which structural and textural considerations seem to be more important than purely mineralogical ones.

*Igneous rocks* A primary subdivision of these rocks is made according to their geological environment (the depth of their occurrence) and the correspondingly different conditions of pressure exerted on them at the time of crystallization. They are thus divided into:

a) *Intrusive* (or plutonic), crystallized slowly at depth, under pressures too great to allow the escape of magmatic gases.

b) *Hypabyssal*, crystallized at medium depths, generally at a pressure sufficient to prevent too great a release of gases, but relatively quickly as evidenced by the limited size of the geological bodies in which they occur (sills, apophyses, laccoliths).

*c) Extrusive* (or volcanic), crystallized on the surface at atmospheric pressures, cooling in very short periods of time and therefore fine grained or even amorphous or glassy.

Intrusive and extrusive rocks are thus classified on modal principles, except for very fine-grained or glassy extrusive rocks, where it is necessary to get the chemical composition and calculate a *norm* which gives the hypothetical percentage of each mineral which would have crystallized if it had not cooled so quickly. Hypabyssal rocks are also classified on the basis of mixed chemical and mineralogical data, referring in the first place to intrusive rocks of similar chemical composition and then giving special names to varieties characterized by certain minerals. International modal classification of fine-grained intrusive and extrusive rocks is based on the *color index* ($M$)—the percentage, by volume, of colored minerals, including those that are opaque and nonsilicates. Rocks are classified in 15 different "fields" according to their color index. Rocks with $M > 90$ (i.e., with more than 90 percent of colored minerals) are called ultramafics (they are allotted a separate field, 16) and are distinguished according to the predominant mineral (hornblendite, pyroxenite, carbonatite, etc.). Those with $M < 90$ are classified according to their relative percentage of three components: alkali feldspar ($A$) (or-

*Horizontally stratified sedimentary rocks incised by the erosion of the Colorado River (USA).*

thoclase, microcline, sanidine, albite, etc.), plagioclase (*P*) (from oligoclase to anorthite) and either quartz (*Q*) or feldspathoid (*F*) (nepheline leucite, analcite, sodalite, etc.), which cannot occur together in equilibrium since they are antithetic to one another. Each group defined in this way is given a name, to which may be added an adjective referring to one or more of the characteristic minerals or a prefix descriptive of the color index: *leuco-* for low values and *mela-* for high Further specifications refer to the texture—equi granular, porphyritic or partially glassy—and the structure, which may be isotropic, oriented and sometimes vacuolar, spherical, zoned, scoria ceous, etc.

*Sedimentary rocks* These rocks are classified not so much according to the process of decomposition of the preexisting rocks from which they are derived as to the mechanism of the deposit which is responsible for their final texture and structure Thus, leaving aside incoherent sediments, they are divided into:

a) *Detrital* (fragmental or clastic) *rocks*, consisting of fragments (clasts) of all sorts of rocks deposited after having been transported for some distance. The environment of the deposit (fluviatile, marine, continental, etc.), which accounts for the type of cementing material (the fine precipitate that makes the rock coherent), is unimportant to their classification, which is based on the size of the clasts. Clastic rocks are thus divided into *rudaceous*, in which the clasts are coarse-grained (minimum diameter 2 mm), *arenaceous* with medium- to fine-grained clasts (diameter between 2 mm and 1/16 mm; .078–.006 in) and *argillaceous* with fine or very fine grain size (diameter less than 1/16 mm). Further subdivisions of these three classes again refer to the dimensions of the grains, and also to their roundness and to the amount of selection exercised on the original material by the transporting agency. Thus the composition of the grains, the percentage of the groundmass and cement present and the mineralogical composition of the cement itself (calcareous, siliceous, etc.) are all important.

b) *Chemical rocks*, consisting for the most part of salts and colloids precipitated from aqueous solutions by evaporation or by an exchange of chemical environment due, for instance, to a sudden mixture with gas or volcanic solutions. Further subdivisions are based on the chemical composition of the precipitates, which may be *calcareous* (calcite, dolomite), *siliceous* (chalcedony, quartz), *ferruginous, manganiferous* and last *saline* (rock salt, gypsum, anhydrite, etc.) generally considered together as *evaporites* because they are derived from the evaporation of water in

432

enclosed basins such as coastal lagoons and salt lakes. Also included among the chemical rocks are *residual* rocks, insoluble parts that remain when preexisting rocks (laterite, bauxite, clay) are washed away, and *metasomatic* rocks, produced by an exchange reaction between sea water and sedimentary rocks of various kinds (dolomites), although this term usually is relegated to metamorphic rocks.

*c*) *Organic* (organogenetic or biochemical) *rocks*, derived from the accumulation of substances of organic origin (shells, skeletons, vegetable matter). These too are divided according to their chemical composition, into *calcareous* (calcite, dolomite), *siliceous, ferruginous, phosphatic* (collophane, apatite) and *carbonaceous* (peat, lignite, subbituminous and bituminous coals and anthracite). In the classification of these rocks, especially of the carbonaceous rocks, consideration is given to the environment inhabited by the organisms from which they are derived (bioherm, benthos, plankton, etc.), to the nature of the organisms concerned and to a possible mixture with clastic materials, and chemical transformation undergone after deposition and before the definitive formation of the rock.

*d*) *Pyroclastic rocks*, including rocks erupted as fragments or as suspensions of lava, glassy and gaseous, in the course of volcanic activity, and deposited in strata consolidated under subaerial conditions (tuff) or subaqueous conditions (tuf-

*Veins of light-colored neosome in dark-colored paleosome, forming folds (ptygmatic folds) in a migmatitic gneiss (Valtellina, Sondrio, Italy).*

fite). These rocks are between igneous and sedimentary in character; they have the composition of the first, a little modified, and are deposited and lithified like the second. As will be seen, the classification of sedimentary rocks is based on criteria quite different from those used for igneous rocks; the genetic-environmental factor, interpreted by the texture and structure of the rocks, plays a primary role, while chemical composition is of minor importance.

*Metamorphic rocks* The classification of metamorphic rocks is also based principally on genetic criteria, according to the process that led to the recrystallization of the parent rock. Some metamorphic rocks · are distinguished according to whether they are derived from igneous rock (prefix *ortho*) or sedimentary rock (prefix *para*). A primary subdivision along genetic lines divides them into *contact metamorphic rocks* (derived from thermal metamorphism), *crystalline schists* (from regional and static metamorphism), *mylonites* (from displacement metamorphism) and *impactites* (from the extremely high temperatures and pressures generated by the impact of a meteorite on terrestrial rocks). The determinant for nomenclature of a metamorphic rock is the *grade of metamorphism*—the extent of recrystallization. This is defined by the presence of particular *mineral indicators* or of special associations of minerals typical of a characteristic equilibrium assemblage (*paragenesis*). In particular, much use is made of the concept of the *facies*, by which all rocks crystallized within a certain range of temperature and pressure are grouped together; there follows from this the concept of the *isograde*, a line joining all points in a geological body or region which have undergone recrystallization to the same extent, recognizable by the appearance or disappearance of a specific mineral indicator or index. These concepts make it possible to group together rocks that are chemically heterogeneous; this aspect comes into consideration more fully when we go on to classify the rocks individually. At least six divisions are made on a chemical basis, each one characterized by the formation of particular minerals as the grade of metamorphism increases:

*a*) Argillaceous rocks, derived from argillaceous (clayey) sedimentary rocks.

*b*) Felsic (quartzo-feldspathic) rocks, from feldspathic arenites and rocks of granitic composition.

*c*) Carbonate rocks, from calcareous rocks and dolomites.

*d*) Mafic rocks, from mafic igneous rocks and tuffs.

*e*) Magnesian rocks, derived from ultramafic igneous rocks and certain minerals in sedimentary rocks, for instance montmorillonite.

*f)* Ferruginous rocks, from iron-bearing sedimentary rocks.

In the course of recrystallization these compositions produce minerals of wide diversity, and the rocks are named after these. In addition to the modal mineralogical criterion (which has not been accepted internationally for metamorphic rocks as it has for igneous rocks), primary consideration in the classification of metamorphic rocks is given to the structure. Thus a specimen composed of quartz, biotite, plagioclase, sillimanite and garnet may be classified as mica schist or gneiss, though the mode is the same, according to whether it has a schistose structure (readily broken into thin, lenticular sheets) or an "augen" gneiss structure (has a layered structure with perhaps plagioclase or garnet concentrated in a few big, round crystals—the *Augen* or eyes). Similarly a rock composed of muscovite, quartz and graphite can be called a phyllite or a mica schist according to the microscopic and macroscopic dimensions of the sheets of mica, with phyllite being finer grained. We are still far from reaching good general agreement on the nomenclature of metamorphic rocks, not least because local names still survive which are known only to researchers. A good knowledge of the terms relating to structure is therefore still indispensable. In this text traditional nomenclature has been followed, classifying the rocks in order of their grade of metamorphism following the principles of metamorphic facies, after first separating them according to the chemical divisions mentioned above and the different types of metamorphic processes.

Finally, it must be pointed out that the nomenclature of migmatites, which are halfway between igneous and metamorphic rocks, the mixed rocks *par excellence* insofar as they are composed of a residual portion (paleosome) and a remelted portion (neosome), is based on textural criteria, i.e., on the extent of penetration of the neosome into the paleosome as it approaches nearer and nearer to complete anatexis.

## COLLECTION AND CONSERVATION OF ROCKS

There are far fewer rock collectors than there are collectors of minerals, but a petrographic collection should nonetheless be a source of some satisfaction. From both the scientific and the aesthetic viewpoint a collection of petrographic specimens may be even more valuable than one of minerals, and it is certainly much easier to organize, at any rate in the early stages. The tools required are simple: mallet, hammer, drills and chisels. The best places for collection are those where the rock has been opened up recently so that there is no coating of alteration on it, such as quarries, mine shafts, manmade trenches, landslides, etc. The specimens must be big enough to display the grain size of the rock and any special features to which attention should be drawn. The specimen is customarily chipped into a parallelopiped, 10 × 5 × 3 cm (4 × 2 × 1.8 in), the two larger surfaces flat and with straight beveled edges, done so that the marks of the hammer blows do not show. It is not advisable to collect specimens and then saw them up into little plates, though some collectors prefer this method, and even polish the surfaces. There are several ways of identifying the specimens on a fragment of the exhibit; it may even be possible to attach a thin section to it. The arrangement of the exhibits should be carried out with the same care as was described for mineral specimens, remembering however that specimens of rock, although sometimes much larger, are also less fragile and can be kept in less delicate cases than those advised for minerals.

## METEORITES

Meteorites are solid bodies that fall onto the Earth from space. They vary very much in size, from fine dust to masses weighing many tons, and as they pass through the atmosphere the friction heats their surfaces until they reach their melting point and produce a thin black, shiny, glassy *fusion crust*. The speed at which they fall to Earth varies considerably, since it depends on their mass and their trajectory, and on factors connected with friction. Some fall at a relatively low velocity, without producing any striking effects or causing any damage; others fall with a violent impact, with effects like those of a bomb spreading for a distance of several kilometers (Meteor Crater in Arizona is an example), and sometimes they explode, bursting into countless fragments or completely vaporizing. A spherical crater, the *impact crater,* is produced on the Earth's surface by the impact; it can be as much as 28–29 km (18 miles) across, as at the Ries Basin in Germany. Larger craters are found in Canada, and craters hundreds of kilometers across must have existed in the early stages of the Earth's history (4–4.6 billion years ago) similar to those now seen on the Moon, Mars and Mercury. The energy released produces rare high-pressure minerals in some of the impacted rocks (e.g., coesite and stishovite, which are polymorphic modifications of quartz), or minerals can develop shock characteristics or become glassy (maskelynite, diaplectic glass from feldspar)

These and many other phenomena are aspects of *impact metamorphism*.

Meteorites are classified by their mineralogical and chemical composition into stones, stony-irons and irons. Some glassy forms known as tektites and impact glasses fall from the sky but they are not considered to be meteorites.

*Stony meteorites* are the most common and are not easily recognized after they fall because they may look and feel superficially like terrestrial rocks. They contain mainly pyroxenes (enstatite, bronzite), olivine, plagioclase, metallic nickel-iron alloys, sulfides (the mineral troilite, FeS) and accessory minerals. Some of them, the *achondrites*, are igneous rocks (often brecciated by impact events), while others are characterized by the presence of *chondrules*—small, partially crystallized droplets consisting of silicate minerals embedded in glass—and so are called *chondrites*. Achondrites are igneous rocks, often brecciated by impact events, formed at or near the surface of small planetary bodies which make up the asteroid belt after these bodies were differentiated. Chondrites are formed as condensation products from the gaseous nebula around the Sun and are the primitive materials for the formation of planets, such as the Earth. The primary source of all the minerals and rocks—as well as the atmosphere and hydrosphere—on the Earth are chondritic meteorites and our Sun, from which all these materials were derived.

*Stony-irons* are composed more or less equally of silicates (olivine, pyroxene, plagioclase) and metallic iron-nickel alloys (the minerals taenite and kamacite) as well as accessory minerals. They are divided into two main types: *pallasites* and *mesosiderites*. Pallasites contain roughly half metallic nickel-iron and half olivine crystals. They form deep within a small planetary body at the interface between the core and mantle. Mesosiderites are roughly half metallic nickel-iron and brecciated basalt. They form near or at the surface of small planetary bodies.

*Iron meteorites* are the easiest to recognize, even when they are not seen to fall, because they differ so much from terrestrial rocks, being composed of metallic nickel-iron alloy minerals called kamacite (low nickel) and taenite (high nickel). They also contain minor amounts of silicates, graphite, diamond, troilite and chromite. In appearance they are black, shiny metallic masses, sometimes with a warty surface. Treatment with nitric acid on a cut and polished surface reveals a striking pattern of bands of taenite and kamacite (*Widmanstätten pattern*) characteristic of many irons. Some iron meteorites represent materials directly condensed from the gaseous nebula around the Sun, while

437

others are the result of the planetary differentiation process whereby the iron sinks to the core of a planet.

*Tektites* are black or dark-green glassy objects of near-granitic compositions sometimes containing tiny metal spheroids. They have pitted surfaces, are discoid or spherical in shape and have conchoidal fracture; they are discovered both on the surface and in sedimentary rocks of relatively recent formation. Superficially they look like obsidian (volcanic glass of granitic composition). They have been found in four main areas: Czechoslovakia, Texas, Ivory Coast and the Australasian field that extends from Indochina across to Java, Malaysia and Australia. There was much controversy as to their origin—whether from the Moon or the Earth—but since the Apollo space program nearly all scientists agree that they could not have originated on the Moon and that they are derived from impact melts of sedimentary rocks which were hurled long distances. They are all relatively recent, only 1–14 million years old.

Scientific interest in meteorites is especially intense at present because they give us valuable information about the early history of the solar system and the early history of planetary formation and differentiation. Chondritic meteorites even contain a few rare grains of pre-solar system particles, possibly derived from supernovae explosions and detected by isotopic studies.

# IGNEOUS ROCKS

## KEY TO THE SYMBOLS

CHEMISTRY OF THE ROCKS

persilicic rocks
($SiO_2$ > 65 percent)

mesosilicic rocks
($SiO_2$, 52–65 percent)

hyposilicic rocks
($SiO_2$ < 52 percent)

MODE OF OCCURRENCE

plutonic rocks
(high pressure and high temperature)

volcanic rocks
(low pressure and high temperature)

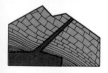

hypabyssal rocks
(medium pressure and high temperature)

MODAL CLASSIFICATION FIELD
following the IUGS modal classification, from
which the nomenclature used here is derived

**Type** Intrusive igneous rock.

**Chemistry** Felsic.

**Components** *Essentials:* potassic feldspar (orthoclase, microcline), quartz. *Accessories:* plagioclase (oligoclase), biotite, pyrite, zircon, monazite, tourmaline, magnetite, ilmenite. The rocks sometimes contain small quantities of orthopyroxene, clinopyroxene, hornblende and garnet. In the alkaline granite variety oligoclase plagioclase is missing, but alkali feldspars such as albite and anorthoclase are present, together with amphiboles and sodic pyroxenes (riebeckite, barkevikite, acmite, acmite-augite) and, as accessory minerals, fayalite, fluorite, cryolite and ore minerals containing rare earths and radioactive elements.

**Appearance** Rosy to dark red in color, sometimes with bluish specks; hypidiomorphic granular texture grading to xenomorphic and porphyritic; massive structure, but with numerous miarolitic cavities filled with crystals of olivine, tourmaline, etc. These granites are also often crossed by veins of quartz, aplite and pegmatite which is rich in rare minerals (cryolite, rare earth-bearing minerals, radioactive minerals, etc.).

**Geotectonic environment** Alkaline feldspar granites form immense homogeneous batholiths in the Precambrian shields, probably derived by anatectic processes. Alkaline granites form as differentiated marginal zones on the edge of minor intrusions, especially those formed at great depth. These are probably products of late differentiation, largely connected with pneumatolytic (gaseous) fluids, and are rich in uncommon chemical elements resulting in rare minerals.

**Occurrence** A typical alkaline feldspar granite is the so-called Swedish red granite, which also occurs in Norway, Finland and Canada. The granite found at Predazzo (Trento) is rich in alkali feldspar and contains tourmaline, slightly altered. Typical alkaline granites are to be found in Portugal, in the Cascade Range of Oregon and in California (USA), in the Cordillera Blanca in Peru and in certain limited areas of the Sardinian pluton (Fonni, Nuoro).

**Uses** These rocks are often highly valued as building material, especially decorative in polished slabs. There are at least three commercial varieties of "Swedish red"—"imperial red," "Ivanovich red" and "orchid red"—differing in shades of color and in grain size. Alkaline granites are used almost exclusively for the extraction of rare earth–bearing minerals.

Granite containing tourmaline (ca. ×1). Predazzo, Trento, Italy.

Granite "Swedish red" (or rose or pink) (ca. ×1). Sweden.

**Type** Intrusive igneous rock.
**Chemistry** Felsic.
**Components** *Essentials:* quartz, potassic feldspar (orthoclase, microcline), plagioclase (albite-oligoclase), biotite mica. *Accessories:* magnetite, ilmenite, apatite, pyrite, zircon, allanite, tourmaline. *Accidentals:* muscovite mica, hornblende, pyroxene, garnet.
**Appearance** Color white, light gray, pink, yellowish, more rarely greenish when altered. Massive structure with medium or fine grain size and occasionally with good mineral orientation. Hypidiomorphic-granular texture; sometimes the feldspar tends to be presént as large crystals which give the rock the appearance of porphyritic texture. Feldspar can be present either as perthite (in some cases microperthite, which is recognizable only under the microscope), an intergrowth of orthoclase and albite, or as an association of separate small crystallites of the two minerals, which are of similar sizes and shapes. Frequent rounded or elongated xenoliths of mafic minerals (melanolites); miarolitic cavities; aplitic and pegmatitic differentiates, either well defined (veins) or grading into the surrounding rock. Contacts of the granite with enclosing rock are either sharp (intrusive type) or gradational (migmatitic type). In some cases circumscribed massifs have an oriented structure produced by elongated and flattened crystals, especially of potassic feldspar; such structure is caused by convective motions, more or less parallel to the margins of the pluton, which occurred in the magma containing previously formed crystals, and not by any metamorphic process. Leucogranites are particularly rich in light-colored minerals.
**Geotectonic environment** Large homogeneous or weakly differentiated batholiths having gradational contacts with host metamorphic rocks in Precambrian shields and in ancient basements. Circumscribed plutons with sharp contacts within metamorphic and sedimentary rocks, with veins and apophyses injected into the surrounding rocks; veins and sills of variable dimensions in sedimentary rocks with sharp contacts and contact metamorphic effects. Granites containing perthites are thought to be derived by rapid cooling of the magma; those with two separate feldspars are thought to be derived by slow cooling at rather low temperatures and in the presence of high water pressures.
**Occurrences** Granites and granitoids (plutonic rocks of similar appearance to granites but difffering from them slightly in the modal proportions of essential minerals) are the most widespread rocks on the surface of the continents of the Earth, the commonest rocks in the great Precambrian shields of Scandinavia, Canada, Russia, Brazil and Africa. The photo opposite shows a typical granite from Baveno (Novara, Italy); its overal color is pink, with pink orthoclase frequently, and typically

▲ Red granite (ca. ×1). Baveno, Novara, Italy.

▲ As above: photomicrograph of a thin section (ca. ×20; crossed Nicols).

twinned; milk-white oligoclase (polysynthetically twinned in thin section), colorless quartz, thin layers of blackish biotite (brown in thin section) and traces of amphibole (showing vivid interference colors in the photomicrograph opposite). Granites are well represented in Great Britain and were intruded essentially in three periods of orogeny (mountain building), excluding the Precambrian granites that are of anatectic origin and are widespread in the Highlands of Scotland. Large areas of Mesozoic granite occur in the West Coast Range of North America (California, British Columbia and Alaska). In the USA Tertiary granite occurs in Montana, where it sometimes shows differentiation toward alkaline types. The occurrence of granite in Germany, Switzerland (Aar-Gotthard), Italy and France (Mount Blanc, Aiguilles-Rouges, Pyrenees) is more limited.

**Uses** Granite is used as a building stone either as polished slabs or as rough. An important source of economically valuable minerals, especially in its differentiated facies as pegmatitic types or in those facies related to the late gases of the magmatic process (greisen). Locally, building stones that are called "granite" are often other rock types, including sedimentary rocks.

▲ White granite (ca. ×1). Montorfano, Novara, Italy.

▲ Granite with two kinds of mica (ca. ×1). Novate Mezzola, Sondrio, Italy.

**Type** Intrusive igneous rock.

**Chemistry** Intermediate.

**Components** *Essentials:* quartz, plagioclase (oligoclase-andesine, often zoned), potassic feldspar, biotite, hornblende. *Accessories:* magnetite, ilmenite, apatite, titanite, allanite, zircon. *Accidentals:* pyroxene, muscovite mica.

**Appearance** Color light gray to dark gray. Massive structure with medium or fine grain size and frequently with flow orientation. Hypidiomorphic-granular texture, rarely with large crystals of dark-green hornblende or white potassic feldspar. It can be confused with granite, so it is often classified together with granite and other similar rocks as a group called granitoids.

**Geotectonic environment** Small circumscribed plutons aligned parallel to a tectonic lineation of regional importance which seems to be derived from a process of subcrustal anatexis. Differentiated zones at the margin or core of large batholiths of granitic composition with gradational contacts. Locally, some granodiorites can be the result of syntexis or assimilation of surrounding or "country" rocks.

**Occurrences** The gigantic "granitic" batholiths of the West Coast Range in North America contain the most extensive development of granodiorite and associated tonalite in the world. The southern California batholith has granodiorite covering a surface area of 9000 sq km (5590 sq mi). The photo opposite illustrates a typical granodiorite. It is light in color, with more phenocrysts of biotite than of hornblende, homogeneous but with many dark-colored inclusions. The plagioclase is twinned and zoned (see photomicrograph), with minute inclusions in the center; the biotite, with vivid interference colors, is sparsely scattered. There are traces of amphibole, sometimes hardly perceptible, in little bright-colored grains. Granodiorites are also widespread in Precambrian batholiths. Other examples include southern Yugoslavia, Romania, the Tauri Massif in Austria, central southern Norway and Japan (Ryoke).

**Uses** Used as a building material in polished slabs or as rough.

Granodiorite (ca. ×1). Valcomonica, Brescia, Italy.

As above: photomicrograph of a thin section (ca. ×20; crossed Nicols).

5

**Type** Intrusive igneous rock.

**Chemistry** Intermediate.

**Components** *Essentials:* plagioclase (oligoclase or ande sine), quartz (in amounts of > 10 percent, i.e., excluding it from the family of granitic rocks), hornblende, biotite. *Accessories* orthoclase, apatite, titanite, magnetite, ilmenite, zircon. *Acci dentals:* allanite, clinopyroxene, orthopyroxene.

**Appearance** Medium-gray rocks with frequent dark inclu sions; massive structure, sometimes with fluidal areas; hypi diomorphic granular texture with local transformations to porphyry, due in particular to xenomorphic hornblende and biotite. Frequent mafic differentiation, less frequent felsi differentiation.

**Geotectonic environment** In granitoid-type, well-defined batholiths and plutons, in which it often forms an igneous core developed during early differentiation. It is found in large masses that have been interpreted to be the result of anatecti remelting at great depths, in veins formed during a stage of compression and in environments with abundant water. As a result it has the same petrological significance as andesite but has remained at depth, crystallizing under plutonic condition and evolving chemically in a felsic sense. Quartz-bearing diorites are virtually identical varieties from a chemical view point but contain no hornblende; some of these are considered to have been derived from gabbro as a result of hybridization with granitic fluids. The term quartz-bearing or quartziferou diorite is often used to indicate all rocks of this type. The lates international agreements have given preference to the term "tonalite," in order to avoid too many attributive forms.

**Occurrence** The type locality is Monte Tonale in the Ada mello complex of the Tyrol. It contains prevalent euhedra plagioclase, easily identifiable in thin section (see photo micrograph) because it is zoned and twinned; with either euhe dral or anhedral amphibole with interference colors tending to brown-green due to masking by the actual color; large sheet of green biotite and light colored interstitial quartz. All thes minerals have a hint of alteration identifiable by the presence of pleats filled with fine-grained minerals. The quartziferou diorite of the Val Masino (Italy) and the Val Bregaglia (Canto des Grisons, Switzerland) is slightly oriented; it is known locall as "serizzo" and contains only orthoclase in the groundmass This rock, which is considerably darker and has a coarse grain size than the above-described tonalite, contains numer ous light and dark colored areas, because it is markedly hybri in origin. In thin section (see photomicrograph) plagioclase ha dense polysynthetic twinning oriented more or less like th biotite with interference colors tending toward green: amph bole has interference colors tending toward yellow: and anhe dral epidote has vivid interference colors tending toward rec violet and orange. In the same pluton there is a transition, b

▲ Tonalite (ca. ×1). Valcomonica, Brescia, Italy.

▲ As above: photomicrograph of a thin section (ca. ×20; crossed Nicols).

enrichment in large crystals (up to 10 cm; 4 in) of orthoclase, to a variety which has an overall granodioritic composition in a dark, biotitic groundmass in which the large orthoclase crystals are conspicuous. Quartziferous diorites and tonalites are very common in the great batholith of the Sierra Nevada in southern California and the Cascade Range in Oregon (USA) and British Columbia (Canada). They are also found in the Caledonian intrusives of Scotland, especially at Loch Awe and in Galloway. In Norway they occur with quartz-rich and rather leucocratic varieties (trondhjemites).

**Uses** A building material, both in the natural state and as polished slabs, sometimes also used to build steps, because of its rigidity. Many mineral deposits, especially of pyrite and copper, are genetically associated with tonalites and quartz-bearing diorites.

▲ Porphyritic tonalite (ca. ×1). Val Masino, Sondrio, Italy.

▲ Orthoclase-bearing tonalite (ca. ×1). Val Masino, Sondrio, Italy.

▲ As above: photomicrograph of a thin section (ca. ×20; crossed Nicols).

## 281 SYENITE

**Type** Intrusive igneous rock.

**Chemistry** Intermediate.

**Components** *Essentials:* potassic feldspar, plagioclase (andesine-labradorite), amphibole. *Accessories:* titanite, orthopyroxene or clinopyroxene, quartz, biotite, magnetite, ilmenite, and sulfides. *Accidentals:* olivine, corundum, nepheline.

**Appearance** pale, gray, pinkish or violet rocks; massive structure, medium grain size, often changing to pegmatitic; frequent fluidal structures and miarolitic cavities; hypidiomorphic granular texture with frequent transitions to porphyry.

**Geotectonic environment** Zones limited to felsic or mafic massifs; they sometimes form the most differentiated part of gabbroic plutons or stratiform intrusions. They are often clearly associated with tectonic environments.

**Occurrence** The type locality in the region of Syene (Aswan, Egypt) contains sufficient quartz for it to be classified as granodiorite and not syenite. The photo opposite shows a typical syenite from the granite to monzonite pluton of Biella (Vercilli, Italy). It is formed of violet-colored orthoclase, due to the presence of fine sheets of ilmenite on cleavage surfaces, which is microperthitic and identifiable in thin section (see photomicrograph) by its almost constant twinning; prismatic hornblende crystals with vivid interference colors, rare plagioclase in polysynthetically twinned crystals; quartz and interstitial albite. A common accessory is titanite. Syenites with biotite but without hornblende are found in the Black Forest (Germany) and in the Oslo region (Norway), where quartziferous syenites pass into nepheline syenites and to saturated (larvikites) and undersaturated (nordmarkites) sodic syenites. Quartziferous syenites are also common in the Adirondacks, New York (USA), where they also pass into undersaturated rocks. The great variety of accessory components leads to a host of local synonyms defining rocks with small variations in mineralogical composition, due to the variety of syenites and the intensity with which they have been studied, by comparison with surrounding rocks.

**Uses** Much used in building as polished slabs, although these are hard to obtain in any size and with a uniform color. In some areas these rocks are associated with important mineral deposits of rare metals (nepheline syenite pegmatites).

▲ Syenite (ca. ×1). Balma, Vercelli, Italy.

▲ As above: photomicrograph of a thin section (ca. ×20; crossed Nicols).

## 282 ALKALINE FELDSPAR SYENITE

**Type**  Intrusive igneous rock.

**Chemistry**  Intermediate felsic.

**Components**  *Essentials:* perthitic potassic feldspar. *Accessories:* quartz, plagioclase (albite), olivine, acmite-angite pyroxene, titanite, allanite, pyrrhotite, magnetite, ilmenite. *Accidentals:* nepheline.

**Appearance**  Pale to dark gray in color, with bluish highlights; hypidiomorphic coarse-grained granular texture; massive structure with frequent fluidal areas.

**Geotectonic environment**  Small masses, often laccoliths or sills, inside or at the edge of blocks of syenite and monzonite. They are derived by local differentiation without any clear tectonic significance.

**Occurrence**  Most characteristic are those of southern Norway (larvikite) and central Norway (nordmarkite). Alkaline feldspar syenites are also present in many parts of Greenland and Canada.

**Uses**  Highly decorative, used as building stone, especially as polished slabs. Larvikite is commercially known as "pale Labrador."

---

## 283 MONZONITE

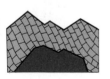

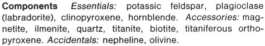

**Type**  Intrusive igneous rock.

**Chemistry**  Intermediate.

**Components**  *Essentials:* potassic feldspar, plagioclase (labradorite), clinopyroxene, hornblende. *Accessories:* magnetite, ilmenite, quartz, titanite, biotite, titaniferous orthopyroxene. *Accidentals:* nepheline, olivine.

**Appearance**  Color dark gray, also greenish or reddish; massive structure with medium grain size, often markedly fluidal; hypidiomorphic granular texture with idiomorphic plagioclase bonded by fine-grained orthoclase.

**Geotectonic environment**  Small intrusive laccolith- or sill-type bodies, usually inhomogeneous, given that within them there are transitions to more mafic and more felsic zones. They often indicate local remelting associated with rift or tectonic zones.

**Occurrence**  The type example is the intrusive shallow plutonic body of Mount Monzoni in the Tyrol. It is also common in the Oslo plutonic mass in Norway. In the USA monzonite is found as laccoliths in the Henry Mountains of Utah and in Colorado. Small intrusions of monzonite are associated with the Caledonian igneous activity in Scotland; on Syke, Tertiary monzonite is found in the Red Hills.

**Uses**  Some use as building stone; it is often associated with mineral deposits.

▲ Larvikite in polished slab (ca. ×0.5). Larvik, Norway.

▲ Monzonite (ca. ×1). Predazzo, Trento, Italy.

**Type**    Intrusive igneous rock.

**Chemistry**    Intermediate.

**Components**    *Essentials:* plagioclase (zoned from bytownite to andesine); in diorites, by definition, the average anorthite component is < 50 percent; hornblende. *Accessories:* magnetite, ilmenite, titanite, allanite, quartz. *Accidentals:* orthopyroxene and clinopyroxene, orthoclase, biotite.

**Appearance**    Overall color is dark gray, blackish-gray. The texture is hypidiomorphic granular to porphyritic by development of tabular plagioclase or hornblende in squat prisms. Massive structure with medium-to-coarse grain size and frequent transition to fluidal texture; differences in color and grain size are common. It sometimes has an orbicular structure. Varieties poor in hornblende and other colored minerals are called leucodiorites; they are similar in appearance to tonalites and granodiorites, but do not contain > 10 percent quartz.

**Geotectonic environment**    Limited blocks in marginal zones of essentially granitic or granodioritic plutons; transition facies at the edge of gabbroic masses. These are rare rocks, often due to the hybridization of more mafic rocks, less frequently due to marginal or upper level differentiation of magmas by means of gravitative separation.

**Occurrence**    Many rocks originally called diorite are now called something else because other names are being used for different types of diorites; thus quartz diorite is now called tonalite. A true diorite is found in the Haute Savoie, France. A thin section of diorite from Sondrio, Italy (see photomicrograph), shows individual prismatic crystals of plagioclase, nearly always polysynthetically twinned and sometimes zoned; prisms of amphibole with vivid interference colors; rare grains of quartz with a tendency to replace amphibole borders; and large laminae of biotite showing reddish and greenish interference colors. Diorites are common in the Hercynian massifs of Central Germany (Thuringia and Sassonia), Romania, Finland and central Sweden and especially in some plutons in the USA (Minnesota). Melanocratic varieties tending toward gabbro are observed in many regions including the Glen Fyne-Garabal Hill Complex in Scotland and in the Channel Islands. A most characteristic melanocratic variety is corsite, an orbicular diorite found in Corsica.

**Uses**    Used as building stone, as polished slabs.

▲ Diorite (ca. ×1). Sondalo, Sondrio, Italy.

▲ As above: photomicrograph of a thin section (ca. ×15; crossed Nicols).

**Type**   Intrusive igneous rock.

**Chemistry**   Mafic.

**Components**   *Essentials:* plagioclase (labradorite-bytown-ite); in gabbro, by definition, the average anorthite component is > 50 percent, clinopyroxene. *Accessories:* magnetite, il-menite, chromite, apatite. *Accidentals:* green and brown amphibole.

**Appearance**   Overall gray, light green or greenish-gray color; very heterogeneous; texture hypidiomorphic granular to por-phyritic, structure massive, coarse-grained but often with visi-ble alternating bands due to grain size or modal variations. A light color, sometimes with emerald-green spots, is due to a low degree of metamorphism that has led to the complete re-placement of the plagioclase by an association of clinozoisite, albite, calcite and sometimes lawsonite, and replacement of clinopyroxene by chlorite.

**Geotectonic environment**   Characteristic of the ophiolitic as-sociation in which it forms intermediate horizons between the ultramafic substratum (probably derived directly from the man-tle) and overlying basaltic lavas. It is presumed that these rocks are formed in a nontectonic environment, during the opening of an oceanic basin, by the separation of two continental plates. Their crystallization must have occurred at depth as a cumulate derived from primary basaltic magma. During the compressive phase that follows they are transformed by metamorphism into amphibole-bearing gabbro.

**Occurrence**   Gabbro (sometimes called euphotide gabbro in European nomenclature) comes from the Alps, where it is well represented in Switzerland and Italy. The gabbro is often al-tered, as at Zermatt (Switzerland), where clinopyroxene is par-tially replaced by smaragdite to form metagabbro. This gives the rock a corona texture of smaragdite around clinopyroxene. In thin section crystals of clinopyroxene are seen to be de-formed, with thin wedges of fine inclusions of opaque minerals set in a mass of coarse-grained crystals of strongly sericitized (altered) plagioclase. Ophiolite bodies containing gabbro are also found in Carrock Fell, Cumberland; New Radnor, Wales; Arran, Scotland; and in California (USA), Cyprus, Greece and Turkey.

**Uses**   This rock type is too fragile to make good construction material. Sometimes it is associated with mineral deposits of copper such as those in Liguria (Libiola), and Cyprus in particular.

▲ Ophiolitic gabbro (ca. ×1). Passo del Bracco, La Spezia, Italy.

▲ As above: photomicrograph of a thin section (ca. ×15; crossed Nicols).

10

**Type** Intrusive igneous rock.

**Chemistry** Mafic.

**Components** *Essentials:* plagioclase (labradorite-anorthite; in gabbros, by definition, the average anorthite component is > 50 percent), olivine, clinopyroxene. *Accessories:* chromite, magnetite, ilmenite, apatite, sulfide, titanite, rutile. *Accidentals:* corundum, brown amphibole, garnet, biotite.

**Appearance** Fairly dark overall, green-gray or brownish-gray and sometimes violet-gray, if rich in titaniferous minerals. Texture typically granular and hypidiomorphic, grain size medium, sometimes with aggregates of colored minerals; sometimes has corona texture around olivine, formed by clinopyroxene and spinel in radiating aggregates; structure massive or in bands, often with cumulate texture due to gravitative settling of the early-formed olivine to the bottom of the magmatic body, depositing itself more or less as a sediment.

**Geotectonic environment** Limited masses in the deep parts of larger intrusions, mainly normal gabbroic or dioritic; intermediate levels in cumulate sequences which are differentiated with an ultramafic lower level and granophyre at high levels.

**Occurrence** Olivine gabbros are common in the large stratiform bodies of the Bushveld Complex (South Africa), the Stillwater Complex and Duluth Complex (USA) and in the Sudbury and Muskox intrusives (Canada), where they are associated with important mineral deposits. The thin section shows (see photomicrograph) an olivine gabbro from Sondrio, Italy. In it can be seen polysynthetically twinned plagioclase, olivine nuclei (high interference colors) surrounded by a corona of pyroxene. Troctolite, which contains mainly plagioclase and olivine, is typical of the Harz Mountains (Germany) and is also found in large stratiform bodies as noted above. Another variety, with an orbicular structure, is found in San Diego County, California (USA), the Kenora District of Ontario (Canada), Romsaas (Norway) and Corsica.

**Uses** Rarely used as building stone, except for some varieties from Sweden which are ilmenite-rich, amphibole-bearing and almost black. Olivine gabbros which are cumulate in origin form the parental rocks for several important ore deposits of chromium, nickel, cobalt, iron and platinum and, in addition, represent an important source of olivine, which is used as refractory material in blast furnaces.

▲ Olivine gabbro (ca. ×1). Sondalo, Sondrio, Italy.

▲ As above: photomicrograph of a thin section (ca. ×15; crossed Nicols).

## 287 NORITE

**Type** Intrusive igneous rock.
**Chemistry** Mafic.
**Components** *Essentials:* plagioclase with an average anorthite content > 50 percent (labradorite-bytownite), orthopyroxene ≥ clinopyroxene. *Accessories:* olivine, magnetite, ilmenite, apatite, chromite, sulfide. *Accidentals:* biotite, hornblende, cordierite.
**Appearance** Color dark gray, texture hypidiomorphic granular, or xenomorphic, often fluidal; structure massive with bands; grain size medium.
**Geotectonic environment** Small differentiated units inside mafic or ultramafic plutonic masses; frequently in stratified horizons.
**Occurrence** It is found in the great stratiform Bushveld Complex (South Africa) and in the Stillwater Complex in Montana (USA). It is also found in great quantity at Sudbury, Ontario (Canada) and is common in southern Norway. Also found in San Diego County, California (USA).
**Uses** Frequently associated with deposits of nickel-bearing pyrrhotite.

## 288 ANORTHOSITE

**Type** Intrusive igneous rock.
**Chemistry** Mafic.
**Components** *Essentials:* plagioclase (labradorite-bytownite). *Accessories:* pyroxene, olivine, chromite, magnetite, ilmenite. *Accidentals:* amphibole, garnet.
**Appearance** Color white or light gray; texture granular, with elongated tabular plagioclase. Structure mainly bands differentiated by grain size and texture.
**Geotectonic environment** Massifs and plutons, generally forming parts of a stratified sequence; sometimes very extensive, especially in Precambrian rocks.
**Occurrence** The largest occurrence in the world is in the Canadian Shield where it covers vast areas. The anorthosite in Newfoundland is highly schillerized. Common in Norway. In the USA anorthosite massifs are found in the Adirondacks (New York) and as cumulate layers in the Duluth Complex (Minnesota) and in the Stillwater Complex (Montana).
**Uses** Labradorite, rather dark but with an iridescent display of colors, is a much-used ornamental building stone, in the form of slabs and blocks. Some types of anorthosite are used for basic refractory materials.

▲ Norite (ca. ×1). Appenine Mountains, Emilia, Italy.

▲ Polished slab of ''labradorite'' or anorthosite (ca. ×1). Norway.

## 289  FELDSPATHOID SYENITE

**Type** Intrusive igneous rock.
**Chemistry** Mafic (undersaturated).
**Components** *Essentials:* alkaline feldspar (with or without albite), feldspathoids (nepheline with sodalite or cancrinite, etc.). *Accessories:* plagioclase (oligoclase-andesine), analcite, acmite, sodic amphibole, aenigmatite, magnetite, biotite, fluorite, eudialyte, astrophyllite, etc.
**Appearance** Color light gray, sometimes also pink or greenish; texture granular and xenomorphic tending toward porphyritic; structure massive with very variable grain size, and frequent transition to fluidal.
**Geotectonic environment** Small masses, seams and lenses associated with syenite and alkaline rocks, probably partly of hybrid derivation (perhaps from contamination with limestone).
**Occurrence** Very common in the Baltic Shield (Finland, Sweden, Kola Peninsula of the USSR) and the Canadian Shield (Ontario, particularly around Bancroft). In the USA it is found near Litchfield (Maine), in the Highwood Mountains (Montana) and Pikes Peak (Colorado).
**Uses** Associated with rare minerals of rubidium, cesium, thorium, uranium, etc.

11

## 290  ESSEXITE

**Type** Igneous rock.
**Chemistry** Mafic (alkaline).
**Components** *Essentials:* plagioclase (labradorite), orthoclase, pyroxene (augite, titanaugite), biotite, amphibole (barkevikite). *Accessories:* olivine, apatite, magnetite, ilmenite, titanite. *Accidentals:* nepheline, sodalite, cancrinite.
**Appearance** Color dark gray to blackish; texture massive with small idiomorphic plagioclase crystals in a mainly hypidiomorphic groundmass, with fine grain size and a tendency toward fluidal texture.
**Geotectonic environment** Small plutons or subvolcanic masses; difffferentiated marginal portions of gabbroic plutons. Sometimes also found as xenoliths in lava flows.
**Occurrence** The name is derived from Essex County, Massachusetts (USA), where the rock contains abundant nepheline. The essexites from Scotland, Québec (Canada) and central Bohemia are more typical. The essexites in Scotland are strongly porphyritic and are known as Crawfordjohn rock since they are used for curling stones. The theralites of Czechoslovakia and the shonkinites of Montana (USA) are analogous in composition to alkaline gabbro, but are more sodic.
**Uses** Sometimes associated with masses of rare minerals.

13

▲ Feldspathoidal syenite (ca. ×1). Bancroft, Canada.

▲ Essexite (ca. ×1). Rongstock, Czechoslovakia.

16

**Type** Intrusive igneous rock.

**Chemistry** Ultramafic.

**Components** *Essentials:* pyroxene (clinopyroxene or orthopyroxene). *Accessories:* olivine, hornblende, chromite, magnetite, ilmenite sulfide, biotite, garnet, apatite.

**Appearance** Color dark green, dark brown, black; texture granular from hypidiomorphic to xenomorphic; structure massive, coarse-grained, sometimes with crystal orientations indicative of gravitative deposition.

**Geotectonic environment** Small independent intrusions, veinlike in appearance, or horizontal units in cumulate mafic sequences, between the lowest ultramafic levels and the upper gabbroic levels.

**Occurrence** This rock is highly differentiated. The commonest variety is websterite, which contains both orthopyroxene and clinopyroxene. Its name is derived from Webster, North Carolina (USA). The variety bronzitite contains only orthopyroxene (variety bronzite) and is about 800 m (875 yds) thick in the Stillwater Complex (USA). It forms large sheets in the Bushveld (South Africa) and Stillwater (USA) Complexes and the Great Dike of Rhodesia. Clinopyroxenite contains > 90 percent clinopyroxene. The thin section (see photomicrograph) shows a clinopyroxenite from Trento, Italy. It consists of large crystals of clinopyroxene showing vivid interference colors. Acmitite is the only pyroxenite of distinctly sodic type. It is found in the Kola Peninsula (USSR), in the Pyrenees and in southern Spain and Morocco.

**Uses** Limited to use as a local building stone.

▲ Pyroxenite (ca. ×1). Malgola, Trento, Italy.

▲ As above: photomicrograph of a thin section (ca. ×20; crossed Nicols).

## 292 HORNBLENDITE

**Type** Intrusive igneous rock.
**Chemistry** Ultramafic.
**Components** *Essentials:* hornblende. *Accessories:* olivine, orthopyroxene and clinopyroxene, magnetite, ilmenite chromite, sulfide.
**Appearance** Color from dark green to black; texture granular and xenomorphic, often feltlike; structure massive, sometimes zoned concentrically around a nucleus of pyroxenite.
**Geotectonic environment** Lenses or masses inside periodotite or at the edges of gabbro complexes; rare veins in nepheline syenite.
**Occurrence** A hornblendite occurs at Garabal Hill in southwest Scotland. It is very coarse-grained, containing individual hornblende crystals up to 5 cm (2 in) in length. Lenses or horizons of hornblendite are found in peridotites, especially in the Pyrenees, the Urals and in Cuba. Pedrosite is a veined alkaline hornblendite in the rocks of the Kola Peninsula (USSR).
**Uses** Of no commercial importance.

16

## 293 DUNITE

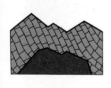

**Type** Intrusive igneous rock.
**Chemistry** Ultramafic.
**Components** *Essentials:* olivine. *Accessories:* chromite, clinopyroxene magnetite, ilmenite, apatite. *Accidentals:* garnet, native platinum.
**Appearance** Color light green; texture granular and xenomorphic, often superficially saccharoid-looking, sometimes blastomylonitic; structure massive, medium to fine grain size, with frequent variations of grain size and color.
**Geotectonic environment** Lenses and masses, sometimes very extensive, at the base of cumulate sequences.
**Occurrence** This rock takes its name from the type locality at Mount Dun (New Zealand), where it is very common. It is also common in the Urals where it contains platinum. In Turkey it contains chromite deposits. Dunite also occurs in the Stillwater (USA) and Bushveld (South Africa) Complexes and as ultramafic nodules in undersaturated basalts.
**Uses** Dunite is associated with important metalliferous mineral deposits of platinum and chromium.

16

▲ Hornblendite (ca. ×1). Valcomonica, Brescia, Italy.

▲ Dunite (ca. ×1). Dun Mountains, New Zealand.

## 294  LHERZOLITE

**Type**  Intrusive igneous rock.
**Chemistry**  Ultramafic.
**Components**  *Essentials:* olivine, clinopyroxene, orthopyroxene, chromite or pyrope garnet. *Accessories:* sulfide, hornblende.
**Appearance**  Color from yellowish green to brownish; texture granular and xenomorphic. Structure massive, sometimes layered.
**Geotectonic environment**  Sometimes large masses, forming the major portion of ophiolitic associations, which are considered to be parts of the mantle pushed to the surface of continents or ocean basins. Also occurs as spinel lherzolite xenoliths in undersaturated basalts and as garnet lherzolites in kimberlite pipes.
**Occurrence**  In California (USA), Norway, Cuba, New Caledonia, Oman, Cyprus, Switzerland and South Africa. The type locality is in the Pyrenees (Lherz, France). Small masses occur in southern Spain and Morocco.
**Uses**  Associated with metalliferous deposits of nickel and chromium.

16

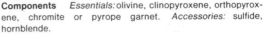

## 295  PERIDOTITE

**Type**  Intrusive igneous rock.
**Chemistry**  Ultramafic.
**Components**  *Essentials:* olivine, clinopyroxene, orthopyroxene. *Accessories:* chromite.
**Appearance**  Color light to dark green; texture granular, with xenomorphic olivine and interstitial pyroxene, often poikilitic; structure massive, sometimes layered with zones enriched with chromite or pyroxene.
**Geotectonic environment**  Limited masses inside ophiolitic complexes; large units at the base of stratiform sequences, differentiated from basaltic magma.
**Occurrence**  Immense zones in the Bushveld (South Africa), Stillwater (USA), Sudbury and Muskox (Canada) complexes and in the Urals (USSR). Also present in the Cortlandt Complex (New York, USA) and the Lizard Complex (Cornwall, England).
**Uses**  Frequently associated with deposits of nickel-bearing pyrrhotite, chromite, platinum and rare metals.

16

▲ Lherzolite (ca. ×1). Lherz, France.

Peridotite (ca. ×1). Levanto, La Spezia, Italy.

16

**Type** Intrusive igneous rock.

**Chemistry** Ultramafic.

**Components** *Essentials:* olivine, clinopyroxene, orthopyroxene, garnet (pyrope). *Accessories:* spinel, diamond, graphite, pyrrhotite.

**Appearance** Color fairly dark green with crimson patches and speckling due to the presence of garnet; texture hypidiomorphic granular with xenomorphic tendency in garnet, which often forms megacrysts; structure massive, or banded by concentration of garnet, but sometimes also clinopyroxene. Garnet frequently has a corona of amphibole plus spinel called a "kelyphitic rim."

**Geotectonic environment** Limited masses in continental shields apparently emerging along deep fractures, or included in metamorphic zones of the granulite facies. They are frequent as nodules in kimberlite pipes, with which diamonds are sometimes associated; they are considered to be fragments of the upper mantle brought directly to the surface by a swift, explosive volcanic event, a kimberlite pipe emplacement.

**Occurrence** Among the most prevalent outcrops of the garnet peridotite in continental shields or granulitic metamorphic zones are those of western Norway, Czechoslovakia (Bohemia and Moravia), Scotland and Canada. In thin section (see photomicrograph) large garnet crystals (surrounded by a fibrous kelyphitic rim mainly of amphibole) are set in an olivine-rich groundmass with vivid interference colors. Those in kimberlite pipes are frequent in South Africa, Yakutia (USSR), Australia and New Zealand (Kakanui). In the Alps garnet peridotite is found at Gorduno (Canton of Ticino, Switzerland) and in Val d'Ultimo (Alto Adige, Italy). Garnet peridotite nodules in kimberlite pipes represent the deepest terrestrial materials brought to the surface of the earth.

**Uses** Parental rock of diamond. Diamond is not only used for jewelry, but is extremely important as an industrial abrasive. Olivine is used as raw material for refractory materials and, more rarely, in jewelry.

Garnet peridotite (ca. ×1). Norway.

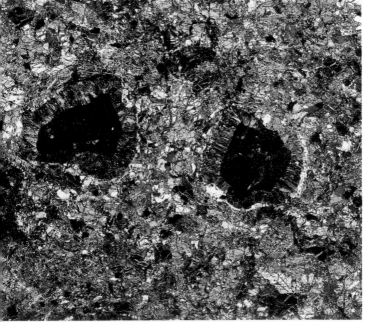

As above: photomicrograph of a thin section (ca. ×15; crossed Nicols).

## 297 KIMBERLITE

**Type** Intrusive brecciated "igneous" rock.

**Chemistry** Ultramafic.

**Components** *Minerals:* serpentinized olivine, phlogopite, pyrope garnet, pyroxene, chromite, ilmenite, melilite, diamond, graphite, calcite, monticellite.

**Appearance** Color black, blue, greenish, yellow when altered; texture granular, almost constantly cataclastic (intrusive breccia type); structure brecciated, grain size very variable, rich in xenoliths.

**Geotectonic environment** Intrusive breccia into pipes or fractures.

**Occurrence** Common and famous are those of South Africa (Kimberley, Roberts Victor) and Lesotho (Taba Putsoa). Also found at Yakutia (USSR) and in the Rocky Mountains (USA). Affinitive rocks are found in Sweden, Canada, Brazil, Zaire and Sierra Leone.

**Uses** Parental rock of diamond and important source of clear pyrope crystals (Cape ruby), used in jewelry.

16

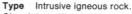

## 298 GRIQUAITE

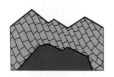

**Type** Intrusive igneous rock.

**Chemistry** Ultramafic.

**Components** *Essentials:* clinopyroxene (chromian diopside), garnet (pyrope). *Accessories:* kyanite, diamond, graphite, orthopyroxene, corundum.

**Appearance** Color from red to green; texture xenomorphic granular, often cataclastic; structure massive, sometimes layered.

**Geotectonic environment** Contained as small nodules in kimberlite pipes; exceptionally, also as lenses in garnet peridotite.

**Occurrence** Always associated with kimberlite in South Africa, Yakutia (USSR) and in some kimberlite breccias in Australia and New Zealand. Griquaitic lenses are described in the peridotites from Bohemia, Moravia and Austria.

**Uses** Discovery of diamond exceptional; however, these rocks are useful as an indicator of diamond.

16

Kimberlite (ca. ×1). South Africa.

Griquaite (ca. ×1.5). South Africa.

**Type** Extrusive igneous rock.

**Chemistry** Felsic (alkaline).

**Components** *Essentials:* quartz, alkaline feldspar (sanidine anorthoclase, albite), sodic clinopyroxene (acmite-augite) sodic amphibole (riebeckite, arfvedsonite). *Accessories:* magnetite, ilmenite, olivine, glass, aenigmatite.

**Appearance** Color fairly variable from light (whitish, greenish, reddish) to quite dark (green or black); texture porphyritic with phenocrysts of essential components set in the ground mass; the variety called pantellerite is characterized by anorthoclase, acmite-augite and aenigmatite phenocrysts; the variety comendite, characterized by sanidine and sodic amphibole is microgranular or micropegmatitic, and only rarely has fairl silica-rich glass.

**Geotectonic environment** Very rare rocks limited to a few flows, marked alkaline-sodic differentiation, perhaps by derivation from anatectic magma, or by removal of some components from a basaltic magma by means of gases.

**Occurrence** The classic alkaline rhyolite is the pantellerite from the island of Pantelleria, southwest of Sicily but also known in Dancalia (Ethiopia), Somalia and the Rift Valley of Central Africa (Uganda and Tanzania). Analogous rocks are found in Colorado (USA) and Oslo (Norway), and are composed of phenocrysts of perthitic microcline and acmite in a groundmass of the same mineralogical composition, along with quartz. The comendites, which derive their name from the locality at Comende, Sardinia, are also known in Dancalia, in the Rift Valley and in Corsica. Very similar is the paisanite from Paisano Pass (Texas, USA), which cuts a syenite; it is a porphyritic rock composed of phenocrysts of alkaline feldspar and quartz and small nodules of amphibole in a microgranular groundmass composed of quartz and alkaline feldspar.

**Uses** No known industrial or commercial use.

Pantellerite (ca. ×1). Pantelleria, Trapani, Italy.

Comendite (ca. ×1). St. Pietro Island, Cagliari, Italy.

# 300 RHYOLITE

**Type** Extrusive igneous rock.

**Chemistry** Felsic.

**Components** *Essentials:* quartz, alkaline feldspar (sanidine). *Accessories:* glass, biotite, albite, magnetite, ilmenite. *Accidentals:* tridymite, cristobalite, amphibole, pyroxene (diopside).

**Appearance** Color very light, except in the glassy varieties, which may even be totally black or other dark colors (obsidian); texture generally porphyritic, with xenomorphic but corroded phenocrysts, groundmass ranging from totally glassy (holohyaline) to totally crystalline (holocrystalline) and, in certain areas, granular, granophyric, dense, etc.; structure very variable, from pimply (pumice) to fluidal, perlitic, spherulitic and massive. Obsidian has a characteristic conchoidal fracture with sharp and often translucent edges.

**Geotectonic environment** Derives from the rapid cooling of a very viscous magma of granitic composition; as a result, it is mainly found in domes, chimneys, dikes and more rarely in proper lava flows. If the flows come into contact with water, they tend to break up and form spherical concentric fractures which give rise to tiny, pearl-shaped bodies (perlite).

**Occurrence** The rhyolites in the Lipari Islands (Italy) were among the first described and the rock liparite is synonymous with rhyolite. Toscanite, a rock midway between rhyolite and dacite, is named after Tuscany (Italy) where it was first described. The photo opposite shows a rhyolite from Padova (Italy). In the thin section (see photomicrograph) large light-colored crystals of idiomorphic sanidine and very small but very clear indented quartz crystals are evident in a microgranular groundmass of the same composition, with a few rare laminae of colored biotite. Rhyolites are very abundant in California and Oregon (USA), North Wales, Devon and Cornwall (England), Hungary, Romania, East Africa, Japan and especially in Dancalia (Ethiopia). A rare obsidian, used in jewelry as an ornamental stone, is the "snowflake" obsidian from Utah (USA) which has tabular crystals of white sanidine gathered in distinct tufts in a black groundmass.

**Uses** When heat treated, perlite becomes an excellent industrial material for acoustic and light thermal insulation used in building. Pumice is used in chemical washing processes, filtration with selective absorption and as a soft abrasive. Obsidian is sometimes used in the production of "rock wool."

▲ Rhyolite (ca. ×1). Euganean Hills, Padua, Italy.

▲ As above: photomicrograph of a thin section (ca. ×20; crossed Nicols).

**Type** Extrusive igneous rock.

**Chemistry** Felsic.

**Components** *Essentials:* quartz, alkaline feldspar (sanidine, orthoclase, albite), biotite. *Accessories:* hornblende, plagioclase (oligoclase-andesine), magnetite, ilmenite, apatite, zircon, molybdenite. *Accidentals:* chlorite, sericite, calcite.

**Appearance** Color light gray, pink, violet, brick-red; texture porphyritic, with large quartz phenocrysts, splintery or corroded, and feldspar phenocrysts (corroded and rounded) in a groundmass from dense to micropegmatitic; partly derived from the devitrification of interstitial glass; texture massive, sometimes zoned or flamed, rich in cavities (miarolitic), with fine crystals of quartz, feldspar, topaz, fluorite, etc. Very rough superficial appearance, especially in varieties which are completely devitrified and thus have a microcrystalline groundmass (felsite).

**Geotectonic environment** They are considered to be the paleovolcanic forms of rhyolite, derived from large flows and explosive units of fissure-type volcanoes. They are consolidated tuff and ignimbrite, partially crystallized with age or as a result of included volcanic gases.

**Occurrence** In Italy large areas of porphyry outcrop in the Paleozoic areas from Biellese to Varesotta (red porphyry of Val Ganna), in the pre-Alps of Bergamasche and Bresciane and especially in the valleys of Aviso, Adige and Isarco. They constitute alternate flows with layers of tuff of the same composition, and are in part derived from them. They vary in thickness from 400 to 1500 m (440–1648 yds) and cover a surface area of more than 3000 sq km (1875 sq mi). In thin section (see photomicrograph) this rock shows large lobate and fractured quartz crystals; some crystals of feldspar are euhedral, but badly altered; small laminae of biotite, having vivid interference colors, have dark segregations and are set in a microgranular groundmass. The "elvan" dikes of Cornwall and Devon (England) are often quartz porphyry, a notable example being the "Prah elvan" in the Prah sands. Quartz porphyries are often associated with felsic intrusions throughout Britain and occur in the Lake District, the Cheviots and the Pentland Hills. Numerous sills and dikes of quartz porphyry are associated with the Tertiary intrusive complexes of Arran, Rum and Mull in Scotland. They are widespread in the western Coast Range of North America, where they are associated with Mesozoic intrusives. A notable occurrence is North Bend in British Columbia (Canada). Other occurrences are found in Switzerland (Luganese) and Germany (Westphalia, Sassonia).

**Uses** Used as building and decorative material, as well as for roadbeds, ballast, grindstones and gravel. European quartz porphyry is associated with uranium deposits and molybdenum deposits. American quartz porphyry is often associated with deposits of chalcopyrite of great economic importance ("porphyry copper ores").

▲ Quartz porphyry (ca. ×1). Bolzano, Italy.

▲ As above: photomicrograph of a thin section (ca. ×15; crossed Nicols).

## 302 DACITE

**Type** Extrusive igneous rock.

**Chemistry** Intermediate.

**Components** *Essentials:* plagioclase (zoned, with compositions from labradorite to oligoclase), quartz, biotite, brown hornblende. *Accessories:* magnetite, ilmenite, sanidine, orthopyroxene, clinopyroxene, glass.

**Appearance** Color medium gray; structure porphyritic, with phenocrysts set in an aphanitic, holocrystalline equigranular groundmass or, more rarely and if glass is present, in a hyalopilitic groundmass; texture from massive to markedly fluidal.

**Geotectonic environment** Lava flows, apophyses, dikes and smallish domes, often associated with localized areas inside or at the edge of plates.

**Occurrence** Very common in Transylvania (Hungary, Romania and Yugoslavia) from which the name (Dacia) comes. It is found in the Old Red Sandstone in Scotland and the Tertiary and Recent lavas of California and Nevada in the USA. It is also found in the Andes, the West Indies, New Zealand, Sardinia and the Puy de Dôme (France).

**Uses** Of no commercial value.

## 303 LATITE

**Type** Extrusive igneous rock.

**Chemistry** Intermediate.

**Components** *Essentials:* sanidine, plagioclase (zoned, with compositions from labradorite to andesine), augite, brown hornblende. *Accessories:* anorthoclase, olivine, feldspathoid, magnetite, ilmenite.

**Appearance** Color gray, sometimes with pink, green or brown hues; texture porphyritic with mainly glassy groundmass which is secondarily equigranular or feltlike; structure massive to fluidal.

**Geotectonic environment** Flows, dikes and small apophyses in tectonically stable environments; possibly derived from anatectic magma in limited areas with high heat flow.

**Occurrence** Not very common, but characteristic, are the latites of the Roman Volcanic Province (Italy) containing the varieties ciminite and vulsinite. Present also in the Puy de Dôme (France), Siebengebirge (Germany), Sydney (Australia) and the Atlantic Ocean islands (Canaries, Azores).

**Uses** Of no commercial value.

▲ Dacite (ca. ×1). Transylvania, Romania.

▲ Latite (ca. ×1). Roccomonfina, Caserta, Italy.

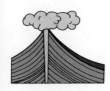

**Type** Extrusive igneous rock.

**Chemistry** Intermediate.

**Components** *Essentials:* sanidine, plagioclase (andesine-labradorite), biotite. *Accessories:* amphibole, pyroxene, magnetite, ilmenite apatite, melanite garnet, zircon, titanite, glass.

**Appearance** Color white, light gray, light brown or greenish; texture markedly porphyritic, expressed by biotite and sanidine phenocrysts set in a groundmass formed of small elongated strips of sanidine and albite, with parallel development, gathered fluidally around the phenocrysts; structure massive to fluidal and, very rarely, vacuolar. The superficial roughness of specimens gives rise to the name of the rock.

**Geotectonic environment** Flows, dikes and limited apophyses, usually closely associated with alkaline basalt of nonorogenic environments, from which these rocks are derived. They are also present in mid-oceanic volcanic islands, again associated with alkaline basaltic lavas and in zones with marked alkalic differentiation.

**Occurrence** Trachytes are the extrusive rocks corresponding to syenites. Two types are recognized, normal or sodic trachyte and potassic trachyte. Sodic trachytes are widespread in Hungary, New Zealand, the Pacific islands (Tahiti) and the Atlantic islands (St. Helena, Ascension). They are also found in the Rift Valley (Central Africa), Drachenfels (Germany) and the Solfatara Volcano, Naples (Italy). They are characterized (see photomicrograph) by large crystals of sieved (or poikilitic) potassic feldspar with glassy borders in a groundmass of sanidine, yellow biotite and opaque minerals. Potassic trachytes are well developed in Peninsular Italy. Undersaturated trachytes contain feldspathoids (sodalite, noselite, hauyne) and in particular leucite (orvietite and vicoitite).

**Uses** Excellent material for paving and flooring in general, because it does not become shiny from rubbing. Is also used in building for facing ("pink trachyte," "classical trachyte," etc.).

▲ Trachyte (ca. ×1). Euganean Hills, Padua, Italy.

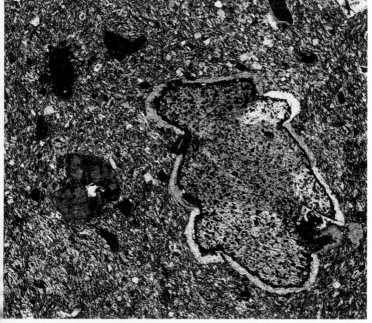

▲ As above: photomicrograph of a thin section (ca. ×15; crossed Nicols).

**Type**  Extrusive igneous rock.

**Chemistry**  Intermediate.

**Components**  *Essentials:* Plagioclase (labradorite-andesine in which, by definition and to distinguish them from basalt, anorthite is < 50 percent), biotite. *Accessories:* magnetite, ilmenite, quartz, hornblende, pyroxene, glass. *Accidentals:* olivine, orthoclase, anorthoclase.

**Appearance**  Color blackish-brown or greenish, especially if the groundmass is mainly glassy; texture markedly porphyritic, in which plagioclase and biotite phenocrysts stand out in a vitreous or fine-grained groundmass, formed by anhedral plagioclase and only locally oriented in fluidal forms; structure massive, sometimes with abrupt changes of grain size and color; xenoliths are fairly common.

**Geotectonic environment**  Lava flows and domes, especially in regions with considerable tectonic activity and in association with basalt. An "andesite line" formed by thousands of volcanoes (consisting of some andesite) circles the entire Pacific Ocean, corresponding to the subduction of an ocean plate with an essentially basaltic composition beneath a continental plate. Andesite is derived by melting these components at great depth, followed by rapid emergence of magma through volcanic pipes. This magma is possibly contaminated by felsic continental material that changes the chemical composition, especially the water pressure. Andesitic volcanoes are also common in the Alpine-Himalayan system.

**Occurrence**  The major andesitic volcanics are in the Andes (hence the derivation of the name). They are also well developed in Central America (Mexico, Guatemala, Costa Rica) and in Martinique, Japan, Indonesia and Melanesia. Those at Elburs (Persia), those from Turkish Tauro extending to the Aegean Sea and those in Romanian Transylvania are also notable. The thin section (see photomicrograph) of an andesite from Padua (Italy) shows visible idiomorphic twinned crystals of plagioclase, laminae of corroded biotite, sometimes with segregations of red hematite at their borders and highly corroded amphibole in a partially glassy groundmass with microlitic plagioclase accentuating the flow texture. A most interesting characteristic of some andesite is formed from the final volcanic gases and is called "propylite." This mineral is dark green in color due to the formation of abundant chlorite, at the expense of hornblende and biotite; it is very widespread in Transylvania, Canada and the USA and is the source of large copper deposits.

**Uses**  Andesitic volcanism is associated with "porphyry copper ores," one of the greatest sources of this metal in the world. The "andesite chain" is an important geological indicator for the reconstruction of ancient geosutures. Locally used as building material.

▲ Andesite (ca. ×1). Euganean Hills, Padua, Italy.

▲ As above: photomicrograph of a thin section (ca. ×15; crossed Nicols).

**Type** Extrusive igneous rock.

**Chemistry** Mafic.

**Components** *Essentials:* plagioclase (labradorite-bytownite: by definition the average anorthite content is > 50 percent), pyroxene (augite, often titaniferous; pigeonite; hypersthene). *Accessories:* magnetite, hematite, ilmenite, apatite, quartz, olivine, glass; *Accidentals:* amphibole (brown titaniferous hornblende), biotite.

**Appearance** Color very dark to black, but changing to brown and even to red in parts of the flow altered by oxidization; texture from holocrystalline to hypocrystalline to completely vitreous; grain size generally fine with rare plagioclase and pyroxene and olivine phenocrysts. The groundmass ranges from intergranular to subophitic. Structure both massive, sometimes with clear-cut columnar fragmentation, and scoriaceous, pimply, with cords, with zones rich in vacuoles (amygdules) often filled with zeolites or carbonates. Also frequent pillow structures (see photo) made up of spherical forms with a massive central part, subophitic texture and amygdaloidal vacuolar surface filled with secondary minerals, and finally a variolitic surface pockmarked with small roundish prominences between 1 mm and 2–3 cm (.025–1.2 in) in diameter. When basalt comes into contact with fresh water or shallow salt water it instantly cools, giving rise to rocks which are completely vitreous, very fractured and reconsolidated, known as hyaloclastites (see photo), formed almost entirely by brown splinters of palagonite, a glass with mafic composition. This glass, when altered, gives rise to fine aggregates composed essentially of chlorite and typically changes to a green color. Many basalts contain fragments of other rocks (xenoliths), partly torn from the walls of the channel as the magma rose, but also partly residual from deep rocks (spinel lherzolite), from which the basalt itself originated by partial melting.

**Geotectonic environment** This is the most common extrusive rock, as enormous subaerial expanses (traps), formed by superposed flows issuing from deep linear fissures and as cones emitted by central volcanos, often intercalated with scoria. Submarine flows are important, very extensive areally, commonly have pillow structures and are emitted from linear fissures which tap the upper mantle (mid-oceanic fractures). Chemically, a distinction is made between the *tholeiitic* type, usually without olivine, containing quartz, and the *alkali basaltic* type, usually olivine-bearing, and richer in potassium and sodium. Tholeiites are more widespread and form pillow basalts and the majority of traps. They are also commonly represented by diabases in small, often stratiform, hypabyssal intrusions. Some basalt is considered to be primary magma, formed by partial melting and separation of the mafic component of the earth's mantle with immediate transfer to the earth's surface, and leaving an ultramafic residue (spinel lherzolite). The abyssal (ocean-bed) tholeiites, especially near mid-ocean fissures, represent the original scars of the breaking-up of two continental plates and are very low in alkalis. The alkali basalts (or alkali olivine basalts) are derived from the melting of only a small part of the upper mantle and are enriched in elements

▲ Olivine basalt (ca. ×1). Etna, Sicily, Italy.

▲ As above: photomicrograph of a thin section (ca. ×15; crossed Nicols).

with larger ionic radii (alkalis). They are formed at greater depths than tholeiites, at somewhat higher temperatures and pressures, in locales with higher heat flow. They vary considerably from one another in composition and give rise to a considerable range of rock types, always tending toward undersaturated types. The detailed nomenclature of these basaltic rocks is extremely complex. Among the most important types are: *oceanites*, olivine-rich basalts, grading into picrites and considered characteristic of ocean islands such as Hawaii; *ankaramites*, augite-rich basalts, probably derived from the same types of processes which gave rise to the oceanites, to which they are related; *mugearites*, basalts containing phenocrysts of oligoclase or andesine, a more albitic plagioclase than in most basalts; and *trachybasalts*, which contain sanidine phenocrysts and are derived from melt which grades into more undersaturated alkaline magma with greater sodic or potassic differentiation. One particular basaltic rock type is spilite, which has basaltic chemistry but contains albite, chlorite and calcite; some authors consider spilites to be basalts which were autometamorphosed by residual magmatic gases, but most workers suggest it is basaltic lava and tuff which was metamorphosed by sea water.

**Occurrence** The largest surficial basaltic units are found in India (the Deccan), the USA (Columbia River, Snake River, Lake Superior), Scotland, Iceland, Greenland and Brazil. Pillow lava is found on the ocean bed, in particular in the mid-Atlantic ridge, and in older beds, which have since emerged. Basalts with differing degrees of differentiation grade toward andesites or toward alkaline magmas in volcanos such as Hekla (Iceland), Demavand (Iran), Mull (Scotland) and Etna (Sicily). A thin section of the lava of this latter rock (see photomicrograph) shows idiomorphic phenocrysts of brightly colored pyroxene, polysynthetically twinned plagioclase, and olivine in violet-colored grains, in an intersertal groundmass consisting of crystals of plagioclase in glass. Those on many ocean islands (the Canaries, St. Helena, Kerguelen and, to some extent, the Hawaiian Islands) are alkali basaltic. The hypabyssal mass of the Skaergaard intrusion (Greenland) is tholeiitic; here the differentiation of the magma has led to great enrichment in iron and titanium, which has been carefully studied.

**Uses** Local material for road paving and crushed stone for road and railroad ballast; also used in a minor way as a building stone and as raw material for the production of rock wool and glass wool (fiber glass).

▲ Pillow basalt (ca. ×1). Mt. Iblei, Sicily, Italy.

▲ Hyaloclastite (ca. ×1). Palagonia, Sicily, Italy.

11

## 307 PHONOLITE

**Type**  Extrusive igneous rock.

**Chemistry**  Intermediate (undersaturated).

**Components**  *Essentials:* potassic feldspar (sanidine), nepheline, clinopyroxene (acmite) or sodic amphibole (riebeckite or arfvedsonite). *Accessories:* anorthoclase, albite, titaniferous augite, aenigmatite, apatite, titanite, magnetite, itmenite, leucite, hauyne, sodalite, analcite, zeolites, melanite garnet, cancrinite.

**Appearance**  Color light gray, greenish, brown or pink, with rather greasy luster; texture from holocrystalline to completely vitreous, with prevalence of pilotaxitic texture passing to fluidal around sanidine and nepheline phenocrysts, though not very common; structure sometimes massive, sometimes fluidal and rather complex; there is a characteristic conchoïdal fracture.

**Geotectonic environment**  Flows and apophyses associated with trachyte, from which they are derived by local differentiation; small-scale differentiation units often associated with nepheline syenites in the upper parts of volcano-tectonic complexes, formed at shallow depths (tinguaites).

**Occurrence**  The extrusive equivalents of feldspathoid syenites thus frequently passing into trachytes. They are very common in association with alkaline basalts, in oceanic islands (Kerguelen, the Azores, the Canaries, the Marquesas) and with nepheline syenites in tectonic volcanic areas (Bearpaw Mountains, Montana, and Colorado and South Dakota [USA]; Tingua Mountains [Brazil]; the USSR). Other occurrences in Swabia and Rieden in the Eifel Mountains (Germany), Bohemia (Czechoslovakia), Cornwall (England) and Bass Rock and Haddington (England). In Italy, varieties with potassic differentiation are frequent, associated with tephrite and basanite in the Roman Volcanic Province. The white pumiceous ash that smothered Pompeii in the eruption of A.D. 79 is of leucite phonolite composition. Hauyne is contained in the phonolites of Vulture, Potenza (Italy), which in thin section (see photomicrograph) show large phenocrysts of zoned sanidine, small pyroxene prisms with vivid interference colors, and hauyne, which appears black, being isotropic, but is nevertheless clearly identifiable because of the square and octagonal outlines, in a conspicuously fluidal groundmass. A variety rich in leucite and melanite and poor in sanidine is leucitophyre, typical of the Leucite Hills, Wyoming (USA) and Rieden (Germany).

**Uses**  Limited use as a building material; attempts have been made to use leucitic concentrations as a mineral fertilizer because of its abundant potassium.

▲ Phonolite (ca. ×1). Bohemia, Czechoslovakia.

▲ Phonolite (Mt. Vulture, Italy): photomicrograph of a thin section (ca. ×20; crossed Nicols).

## 308 TEPHRITE

**Type** Extrusive igneous rock.
**Chemistry** Mafic (undersaturated).
**Components** *Essentials*: plagioclase (labradorite-bytownite), clinopyroxene (augite, acmite), feldspathoid (nepheline or leucite). *Accessories*: brown hornblende, biotite, sodalite, analcime, magnetite, ilmenite, glass.
**Appearance** Color fairly dark gray; texture porphyritic with prevalently intergranular crystalline groundmass; structure pimply or vacuolar, sometimes with zeolitic fill.
**Geotectonic environment** Flows and dikes in nonorogenic zones, mainly derived by assimilation of carbonate by alkaline basaltic magma.
**Occurrence** Tephrite is common in the mid-Atlantic islands. Characteristic leucite tephrites are the lavas of Vesuvius and the Roman Volcanic Province (Italy) and Uganda. Nepheline tephrite is common in Germany and tephrite containing hauyne is found at Vulture, Potenza (Italy).
**Uses** Limited to local use as building stone.

14

## 309 BASANITE

**Type** Extrusive igneous rock.
**Chemistry** Mafic (undersaturated).
**Components** *Essentials*: plagioclase (labradorite-bytownite), clinopyroxene (augite, acmite), feldspathoid (nepheline or leucite), olivine. *Accessories*: brown hornblende, biotite, sodalite, analcime, magnetite, ilmenite, glass.
**Appearance** Color fairly dark gray; texture porphyritic with intergranular prevalently crystalline groundmass; structure pimply or vacuolar, with frequent zeolitic fill.
**Geotectonic environment** Flows and dikes associated with alkaline basalts, especially in zones that are being stretched to the point of separation between continental plates or at the intersection of plates with local heat concentrations. They grade almost continually into tephrite. Commonly contain spinel lherzolite xenoliths, which are rocks derived from the Earth's upper mantle.
**Occurrence** Nepheline basanites are found at Eifel (Germany); in the mid-Atlantic islands (the Canaries, Azores, Madeira, Ascension, St. Helena); at Dunedin (New Zealand) and New South Wales and Tasmania (Australia). Leucite basanites are well known in the Roman Volcanic Province (Italy), especially in Monte Somma and Vesuvius. Basanites are found in the Sahara (Tassili), in Uganda and in Montana and San Carlos, Arizona (USA).
**Uses** Limited to local use as building stone.

14

▲ Leucite tephrite (ca. ×1). Lazio, Italy.

▲ Leucite basanite (ca. ×1). Viterbo, Italy.

## 310   NEPHELINITE

**Type**   Extrusive igneous rock.
**Chemistry**   Mafic (undersaturated).
**Components**   *Essentials*: nepheline. *Accessories*: olivine, augite, titanite, perovskite, melilite, noselite, sodalite, hauyne, plagioclase. *Accidentals*: leucite, zeolites.
**Appearance**   Color light gray, sometimes pinkish or greenish; texture porphyritic with generally and often feltlike holocrystalline groundmass; structure from massive to porous.
**Geotectonic environment**   Small flows, associated with alkaline basalts and formed in zones with low heat flow.
**Occurrence**   Typical examples are those from Central Germany (Odenwald, Katzenbuckel), Czechoslovakia (Mittelgebirge) and France (Auvergne). Those containing leucite are found in the Rift Valley of Central Africa (Ruwenzori, Mt. Meru, Oldonyo Lengai). Nephelinite also occurs in the Dunedin district (New Zealand) and near Venice (Italy).
**Uses**   Local use as building stone.

## 311   LEUCITITE

**Type**   Extrusive igneous rock.
**Chemistry**   Mafic (undersaturated).
**Components**   *Essentials*: leucite. *Accessories*: augite, often titaniferous; olivine, hauyne, melilite, melanite, brown hornblende, titaniferous biotite. *Accidentals*: nepheline, glass.
**Appearance**   Color whitish or light gray; texture porphyritic, with holocrystalline granular groundmass; glass-bearing varieties are more compact, dark gray in color; structure from massive to porous, with zeolite-rich cavities.
**Occurrence**   Known on all continents, although the most famous occur in Italy in the Roman Volcanic Province, including Vesuvius and Monte Somma. In the USA they are found at Leucite Hills, Wyoming, and in the Bearpaw Mountains, Montana. Other notable occurrences are found in the Kimberley Range in Western Australia; Celebes and Java, Indonesia; and Central Africa (Zaire, Uganda, Tanzania).
**Uses**   Locally used as building stone and for curbstones.

▲ Nephelinite (ca. ×1). Bohemia, Czechoslovakia.

▲ Leucitite (ca. ×1). Acquacetosa, Rome, Italy.

## 312 MELILITITE

**Type** Extrusive igneous rock.

**Chemistry** Ultramafic.

**Components** *Essentials*: melilite, acmite-augite, olivine. *Accessories*: perovskite, chromite, picotite spinel, melanite, phlogopite, apatite, titanite. *Exceptional*: leucite.

**Appearance** Color gray or pale brown; texture porphyritic with granular holocrystalline groundmass; structure from massive to finely porous, often with xenoliths.

**Geotectonic environment** Related to some kimberlite events and locally derived by advanced contamination of basaltic magma with carbonate rocks.

**Occurrence** Common among the alkali-rich lavas in central Africa (Uganda, Zaire, Kenya). In Sweden (Alnö) it is associated with kimberlite and carbonatite. Also found in the Swabian Alps (original occurrence), Madagascar, Tasmania and at Isle Cadieux in eastern Canada.

**Uses** Sometimes associated with rare-earth minerals, niobite and columbite-tantalite.

**16**

## 313 CARBONATITE

**Type** Intrusive and extrusive igneous rock.

**Chemistry** Ultramafic.

**Components** *Essentials*: calcite or dolomite. *Accessories*: complex carbonates of manganese, calcium, and iron, nepheline, phlogopite, olivine, apatite, monazite, barite, pyrochlore, fluorite, perovskite, magnetite, ilmenite, pyrite, titanite, niobite, etc. High concentrations of minor elements with large ionic radii (zirconium, strontium, barium, etc.).

**Appearance** Color light, grayish or yellowish; texture xenoallotriomorphic, usually without phenocrysts; structure massive, superficially cavernous due to being dissolved by water.

**Geotectonic environment** Associated with nepheline syenite intrusions and formed at great depth. They are found as veins and dikes, often in groups, and as circular intrusions with radial or concentric development. Observed in exceptional cases as flows (Oldonyo Lengai, Ruanda).

**Occurrence** Most common in central and southern Africa, which contains half of the world's occurrences. A famous occurrence is that of Alnö (Sweden). Also found in Norway (Telemark), the USSR (Kola Peninsula) and Canada (Oka). Small outcrops have been noted in the USA and France.

**Uses** Important for the extraction of certain minerals which contain rare elements.

**16**

▲ Melilitite (ca. ×1). Lazio, Italy.

▲ Carbonatite (ca. ×0.5). Oka, Canada.

## 314 OBSIDIAN

**Type** Extrusive igneous rock.

**Chemistry** Variable (usually felsic).

**Components** *Essentials*: glass. *Accessories*: magnetite, ilmenite, oxides and other minerals, such as feldspar, as microphenocrysts.

**Appearance** Color shiny black, with clear conchoidal fracture; texture glassy with rare microphenocrysts and abundant opaque fine dust scattered in concentric zones. Structure massive with, in some cases, the presence of latent concentric fractures (perlitic fractures).

**Geotectonic environment** Derived from the rapid cooling of a viscous granitic magma. Found as volcanic units, as fragments thrown into the air by erupting volcanos, sometimes as long filaments ("Pele's hair") and as the outer covering of rhyolitic and dacitic domes.

**Occurrence** Very common in recent lava flows, mostly from felsic volcanos (Japan, Java). Found in the Tertiary lavas of the Lipari Isles (Italy). The most striking occurrence is in Obsidian Cliff in Yellowstone National Park, Wyoming (USA). Also found in Hungary.

**Uses** Used in prehistoric times for tools and sculptures. Currently used industrially as raw material for rock wool.

## 315 PUMICE

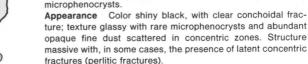

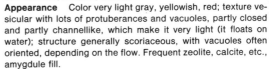

**Type** Extrusive igneous rock.

**Chemistry** Variable (usually felsic).

**Components** *Essentials*: glass. *Accessories*: crystals of various silicates, zeolites, calcite.

**Appearance** Color very light gray, yellowish, red; texture vesicular with lots of protuberances and vacuoles, partly closed and partly channellike, which make it very light (it floats on water); structure generally scoriaceous, with vacuoles often oriented, depending on the flow. Frequent zeolite, calcite, etc., amygdule fill.

**Geotectonic environment** Material expelled violently upward into the air, in the initial gas-rich phases of volcanic explosions. Common in felsic volcanos and also on the outermost crust of some flows.

**Occurrence** The Lipari Isles (Italy), and most of the volcanos of Sunda (Indonesia) and Japan are rich in pumice. Vesuvius has had emissions of pumice and volcanic ash at various times, in particular at the start of the eruption of A.D. 79, which destroyed Pompeii.

**Uses** Reasonably good abrasive, especially for soft metals; light insulation in building.

▲ Obsidian (ca. ×1). Aeolian Islands, Sicily, Italy.

▲ Pumice (ca. ×1). Aeolian Islands, Sicily, Italy.

## 316 GRANITE PORPHYRY

**Type** Hypabyssal igneous rock.
**Chemistry** Felsic.
**Components** *Essentials*: potassic feldspar (orthoclase, microcline), quartz. *Accessories*: biotite, plagioclase (albite-oligoclase), muscovite, amphibole, magnetite, ilmenite, apatite, zircon, xenotime, cassiterite, molybdenite, etc.
**Appearance** Color light gray, pink, often red; texture porphyritic with quartz and idiomorphic feldspar phenocrysts but often corroded, set in a microgranular and microfelsitic groundmass; structure from massive to zoned with flow alignment, frequently with miarolitic cavities.
**Geotectonic environment** Veins and apophyses in granite or in nearby rocks; local marginal zones of a shallow, intrusive, granitic pluton.
**Occurrence** Present in almost all granitic plutons. It is most characteristic along the margins of the Canadian Shield and at Erzgebirge (Germany).
**Uses** Local building stone; sometimes contains concentrations of rare sulfides (e.g., molybdenite).

## 317 GRANOPHYRE

**Type** Hypabyssal igneous rock.
**Chemistry** Felsic.
**Components** *Essentials*: potassic feldspar (orthoclase, microcline), quartz. *Accessories*: plagioclase (albite-oligoclase), biotite, amphibole, sodic pyroxene, magnetite, ilmenite, muscovite, apatite, zircon, molybdenite, topaz, fluorite, etc.
**Appearance** Color light, often reddish; texture porphyritic with quartz and feldspar in a medium-grained groundmass formed by a graphic or micropegmatitic intergrowth of the same; frequently zoned, and containing miarolitic cavities lined with minerals in the form of fine crystals.
**Geotectonic environment** Veins, apophyses, indistinct masses in the peripheral parts of shallow intrusive granitic plutons.
**Occurrence** Present in almost all plutons, especially in association with granite porphyry. It is even present in intrusions of basaltic magma as the final product of differentiation; found in the upper parts of basaltic sills such as the Palisades sill (New Jersey, USA).
**Uses** Building stone for curbstones or as ornamental gravel for parks and gardens; sometimes has concentrations of rare minerals that are the source of rare chemical elements.

▲ Granite porphyry (ca. ×1). Erzgebirge, Germany.

▲ Granophyre (ca. ×1). Cuasso al Monte, Varese, Italy.

## 318  DIORITE PORPHYRY

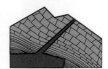

**Type**  Hypabyssal igneous rock.

**Chemistry**  Intermediate.

**Components**  *Essentials*: plagioclase (andesine-labradorite), hornblende, biotite, quartz. *Accessories*: clinopyroxene, epidote, magnetite, ilmenite, apatite, titanite, rutile, zircon, monazite, pyrite, orthoclase. *Secondaries*: uralitic-type amphiboles, hematite, etc.

**Appearance**  Color gray or green, sometimes also reddish by alteration, or bluish; texture markedly porphyritic with tabular phenocrysts of zoned plagioclase and hornblende or biotite in a holocrystalline groundmass, feltlike or intergranular, sometimes with flow traces; structure massive or fluidal parallel to vein walls, with frequent xenoliths or xenocrysts, especially of quartz. In some veins there may be calcareous blocks, torn from the encasing rocks and containing minerals typical of contact metamorphism.

**Geotectonic environment**  Veins, of varying size, within plutonic rocks or in their contact aureole, even some at some distance. To a large extent they are squeezed into fractures in dioritic and tonalitic rocks but may also, in part, represent crystallization from andesitic magma that intruded into a geosuture and emerged in veinlike form (abortive volcanos).

**Occurrence**  Very common in association with all the large granite-granodiorite plutons, sometimes also in dike swarms, especially the numerous dike swarms of Scotland. Dark gray diorite porphyry with large white phenocrysts of plagioclase (see photo) is found in Valtellina, Italy. In thin section this rock shows idiomorphic, zoned and twinned plagioclase (see photomicrograph) and prismatic amphibole with vivid interference colors, in a microgranular groundmass predominating in plagioclase.

**Uses**  Very tough rock used in blocks and slabs for pavements and facings (e.g., ''Val Camonica porphyrite'' and ''Egyptian red porphyry'').

▲ Diorite (ca. ×1). Sondalo, Sondrio, Italy.

▲ As above: photomicrograph of a thin section (ca. ×15; crossed Nicols).

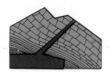

10

**Type** Hypabyssal igneous rock.

**Chemistry** Mafic.

**Components** *Essentials*: plagioclase (labradorite-bytownite), clinopyroxene. *Accessories*: orthopyroxene, olivine, brown hornblende, magnetite, ilmenite, apatite, calcite, pyrrhotite, chalcopyrite, sometimes glass. *Secondaries*: serpentine, chlorite, calcite.

**Appearance** Color dark, from green to blackish, brown on altered surfaces; texture often porphyritic, characterized by an ophitic or subophitic groundmass in which intersecting laths of plagioclase are enclosed by allotriomorphic pyroxene; structure massive, sometimes with columnar fissuring and conchoidal fracture. Some layering due to gravitative settling. A typical feature is the alteration of pyroxene to chlorite, which gives the rock a greenish color.

**Geotectonic environment** Sills and dikes in large swarms extending for kilometers. Some sills have cumulate structures, both macroscopically and microscopically. Their tholeiitic composition suggests derivation from the upper mantle by intrusion into the crust, although they may be modified by differentiation processes before emplacement. Some take the form of fissure-filling or feeder dikes for large basaltic fields. Massive diabase, with an ophitic structure but fairly fine grain size, forms the lower slowly crystallized parts of particularly thick basaltic flows.

**Occurrence** Large seamlike systems of different ages are found in various places: the largest are those of the Karroo (southern Africa), Tasmania, and Nipissing (Canada). The Triassic diabase in New Jersey (USA) is very important, even though more limited; part of this is formed by the Palisades Sill, which is famous for its variations of composition and texture from bottom to top, due to differential conditions of cooling accompanied by gravitative settling. Those of Scotland, Germany and Dillsburg, Pennsylvania (USA), are also important because they are differentiated, with large excesses of iron and titanium resulting in a rock type called ferrodiabase. A typical diabase shown in thin section (see photomicrograph). It has evident ophitic texture: pyroxenes are deeply altered to amphibole and chlorite and are set between twinned divergent idiomorphic crystals of plagioclase; rare crystals of unaltered pyroxene retain their vivid interference colors. The term diabase has sometimes been incorrectly applied to basaltic lava with pillow structure produced by submarine extrusion and showing marked spilitic characteristics (formation of chlorite, hematite and albite at the expense of plagioclase and pyroxene).

**Uses** Frequently associated with economic concentrations of copper minerals; locally used for building stone.

▲ Diabase (ca. ×1). Edolo, Brescia, Italy.

▲ As above: photomicrograph of a thin section (ca. ×20; crossed Nicols).

## 320  PEGMATITE

**Type**  Hypabyssal igneous rock.
**Chemistry**  Felsic.
**Components**  *Essentials*: quartz, alkaline feldspar (orthoclase, microcline, albite), mica (muscovite, biotite, lepidolite). *Accessories*: tourmaline, beryl, topaz, zircon, apatite, rare minerals (e.g., cassiterite, columbite-tantalite, etc.).
**Appearance**  Color very light but varying with content and type of accessories; texture produced by intergrowth of large, almost idiomorphic crystals, more rarely graphically intergrown; structure sometimes zoned with many cavities and druses lined with crystals growing into them.
**Geotectonic environment**  Associated with granitic and syenitic plutons as large veins. Magma was rich in volatile elements resulting in large crystals, containing abundance of elements not used up in earlier crystallization history. Indistinct patches in enclosing rocks.
**Occurrence**  Pegmatite is well known because it contains beautiful crystals of many different minerals. In the USA the most famous mineral occurrences are in pegmatites and include Stoneham and Hebron, Maine, Chesterfield, Massachusetts, Haddam, Connecticut, Pikes Peak, Colorado, and Harney's Peak in the Black Hills of South Dakota. Other notable occurrences are in the Minas Gerais area (Brazil), the Kola Peninsula (USSR) and Malagasy Republic.
**Uses**  Rocks of considerable economic importance for the extraction of minerals such as orthoclase, muscovite, beryl, etc.

---

## 321  APLITE

**Type**  Hypabyssal igneous rock.
**Chemistry**  Felsic.
**Components**  *Essentials*: quartz, alkaline feldspar (orthoclase, microcline, albite) and mica (muscovite, biotite). *Accessories*: tourmaline and a few other minerals.
**Appearance**  Color white or light gray; mosaiclike texture with fine or very fine grain size; structure massive or zoned.
**Geotectonic environment**  Fracture filling in felsic intrusive and veined rocks, both by lateral secretion and by the intrusion of melt within the fracture. Zonation is the result of several successive fillings of the same fracture.
**Occurrence**  Aplites are common in all intrusive masses, mostly toward their borders or in the contact zone. They are often found in fractures due to the differential movement between masses.
**Uses**  Of no commercial or industrial importance.

Pegmatite with tourmaline (ca. ×0.5). Piona, Como, Italy.

Aplite vein in orthoclase-bearing tonalite (ca. ×1). Val Masino, Sondrio, Italy.

**13**

**Type** Hypabyssal.

**Chemistry** Mafic.

**Components** *Essentials*: feldspar (orthoclase, labradorite plagioclase), biotite. *Accessories and accidentals*: augite, olivine, hornblende, calcite, titanite, barkevikite, magnetite, ilmenite, siderite.

**Appearance** Color dark gray, fairly reddish or brownish, sometimes also shiny black; texture porphyritic or granular with large phenocrysts, usually of biotite often packed· or aligned in a finer, mainly feldspar groundmass; structure zoned; vacuolar, contains carbonate concentrations, and often contorted.

**Geotectonic environment** Sills and small masses sometimes associated with granitic-syenitic massifs, both as early differentiation products and, more commonly, as late derivatives of magmatic activity in zones at the margin of the main intrusive body.

**Occurrence** Lamprophyre is distinguished by the magmatic pluton to which it is related. With diorite it is characterized by the presence of hornblende associated with plagioclase, and with syenite by prevalent biotite associated with orthoclase. Each group contains numerous varieties that are differentiated by the prevalent mineral among the phenocrysts and sometimes by the prevalent mineral in the groundmass if it is recognizable. Therefore, among the dioritic lamprophyres the varieties spessartite, ondite, malchite and kersantite are distinguished; among the syenitic varieties durbachite, vogesite, cascadite and minette are recognized. Undersaturated lamprophyres also exist: alnöite, containing melilite in the groundmass; monchiquite, with analcite, and a sodic differentiate camptonite, containing barkevikite. Dioritic lamprophyres are very common in the Hercynian plutons of Central Europe and especially in Brittany and the Vosges (France) and Spessart, Odenwald and ·Schwarzwald (Germany). Syenitic lamprophyres are common in France, Germany, Norway, Finland, Arizona and California (USA), and Jersey (Channel Islands) and the Lake District (England). The classic alnöite is found in Sweden and monchiquite was first described from the Caldas de Monchique in southern Portugal.

**Uses** Of no commercial value.

▲ Kersantite (ca. ×1). Kersanton, France.

Minette (ca. ×1). Vosges Mountains, France.

# SEDIMENTARY ROCKS
## KEY TO THE SYMBOLS

FORMATIONAL ENVIRONMENT

alluvial; detrital rocks deposited by flowing rivers

lacustrine; detrital, chemical and organogenetic rocks formed in lakes

marine; detrital, chemical and organogenetic rocks formed in the ocean

lagoonal; mainly chemical and organogenetic rocks formed in lagoons

chemically formed rocks formed in a cave or karst environment

morainic; detrital rocks formed in glacial and fluvioglacial environments

aeolian; detrital rocks formed in hot and cold deserts

pyroclastic rocks; formed from volcanic materials (could also be classified as igneous rock)

GRAIN SIZE OF ROCKS

coarse          medium          fine          very fine

## 323 CONGLOMERATE

**Type** Clastic sedimentary rock.
**Class** Coherent rudite.
**Appearance** Color variable and irregularly distributed; texture characterized by rounded pebbles (diameter larger than 4 mm; .16 in) and by blocks and masses scattered over a finer grained matrix, but with grains larger than 2 mm (.08 in); structure rather irregular: the bedding is absent or poorly developed, but the pebbles may have a preferential orientation. Fossils are rare.
**Components** The pebbles may be of the most varied rocks and are either all of one type (monogenetic conglomerate) or of several types (polygenetic conglomerate); the matrix is usually sandy with calcareous-clayey bonding.
**Geological environment** Consolidation of gravel associated with deposition in shallow and turbulent water, often indicating marine transgressions. Conglomerates also represent coarse fluvial deposits in torrential environments.
**Occurrence** Widespread in the Permian and Triassic periods overlying the metamorphic basement. Characteristic conglomerates in Great Britain occur in the Torridonian and Old Red Sandstone of Scotland. The Triassic conglomerates of the eastern USA are thought to be the accumulation of gravels from the Appalachian Mountains. Another example is the gold-bearing conglomerate of the Witwatersrand (South Africa).
**Uses** Used locally as a building stone.

---

## 324 PUDDING STONE

**Type** Clastic sedimentary rock.
**Class** Coherent rudite.
**Appearance** Color variable, from whitish to dark gray; texture defined by fairly roundish pebbles (diameter > 4 mm; .16 in), relatively unsorted by size, striated and compressed; structure without bedding except in very thick sequences. A type of conglomerate.
**Components** The pebbles may be of rocks of every type, the matrix is sandy or muddy, the bonding usually calcitic.
**Geological environment** Consolidation of fairly random gravel, of glacial, periglacial and fluvioglacial origin, associated with particular areas and geological eras.
**Occurrence** Pudding stone, an unusual variety of siliceous, strongly cemented conglomerate, is found in the Eocene rocks of the London Basin (England) and Post-Cretaceous strata of South Africa, the USA and Australia. Nagelfluh, a type of pudding stone, is widespread around Lucerne (Switzerland) at the base of the Miocene marine *molasse*. Often used for glacial and fluvioglacial deposits (*tillitos*).
**Uses** Often used as a building-stone facing.

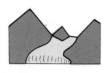

▲ Conglomerate (ca. ×1). Valcamonica, Brescia, Italy.

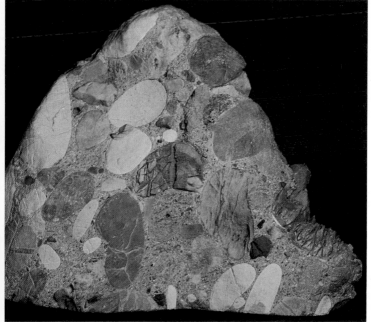

▲ Pudding stone (ca. ×0.5). Montorfano, Brescia, Italy.

## 325 BRECCIA

**Type**  Clastic sedimentary rock.
**Class**  Coherent rudite.
**Appearance**  Color variable; texture defined by angular rock fragments in a fine- or very fine-grained matrix; structure often fairly well bedded in the matrix. Fossils are rare.
**Components**  The fragments may be rocks of every type (polygenetic breccia) but, more frequently, all of the same type (monogenetic breccia); the matrix and bonding are generally clayey, siliceous, calcareous and also limonitic.
**Geological environment**  Consolidation of landslides, detrital strata, cave-ins in karst areas, mainly in subaerial environments. Other breccia types are due to fracturing during the folding of rock masses (friction breccia) or to the dislocation of sediments which are still incoherent before permanent consolidation (intraformational breccia).
**Occurrence**  Found in limestone and dolomite formations such as the Beekmantown of Pennsylvania and the Siyeh Limestone of the Rocky Mountains (USA). It is also found in the Cotham Marble (Bristol) and the Eilean Dubh Dolomite (Sutherland) in Great Britain. Breccias are fairly common but restricted in extent.
**Uses**  Under various commercial names used in building for decorative purposes or for inside paving.

---

## 326 OPHICALCITE

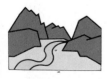

**Type**  Clastic sedimentary rock.
**Class**  Coherent rudite.
**Appearance**  Color fairly dark green, red or violet; texture defined by angular fragments of varying size, bound by a white cement of various sorts; structure nonbedded. No fossils.
**Components**  The fragments are formed by serpentine and gabbro, often with red alteration from formation of hematite. The cement may be calcite, dolomite, magnesite or rarely quartz.
**Geological environment**  Derived from the consolidation of landslides, detritus or friction breccias due to tectonic movements in areas with ultramafic rocks.
**Occurrence**  Common in the Alps and Appennines. The best known varieties include Verde Alpi and Cipollino. A translucent variety, called the "onyx of Châtillon," is found in the Valle d'Aosta (Italy). It also occurs at the Lizard in Cornwall (England) and in Montana (USA).
**Uses**  Highly prized as decorative material in slabs for facings and paving.

▲ Breccia in polished slab (ca. ×0.5). Camaiore, Lucca, Italy.

▲ Ophicalcite (ca. ×1). Levanto, La Spezia, Italy.

# 327 ARENITE

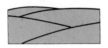

**Type**  Clastic sedimentary rock.
**Class**  Coherent arenite.
**Appearance**  Color very variable, from white to gray, yellow, green, red and brown; texture clastic, grain size very fine to very coarse but less than 2 mm (.08 in), normally with grains that are sorted and somewhat rounded; matrix much less sorted and very fine cement; usually well stratified, often with regular reduction of the grain size from the bottom to top of the stratum (graded bedding). Imprints of flow and waves (ripple marks) are common. Fossils are frequent as are concretions of various shapes and sizes.
**Components**  The grains may be rock fragments or individual minerals, especially quartz, feldspar, mica and calcite. Heavy minerals, fairly rounded, are common (zircon, apatite, olivine, monazite, magnetite, pyrite). The cement is siliceous (quartz, opal, chalcedony), calcitic, dolomitic, clayey (graywacke), limonitic. Pyrite, gypsum and glauconite form as diagenetic minerals.
**Geological environment**  Accumulations of clastic matter transported by wind and river water and sea water (torrents of alluvial matter), often distinguishable on the basis of the shape and superficial imprints of the grains. In marine deposits, commonly found at the base of transgressive conglomerates grading upward to clays and limestones.
**Occurrence**  The purest arenite is composed of quartz sand cemented by quartz. It is called orthoquartzite as in the Fontainebleau formation (France). Some arenites have a kind of flexibility and are bendable in thin units; these are called itacolumite, found in Brazil and India. Many arenites have a calcareous cement; calcareous arenites in the USA occur in Nebraska, Wyoming, California and in the sandstones of the White River Badlands of South Dakota. The *molasse* of the Alpine foothills contain widespread calcareous arenites in Switzerland and Bavaria; in north Italy they contain abundant clay minerals as well. The thin section (see photomicrograph) shows a calcareous-argillaceous-micaceous arenite from Val Camonica (Italy). It shows numerous fragments of gray quartz with very fine calcite, clay, mica cement which appears almost isotropic under crossed nicols. A glauconite cement is particularly common in marine arenites. These include the greensands which are strongly developed in the lower Cretaceous of Europe and of North America. Cambrian greensands are found in Wales and Scandinavia. Modern greensands are found along the south and east coasts of Africa, the south and west coasts of Australia; the west coast of Portugal; the west and east coasts of North America. If muscovite is present, it tends to accumulate in layers producing flagstones as in Brazil and India.
**Uses**  Frequently used as a building stone, also for exteriors, but sometimes with bad results because the cement degrades easily; for millwheels and lapping wheels, used by artisans; as crushed stone.

▲ Arenite (ca. ×1). Brianza, Como, Italy.

▲ Arenite (Valcamonica, Italy): photomicrograph of a thin section (ca. ×20; crossed Nicols).

## 328 ARKOSE

**Type**  Clastic sedimentary rock.
**Class**  Coherent pelite.
**Appearance**  Color gray, pink or reddish; texture clastic with fairly coarse grain size, not well sorted and with angular grains, structure defined by poor stratification, with frequent current imprints. No fossils.
**Components**  The grains are mainly feldspar, with a little quartz, biotite, muscovite and other minerals derived from igneous rocks and lithic fragments. The cement may be silicate (quartz, illite), calcitic or limonitic.
**Geological environment**  Degradation of granitic and gneissic rocks, followed by movement over a short distance, and deposition in a fluvial, lacustrine or shallow marine environment.
**Occurrence**  Large deposits of arkose are widespread throughout all geological periods. Precambrian examples are the sparagmites of Scandinavia and the Torridon Sandstone of Scotland. Paleozoic examples occur in the marine-subaerial sedimentary series around the Hercynian Chain (the Vosges, France; the Black Forest, Germany; Wales; Bohemia). The arenaceous rocks in the Californian Coast Ranges are predominantly arkose from the Jurassic to the present day.
**Uses**  Local use in building, like sandstone.

---

## 329 GRAYWACKE

**Type**  Clastic sedimentary rock.
**Class**  Coherent arenite.
**Appearance**  Color fairly dark, from gray to brown, sometimes greenish; texture clastic and not well sorted, with very angular grains, in a muddy or clayey matrix; the stratification is often in very thick banks (or bars) and is characterized by clear-cut gradation; frequent slip structures, curling of individual strata, grooves from dragging, and inorganic imprints. Fossils very rare and fragmentary.
**Components**  The clasts consist of quartz, feldspar, and rock fragments (mafic igneous, schist, etc.) in almost equal amounts; the matrix and the cement are clayey or chloritic, with iron oxides and hydroxides.
**Geological environment**  Produced by currents of submarine silt, in deep water; they slip over the edge of the continental slope and are then gradually redeposited. Typical of the initial phases of tectonic movements.
**Occurrence**  Very common, especially in the Silurian Tuscarora Sandstone in West Virginia (USA); the Lower Paleozoic grits of England and Wales; the Rhine Valley, Germany; the Northern Appennines of Italy; and the Cretaceous rocks of Kyushu, Japan.
**Uses**  Local use as construction stone.

▲ Arkose (ca. ×1). Vosges Mountains, France.

▲ Graywacke (ca. ×1). Rhine Valley, Germany.

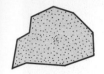

## 330  ARGILLITE

**Type**  Clastic sedimentary rock.
**Class**  Coherent pelite.
**Appearance**  Light to dark gray in color, sometimes black (carbonaceous substances), red or green. Texture clastic with very fine grain size (< 1/256 mm; 1/1024 in; grains not distinguishable even through the microscope). Stratification absent; good cleavability in laminae. Frequent structures of drying, raindrops, fossil traces, and concretions.

**Components**  Mixture of clay minerals (illite, kaolinite, montmorillonite, etc.) with quartz, feldspar, carbonate, mica. Often has a carbon residue, iron oxides (limonite, hematite), nodules of gypsum and pyrite crystals.

**Geological environment**  Lacustrine, lagoonal and marine deposits of materials derived from lengthy movement mainly in water. Some are residual, i.e., formed in situ by chemical and physical alteration of preexisting rocks. Glacial clays are called varves, with very thin seasonal layers.

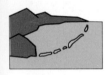

**Occurrence**  Widely distributed throughout the world. Of special interest are the black shales that are rich in graptolites in the Lower Paleozoic of England, Wales and Germany. The Kupferschiefer is a black shale extending from Poland to England containing localized high concentrations of copper and zinc. The Precambrian Nonesuch Shale of Michigan contains up to 10 percent copper in places.
**Uses**  Raw material for the ceramic industry, brick manufacture and refractory materials. Oil- and uranium-bearing shales are being carefully studied for their economic potential.

## 331  MARL

**Type**  Clastic or mixed sedimentary rock.
**Class**  Calcareous-pelitic.
**Appearance**  Color from light to dark gray, brownish, greenish; texture clastic with very fine or fine grain size, few grains distinguishable with the naked eye. Medium to fine stratification often only barely evident, but generally with good cleavability. Frequent sedimentary structures, fossils and concretions.

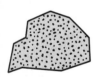

**Components**  Mixture of calcite (more rarely dolomite) and clay minerals, with traces of quartz, mica and carbonaceous residue. Frequent gypsum, calcite and pyrite nodules.

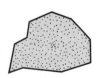

**Geological environment**  Marine or lacustrine deposits of clastic material that has undergone lengthy movement and has mixed with products of chemical precipitation or with organogenetic residue (calcite).

**Occurrence**  Common in the Tertiary Piedmontese Basin in Italy and in southern England. It is also found in the Permo-Triassic basins of the Northern Hemisphere.
**Uses**  Raw material for the cement industry.

▲ Argillite (ca. ×1). Sassuolo, Modena, Italy.

▲ Fucoidal marl (ca. ×1). Induno Olona, Varese, Italy.

## 332 TRAVERTINE

**Type**   Chemical sedimentary rock.

**Class**   Carbonatic rock.

**Appearance**   Color light, yellowish or pinkish; texture fibrous and concretionary, grading to vacuolar; no stratification but small thin bands differentiated by concentrations of impurities or by grain size.

**Components**   Calcite or aragonite, with limonite impurities; frequent fossils and imprints, especially of plants.

**Geological environment**   Derived from the evaporation of spring water, rich in calcium carbonate, both in rivers near waterfalls and in caves. Similar to the much more porous, light and less concretionary "calcareous tufa," and alabaster, formed mainly by deposits of thermal water in subterranean cavities.

**Occurrence**   Quaternary deposits on a large scale occur in Italy, especially in Tuscany and near Rome. Famous deposits occur at Tivoli. Travertine occurs in the USA at Dubois, Wyoming, and in the Great Basin region around Nevada.

**Uses**   Travertine is the classic construction stone, for facing and paving. Calcareous alabaster is used in polished slabs and as ornamental objects.

## 333 GYPSUM

**Type**   Chemical sedimentary rock.

**Class**   Saline evaporite.

**Appearance**   Color white, gray, red, green or brown; texture massive, grain size fine (mealy) to coarse, also earthy and crumbly; stratification present only in the less coarse varieties, with frequent folds, distortions and boudinage phenomena.

**Components**   Mainly gypsum (selenite variety), associated with anhydrite, rock salt and other marine salts, calcite, dolomite, clays and limonite.

**Geological environment**   Banks, lenses and pockets intercalated with clays, limestone and various salts in evaporite sequences. It is thought that gypsum is derived from hydration of anhydrite during diagenesis. Alabaster, which is waxy-looking, is a compact form of gypsum.

**Occurrence**   Gypsum is particularly common in the USA in practically all ages from the Ordovician to the present day. Occurrences include the Carboniferous of Utah, the Permian basin of New Mexico and west Texas and the Cretaceous of central Florida. Alabaster of great beauty is found at Volterra in Tuscany (Italy), in Iran and in Pakistan ("onyx").

**Uses**   Alabaster is used as polished slabs, sometimes translucent, and in decorative craft objects; gypsum deposits are mined for the manufacture of plaster of Paris, certain types of stucco and filler materials by the paper, rubber and other industries.

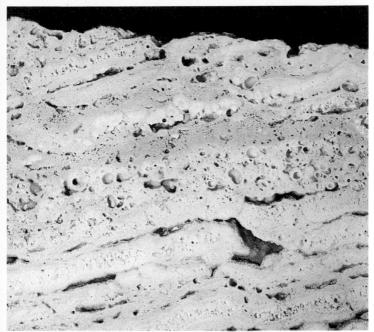

▲ Travertine in polished slab (ca. ×1). Lazio, Italy.

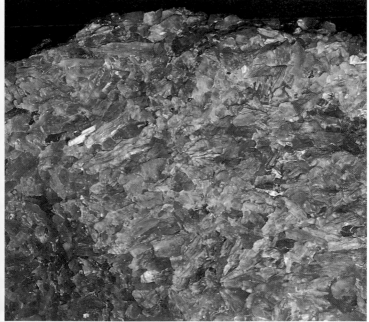

▲ Gypsum rock (ca. ×1). Casteggio, Pavia, Italy.

## 334  ANHYDRITE

**Type**  Chemical sedimentary rock.
**Class**  Saline evaporate.
**Appearance**  Color bluish-gray; texture from massive to striped, with a grain size ranging from medium to saccharoidal; usually with indistinct stratification, but sometimes in thin strata with complex folds or rounded nodules.
**Components**  Mainly anhydrite, associated with gypsum, limestone and various sea salts.
**Occurrence**  The Permian Zechstein Basin, which extends from Russia westward to the north of England, contains many thick sequences of anhydrite. It is best developed in the Hanover region (Germany) and less so in Yorkshire (England). In the USA thick sequences are present in the Paleozoic rock of the Great Lakes region, Utah, New Mexico and Texas. It is also present in the Cretaceous of Florida. The oldest anhydrites are Cambrian in age and occur in Pakistan, Iran and Siberia.
**Uses**  Rocks are used as polished slabs for interior facings; its chief use is in the paper industry, as filler.

---

## 335  PISOLITIC LIMESTONE

**Type**  Chemical sedimentary rock.
**Class**  Carbonatic rock.
**Appearance**  Color white, yellowish or brown; texture defined by a dense compaction of spnerules (diameter > 2 mm; .08 in), sometimes slightly flattened and with an inner concentric structure; matrix scarce. Frequently in banks ranging from thin to average thickness, with scattered fossil fragments.
**Components**  Normally composed of calcite, with dolomite, siderite, hematite or chalcedony as accessories.
**Geological environment**  Pisolites and oolites (smaller spheres always with internal concentric or fibrous-radiated structure) are formed by the precipitation of calcite or aragonite around quartz or carbonate nuclei in warm, turbulent water, probably rolling along the bottom during growth.
**Occurrence**  Forms the Bahamas and the Great Barrier Reef (Australia). Also found around certain hot springs (Karlovy Vary, Czechoslovakia; Karlsbad, Bohemia; Vichy, France). Ancient pisolitic limestones occur in the Lower Carboniferous of Edinburgh, the Isle of Man and in the Lake District (England), in Pennsylvania (USA) and in Germany. Currently forming oolitic sands are found in the eastern Mediterranean and in Great Salt Lake, Utah (USA).
**Uses**  No industrial applications.

▲ Anhydrite (locally called "volpinite") (ca. ×1). Costa Volpino, Italy.

▲ Pisolitic limestone (ca. ×1.5). Germany.

## 336 JASPER

**Type**  Chemical sedimentary rock.
**Class**  Siliceous rock.
**Appearance**  Color very variable; white, gray, red, brown or black; texture compact or microfibrous, with splintery or conchoidal fracture; stratification thin, often with variously colored zones with folds and constricted areas.
**Components**  Mainly chalcedony and quartz with hematite, pyrolusite, clay and sometimes calcite.
**Geological environment**  Deposition of silica directly by precipitation from water, possibly enriched by volcanic solutions, at great depth and in mid-ocean. It is also probable that there is diagenesis at the expense of organogenetic siliceous sediments.
**Occurrence**  In the USA jasper is found in the Lake Superior and Mesabi Range iron ore deposits (jaspilites). It is also found in the Jurassic of California, and the variety novaculite, from Arkansas, is used for making hones and whetstones. The variety containing volcanic material (porcellanite) is found in England and Michigan (USA).
**Uses**  Sometimes used as a decorative stone.

## 337  MANGANESE NODULE

**Type**  Chemical sedimentary rock.
**Class**  Siliceous-metalliferous rock.
**Appearance**  Color brown or black; shape roundish, lenticular or concretionary, often with growth nuclei and concentric zones, grain size fine or very fine.
**Components**  Iron and manganese oxides and hydroxides, partly amorphous (pyrolusite, manganite and psilomelane): small amounts of clay, quartz, calcite and detrital minerals.
**Geological environment**  Deposits formed deep in the ocean, especially in the vicinity of submarine volcanos, from which they are derived by means of solutions which are precipitated around bits of fossil shells, detrital grains or lava nodules.
**Occurrence**  Scattered over the sea floor of the Pacific, Indian and Southern Atlantic oceans in great quantity. They are also common in many Mesozoic and Tertiary fine-grained sediments.
**Uses**  Potential source of metals (manganese, cobalt, nickel and zinc) not exploited to date due to the great difficulty of extraction from the ocean depths.

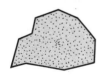

▲ Diaspore (ca. ×1). Sardinia, Italy.

▲ Manganese nodule (ca. ×1.5). Pacific Ocean.

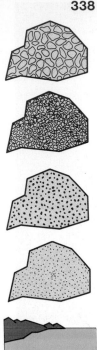

**Type**   Organogenetic sedimentary rock.
**Class**   Calcareous rocks.
**Appearance**   Color very variable: whitish, yellowish, light brown, pink, red, sometimes also dark brown or almost black if it contains bitumens due to decomposition and transformation of the soft parts of organisms which have given rise to them. Texture very variable, from compact and almost porcelainlike, with very minute grain size, to saccharoidal, to very coarse grain size. The abundance and distribution of fossils can be a determining factor for the texture. The structure is also variable but, as a rule, the stratification is distinct with strata of differing coherence, from just a few millimeters to several meters thick. The fossils are frequently concentrated on the surfaces of the strata. Fairly frequent occurrence of cavities due to the dissolution of parts of organisms, in general as negative imprints, or covered by calcite or dolomite crystals. In reef limestones the structures of the living reef are preserved *en bloc* even when the entire rock mass is no longer in its original position, i.e., even when it has suddenly undergone folding and sometimes overturning due to tectonic movements following diagenesis. There are transverse veins of calcite of diagenetic origin and sometimes mineralized veins and zones of metasomatic substitution, also occurring in a diagenetic environment (galena, sphalerite, etc.). It is effervescent with dilute hydrochloric acid.
**Components**   Largely calcite, with secondary dolomite and aragonite. Small quantities of chalcedony, quartz and silicates, especially clay minerals, partly of detrital origin, partly from chemical or biochemical precipitation. Locally rich in bitumens, sulfides (pyrite, marcasite) and iron and manganese hydroxides, either crystalline or amorphous.
**Geological environment**   Formed from the accumulation of calcareous skeletons of marine organisms and therefore to be considered as autochthonous. They can be subdivided, on the basis of the conditions of the formational environment, into *biohermal* limestones (reef), *biostromal* limestones (shells), or *pelagic* limestones. A bioherm is a well-defined mass which is elongated in a vertical sense (formed by living organisms which are in situ) and surrounded by rocks of another type. Bioherms may be due to accumulations of algae (algal limestones), coral colonies (coralline limestones), the remains of crinoids, the shells of brachiopods, etc. Usually they are rather porous and therefore easily subject to chemical reactions during diagenesis, with absorption of magnesium from sea water (dolomitization of aragonite and, to a lesser extent, of calcite). As a result they are rich in druses and cavities lined by crystals or filled with residual clay. These frequently have brecciated structures (intraformational breccias) and develop stylolites, forms of chemical dissolution with redeposition of the insoluble part (especially clays and iron oxides) which, in cross-section, appear as sinuous or broken lines similar to cranial sutures. Fossil remains are not very abundant because they have been partly obliterated during diagenesis, except in particularly well protected zones, which are very rich in these remains. Biostromal limestones are derived from the accumulation of organisms

▲ Fossiliferous limestone (ca. ×1). Sicily, Italy.

▲ Cerithium-bearing limestone (ca. ×1.5). Chàtillon, France.

in flat, extensive beds, or in the form of pillows, poorly stratified, reasonably porous, intercalated with clayey beds or beds
of detrital limestones (calcilutites). They are formed in particular by the shells of living organisms lying on the bottom (lamellibranchs, brachiopods, gastropods and other animals making
up the *benthos*). They are formed either during long periods
without detrital sedimentation or in very short periods because
of the sudden death of all the organisms by pollution or poi-
soning of the water or by sudden environmental changes
(thanatocenosis). They are frequently dolomitized and may
contain large amounts of bitumens. Pelagic limestones are
formed by the accumulation of skeletons of floating organisms,
usually microscopic (plankton), in mid-ocean, where there is
minimal detrital sedimentation. Generally they have fine grain
size and contain microfossils in particular. They often alternate
with siliceous beds (jaspers) or contain flint nodules. Because
of their fine grain size they are not easily dolomitized and are
often particularly well suited to the preservation of fossil remains, down to the very smallest anatomical details.

**Occurrence**   The Coquinas are typical biostromal limestones,
very rich in fossil shells bounded by shell fragments and by
small amounts of fairly bituminous calcitic cement. Typical examples can be found on the beaches of the Bahamas and Florida (USA). Among ancient examples, sometimes used as
decorative rocks, are those of Carinthia (Austria), Purbeck
marble (Dorset, England), Branzi (Bergamo, Italy) and the Eden
and Maysfield beds of southwestern Ohio (USA). The stromat-
olites of the Precambrian and Paleozoic periods, found in particular in the Canadian, African and Australian shields, are also
considered to be biostromal limestones, formed by a large
number of parallel sheets probably derived from algal deposits.
Examples in the USA include the Cambro-Ordovician rocks of
the Appalachians and the Arbuckle Mountains of Oklahoma
(McLish Limestone). A typical biohermal limestone is coralline
or madrepore limestone, formed almost entirely from the skeletons of colonial corals. These rocks are well developed in the
Carboniferous of Great Britain and northwest Europe. The
reefs of Lancashire are a good example. Probably the best exposed reef structures are found in Durham and represent the
western barrier of the Permian Zechstein Sea. These rocks are
also well exposed in the Italian Tyrol in Triassic formations and
are very similar to the Permian madrepore limestones of Texas
and Louisiana (USA), which are associated with petroleum deposits. Those of Arabia and Iran, which are partly dolomitized
and form the reservoir rock for the world's largest hydrocarbon
deposits, are Jurassic-Cretaceous. Coralline limestone are
currently being formed in all the tropical seas, in particular the
Great Barrier Reef of Australia in the Pacific Ocean. However,
many of the fossil bioherms are formed from algal remains,
such as the Tertiary and present day ones formed by the gen-
era *Lithothamnium* and *Halimeda* in the Bahamas, Yucatan
and the whole Caribbean region. The Rudiste limestones in
southern England are also important. Pelagic limestones,
which are usually thinly stratified and often yellow or red in
color, may contain macrofossils perfectly preserved in a very
fine matrix. The earliest pelagic limestones are found in the

▲ Coralline limestone (ca. ×1.5). Dolomites, Italy.

▲ Ammonite-bearing limestone (ca. ×1). Great Britain.

Cretaceous, since this is when foraminifera began secreting a calcareous test. The most famous example is the Cretaceous Chalk of England and northern France, exemplified by the white cliffs of Dover. The Kansas Chalk (USA) is very similar. Pelagic limestones are also found in the Cretaceous of Alabama, Mississippi, Tennessee and Nebraska and adjoining states (USA). They are also found in geosynclinal belts in the Alps. In Italy foraminiferal limestones (orbitolites, nummulites, *Globigerina*) are found, especially in the Eocene period at Veneto, where they are known, in the pinkish-yellow and often variegated variety, by the trade name of "Chiampo marbles."

In thin section (see photomicrograph) one can identify numerous lenticular chambered sections of nummulites and orbitoids and spiral sections of *Globigerina* alongside shell fragments of macrofossils (lamellibranchs), the whole contained in a very fine groundmass. Also pelagic in origin are the Carboniferous fusiline limestone of southern England and the nummulitic limestone of the Nile Valley (Egypt), which was used in the construction of the pyramids.

**Uses**  For construction work, and as crushed stone for road and railroad ballast. Some varieties are used decoratively in the form of polished slabs with commercial names derived from either color, zonation, veining, etc., or from the place or region in which they are found. The Indiana limestone (USA) is one of the most important building stones.

▲ Limestone with a section of pyritized ammonite (ca. ×1). Great Britain.

Polished slab of limestone with nummulites (ca. ×1.5). Veneto, Italy.

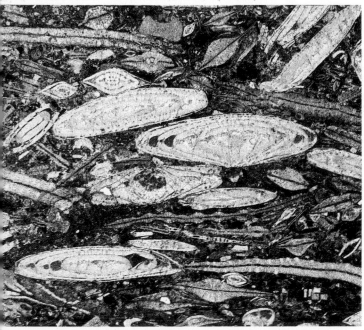

Limestone with foraminifera (Veneto, Italy): photomicrograph of a thin section (ca. ×20).

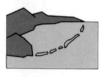

**Type**   Organogenetic sedimentary rock.
**Class**   Calcareous rock.
**Appearance**   Color very variable: white, yellowish, pink, red because of the presence of limonitic or hematitic ochre, or gray or black because of the presence of carbon or bitumen pigments; grain size fine, and texture compact, grading to brecciated (intraformational breccia) or stylolitic (i.e., with forms of dissolution which, on the surface, are translated into lobate or sawtoothed veins). Stratification usually indistinct, or in medium coherent strata on the surface of which one can see various types of fossils. Fracture sometimes conchoidal. Effervescent with cold dilute hydrochloric acid.
**Components**   Calcite very dominant, sometimes with dolomite and minutely subdivided hematite, limonite, carbonaceous pigments, and, on the stylolitic veins or stratum surfaces, traces of clay. In some compact limestones there are siliceous zones of chemical deposits in the form of lenticles, kidneys, and strata of various colors (flints).
**Geological environment**   Limestones of biostromal and biohermal origin, compacted during diagenesis with the elimination of small fossil remains, but with the occasional preservation of larger fossils, also in the form of imprints that are sometimes filled with calcite crystals. Pelagic limestones are less widespread, often very silicified, formed by the shells of microscopic planktonic organisms. Phenomena of diagenesis are common: the originally aragonitic parts of the organisms are replaced with dolomite as a result of the action of moving sea water, and stylolites are also often formed. These are superficial where there has been a dissolution of the rock and, either subsequently or simultaneously, a redeposition of insoluble material (clays or iron oxides); in cross section they are thin, flat or uneven parallel lines usually on the stratification surface, pinkish or brownish in color. Other fossils may be replaced by quartz or chalcedony, or by pyrite (the pyritized ammonites of Lyme Regis, England, are well known). In many cases compact limestones are subject to farreaching chemical solution phenomena—Karst topography—made simpler by their wide-scale homogeneity, which gives rise to furrowed areas, dolinas and sink-holes. In such cases the surface of the limestone is covered by red limonitic or reddish-yellow bauxitic crusts of residual origin.
**Occurrence**   A typical compact limestone with a very fine grain size, of pelagic origin, is the "lithographic stone" of Solenhofen (Bavaria, Germany), which contains magnificent fossil remains preserving very fine details, even of the softer parts. This rock can be classified as a calcilutite. Calcilutites have very fine grain size, are usually pale gray in color (but are often darker) and have an almost conchoidal fracture. They are very well developed in the Ordovician of the Appalachians (USA) and include the Lowville Limestone and the Mosheim Limestone. Oolitic limestone, which is made up of tiny concentrically

▲ Compact limestone (ca. ×1). Varese, Italy.

▲ Polished slab of red compact limestone with stylolites ("Holla red") (ca. ×1). Vicenza, Italy.

layered spheres resembling fish roe (if the spheres are pea sized the rock is called pisolitic limestone), is very well developed in the Jurassic of England. It is represented by the Middle Jurassic Inferior Oolite and Great Oolite Series, which is very common in the Cotswolds. Pellet limestones, which are thought to be fecal in origin, are very well developed in the Devonian Nubrigyn limestones of New South Wales (Australia). They are well developed in the Ordovician limestones of the Appalachians. Coarser grained compact limestones are calcarenites. In the USA they are common and are found in the Ordovician and Pennsylvanian limestones of New Mexico and the Upper Ordovician of Idaho. They are particularly common in late Precambrian and early Paleozoic rocks, e.g., Beltian rocks of the Rocky Mountains and Beekmantown Limestone and equivalents in the Appalachian Mountains. The coarsest grained limestones are the limestone conglomerates. In the New Red Sandstone of Britain the marginal deposits are formed by the erosion of the underlying Carboniferous Limestone Series (South Wales and the Mendips).

**Uses** Although from a petrographic point of view the term "marble" only indicates metamorphosed calcareous rocks, i.e., recrystallized in conditions differing from the depositional conditions, commercially speaking this term is used for all calcareous rocks which can be turned into slabs and polished. This is perhaps the widest use to which compact limestone is put, in addition to its use as a building stone in the form of unfinished blocks. Its use as lime and crushed stone is becoming increasingly rare, these being more easily obtained from incoherent deposits of fragments of limestone (talus) or other rocks. Some varieties which have beautiful arborescent and ruin-shaped forms are used as ornamental stones in jewelry.

▲ Polished slab of compact limestone ("Siena yellow") (ca. ×1). Val d'Elsa hills, Siena, Italy.

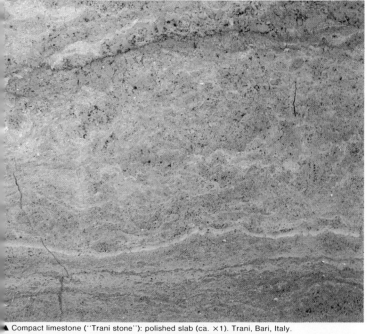

▲ Compact limestone ("Trani stone"): polished slab (ca. ×1). Trani, Bari, Italy.

## 340 DOLOMITE

**Type**  Organogenetic sedimentary rock.
**Class**  Calcareous rock (dolomitic variety).
**Appearance**  Color light, grayish, yellowish, pinkish; smell sometimes fetid when struck; grain size from minute to saccharoidal, grading to vacuolar in structure; stratification absent or very irregular. Fossils frequent. Does not effervesce with cold dilute hydrochloric acid.
**Components**  Dolomite and calcite, with traces of clay, quartz, pyrite, marcasite and bitumen.
**Geological environment**  Primary dolomites are rare, formed by direct precipitation from sea water; most are due to metasomatic substitution during the diagenesis of aragonite, present mainly in fossil shells, by magnesium in solution in sea water and in brackish lagoons or lakes in inland continental areas.
**Occurrence**  The classic occurrence is the coral and algal reefs of the Dolomites in northeast Italy. They are widespread in the Cambrian throughout the world.
**Uses**  Important as reservoir rocks for petroleum (Algeria, Iran); locally used as construction stone, less commonly in the production of crushed stone.

---

## 341 DOLOMITE (WEATHERED)

**Type**  Chemico-organogenetic sedimentary rock.
**Class**  Calcareous rocks (mixed with evaporites).
**Appearance**  Color light, ochre (yellow) or reddish; sometimes smells fetid when struck; grain size quite coarse with clear vacuolar porous structure and high permeability; stratification indistinct because of obliteration by diagenetic phenomena. Fossils rare.
**Components**  Dolomite and calcite, with residue of anhydrite and gypsum, which formed small nodules that have now been dissolved to a large extent; quartz, micas, bitumens and clays are frequent as accessories.
**Geological environment**  Mixed chemico-organogenetic origin, probably with anhydrite as the primary chemical mineral, redissolved and only partly transformed into gypsum during diagenesis; forms limited zones in reefs and coral barrier reefs, where the water is nearly stagnant.
**Occurrence**  Found in the Carboniferous Limestone of England and Wales and the Sailmhor Dolomite of Scotland. North American examples include the Bighorn Dolomite of Wyoming and the Niagara Dolomite. It is found associated with gypsum in England, the Western Alps and the Zechstein Basin in Germany.
**Uses**  Of no commercial importance.
**Note**  A term commonly used in Alpine geology.

▲ Dolomite (ca. ×1). Arona, Novara, Italy.

▲ Dolomite (weathered) (ca. ×1). Piedmont, Italy.

## 342 DIATOMITE

**Type**  Organogenetic sedimentary rock.
**Class**  Siliceous rocks.
**Appearance**  Color white, yellowish, gray: they are incoherent, mealy and extremely porous, light, and very fine grained; they form banks and limited units intercalated with clays.
**Components**  Frustules of diatoms (single-celled algae) sometimes mixed with shells of radiolarians, spicules of sponges and foraminifera; traces of sulfides.
**Geological environment**  Lacustrine and marine deposits of single organisms developed in silica-rich water, probably of volcanic origin, and far removed from any kind of clastic deposition.

**Occurrence**  "Fossil flour" is found in Denmark, Germany ("Kieselguhr"), Brazil and the USA. Diatomaceous earth and Tripoli-powder (Italy) is analogous. "Spongilite," formed from the accumulation of sponge spicules in compact dark-gray banks, is found in Great Britain among siliceous rocks of the Carboniferous, Jurassic and Cretaceous (Lower and Upper Greensands). The Miocene rocks of the California Coast Range (USA) contain abundant diatomite.

**Uses**  Fossil flour is used as a filter to remove impurities and discoloration from liquids, as an abrasive to polish soft metals, and, when mixed with nitroglycerine, as an absorbent in the production of dynamite.

---

## 343 FLINTS AND RADIOLARITES

**Type**  Organogenetic sedimentary rocks.
**Class**  Siliceous rocks.
**Appearance**  Flints are colored white, gray, red, black and sometimes zoned in color; texture fine and compact with conchoidal or shiny scaly fracture; they form kidneys, lenses, and thin beds in compact calcareous rocks. Radiolarites are hard, very compact rocks within thin-bedded strata, and are red, black or green in color.
**Components**  Chalcedony, quartz, opal with residue of radiolarians, sponges, and diatoms: traces of calcite, hematite and various iron oxides, bituminous or carbonaceous substances and sometimes gypsum.

**Geological environment**  Derived from the diagenesis of organogenetic siliceous rocks, especially with radiolarians, deposited in marine pelagic or lacustrine environments.
**Occurrence**  Flint is ubiquitous throughout the Upper Cretaceous White Chalk of western Europe and particularly of southeast England. Radiolarites are forming at the present day in the abyssal depths of the Indian and Pacific oceans (radiolarian ooze). Carboniferous radiolarians are found in the Millstone Grit (Wales) and in Devon (England) and the Rhine Valley (Germany). Jurassic radiolarites are represented by the Franciscan cherts of the Coast Range of California and Oregon (USA).

**Uses**  Radiolarites with medium to fine grain size are sometimes used as whetstones.

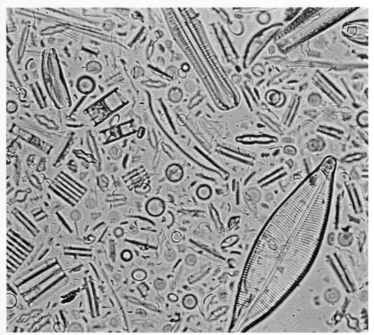

▲ Diatomite: photomicrograph (ca. ×200). Mt. Amiata, Grosseto, Italy.

▲ Flintstone in limestone (ca. ×1). Gavarno, Bergamo, Italy.

## 344 PHOSPHORITE

**Type** Organogenetic sedimentary rocks.
**Class** Phosphatic rocks.
**Appearance** Color yellowish, reddish, brown; texture vacuolar and porous, often oolitic, sometimes with carbonatic cement and flint nucleus; they form grains, pockets and nodules in clayey, sandstone, calcareous and glauconitic rocks; more rarely stratified horizons, several meters thick.
**Components** Apatite, sometimes with fibrous and semi-amorphous varieties (collophane), calcite, clays, limonite, bitumen, shells of foraminifera and various types of fossil remains; accessories include many phosphates, arsenates and molybdates, sometimes brightly colored.
**Geological environment** Continental and marine deposits due to the accumulation of coprolites (guano) or the bones of fishes and vertebrates. Formation by the precipitation of phosphorus salts directly from sea water (nodules of phosphates currently in the oceans) is much rarer.
**Occurrence** Large stratified deposits are found in Morocco, Algeria, Tunisia and the Spanish Sahara and especially the Permian Phosphoria Formation of Idaho (USA). Other deposits are found in Tennessee and California (USA), France, China and Siberia. Deposits containing bones are often rich in phosphate as are those of Florida (USA) and the Rhaetic Bone Bed of southwestern England.
**Uses** Fairly important for the production of phosphatic fertilizers.

## 345 BOG IRON ORE

**Type** Organogenetic sedimentary rock.
**Class** Ferriferous rocks.
**Appearance** Color reddish, blackish, yellowish; compact with fine grain size or, more commonly, oolitic porous texture in which the spherules are consolidated by clays or limonite. Frequently contains animal and vegetable fossil remains.
**Components** Limonite, siderite, chamosite (ferriferous chlorite), goethite, hematite are the main minerals, associated with phosphates, calcite, montmorillonite, illite, etc.
**Geological environment** Deposited by lake water, lagoon water or sea water with restricted movement by bacteria and subsequently profoundly transformed during diagenesis.
**Occurrence** The true bog iron ores are found in terrestrial sediments in Scandinavia. Other sedimentary iron ores are the important Mesozoic economic deposits of Alsace-Lorraine (France) and Luxemburg, which are predominantly chamositic (minette).
**Uses** Important raw material for extraction of iron.

▲ Phosphorite (ca. ×1). Algeria.

▲ Bog iron ore (ca. ×1). Lorraine, France.

## 346  PEAT

**Type**  Organogenetic sedimentary rock.
**Class**  Coal.
**Appearance**  Color light, brownish, very variable from place to place; structure feltlike in which plants are still clearly visible; forms stratified deposits alternating with clays and sand, mainly in very limited depressed continental areas.
**Components**  Vegetable remains, slightly transformed (50–60 percent carbon), mixed with various detrital material (quartz, calcite), animal shells, diagenetic minerals and solid hydrocarbons (simonellite, amber).
**Geological environment**  Fill and compacting of marshes, ponds, inframorainic lakes or lakes carved out by Quaternary glaciers.
**Occurrence**  Peat is most common in northern Europe (Scotland, Ireland, Germany, Denmark, Poland) and North America (Canada, USA). It is restricted to areas having a temperate or subarctic climate.
**Uses**  A poor fuel (heat output is very low); requires preliminary drying and produces annoying ammoniac fumes. Mixed with loam it is used in horticulture.

---

## 347  ANTHRACITE

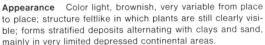

**Type**  Organogenetic sedimentary rock.
**Class**  Coal.
**Appearance**  Color dark brown to black, shiny, hard, scaly and compact, sometimes reticular in appearance produced by opaque and fine scaly zones intersected by glassy-looking or horny-looking veins.
**Components**  Fixed carbon (93–98 percent), with scattered volatile substances (light hydrocarbons: 0–6.5 percent); small amounts of pyrite and quartz.
**Geological environment**  Strata forming large elongated basins corresponding to lagoons and coastal lakes, in slowly subsiding zones. Often part of a repeated series, alternating with sandstone and clay (cyclothems). Originally produced as bituminous coal, and later transformed to more carbon-rich anthracite.
**Occurrence**  Restricted to Paleozoic rocks, in particular the Carboniferous of North America (Appalachian Mountains, Pennsylvania) and Europe (England, France, Belgium, Germany) and the Permian of China, Siberia and Australia.
**Uses**  Good fuel (7000–8000 calories per kilogram); it is used in the production of coal gas, in the iron and steel industry, the synthetic rubber industry, the dye industry, etc.

▲ Peat (ca. ×1). Leffe, Bergamo, Italy.

▲ Anthracite coal (ca. ×1). Wales.

## 348  VOLCANIC BOMB

**Type**  Pyroclastic rock.
**Chemistry**  Intermediate or mafic.
**Components**  Glass; plagioclase (labradorite-anorthite), clinopyroxene, olivine, magnetite, ilmenite.
**Appearance**  Color dark, brownish, greenish or black; texture from granular to hyaloclastic; structure massive, obsidianlike externally, vacuolar internally. Frequently spindle-shaped in appearance, due to rotation during flight, or with a "bread crust" surface (see photo), grooved with cracks due to sudden degasification.
**Geotectonic environment**  Produced by the violent expulsion of large blobs of lava during the initial phases of an explosive eruption. When consolidated forms coherent rocks called volcanic breccia and tuff-breccia.
**Occurrence**  The tapered bombs are common from volcanos with fluid lava (Etna, Vesuvius). Those resembling a crust of bread are typical of more felsic and hence more viscous lavas (Vulcano, Stromboli, Mt. Pelée).
**Uses**  Of no practical importance.

---

## 349  IGNIMBRITE

**Type**  Pyroclastic rock.
**Chemistry**  Felsic.
**Components**  Crystals of quartz, alkaline feldspar (sanidine, albite), biotite and sometimes feldspathoids in a matrix composed mainly of consolidated glass (in part devitrified). Often contain fragments of sedimentary rocks.
**Appearance**  Color light, gray or brownish, sometimes reddish or violet due to oxidization; texture clastic, very consolidated; structure massive, with columnar fissuring and abundant pneumatolytic druses, sometimes with flow phenomena made evident by alignment of bubbles or glass shards.
**Geotectonic environment**  Deposits of hot incandescently glowing tuff (ash), having rolled down slopes and been derived from highly explosive volcanic activity ("eruption clouds" or "nuée ardente"). Consolidated on the spot because of the plasticity of the glass and the presence of residual gases and heat.
**Occurrence**  Widespread, especially in New Zealand, Asia (Kamchatka Peninsula), Armenia and the Caribbean. Forming at the present day from the active volcanos Bezymianny (Siberia), Katmai (Alaska) and Mt. Pelée (Martinique).
**Uses**  Locally used as building material, especially when compact.

▲ "Bread crust" bomb (ca. ×1). Vulcano, Aeolian Islands, Italy.

▲ Rhyolitic ignimbrite (ca. ×1). Val Trompia, Brescia, Italy.

## 350 PORPHYRITIC TUFF

**Type**  Pyroclastic rock.

**Chemistry**  Felsic or intermediate.

**Components**  Together with individual crystals of augite, plagioclase and olivine, they often contain bubbly lava fragments (lapilli), volcanic ash and sedimentary material which helps cementation (zeolites, calcite) or glass.

**Appearance**  Color light gray, pink, greenish, yellowish or brown; texture markedly clastic, with lapilli and sometimes also sedimentary rocks torn from the volcanic pipe or dislocated during rolling downslope following the volcanic explosion, mixed with ash and partly cemented by authigenic minerals (zeolites) and in part by sedimentary material (clays and calcite).

**Geotectonic environment**  Interstratified with lava in volcanos of a mixed explosive-eruptive type; large deposits also present at some distance from volcanic centers, in explosive zones.

**Occurrence**  Associated with the lavas of Vesuvius (Italy). Also present in ancient lavas as in the Ordovician volcanics of the Lake District (England) and North Wales. It is also found in Greece, Turkey, Germany, Indonesia, Japan and the circum-Pacific belt of the Americas.

**Uses**  Some tuffaceous coherent rocks are used locally as construction stone and as material for the production of special cements.

---

## 351 PEPERINO TUFF

**Type**  Pyroclastic rock.

**Chemistry**  Intermediate.

**Components**  Prevalence of black crystals of augite and biotite, with scattered feldspars and abundant leucite, associated with lapilli of varying composition and calcareous small pebbles, in a scanty cement which is mainly authigenic (zeolites).

**Appearance**  Color gray or brown, speckled with black (peppery); texture clastic and well consolidated, wtih medium to coarse grain size; often grading to breccia because of the presence of "projectiles" containing well crystallized minerals or dissolved crystals.

**Geotectonic environment**  Extensive units produced by the explosive activities of trachytic or tephritic-leucitic volcanism.

**Occurrence**  Characteristic of central Italy (Roman comagmatic region) and in particular of Colli Albani.

**Uses**  Stone for light construction, solid and very easy to work.

▲ Porphyritic tuff (ca. ×1). Arona, Novara, Italy.

▲ Peperino tuff (ca. ×1). Lazio, Italy.

# METAMORPHIC ROCKS

## KEY TO THE SYMBOLS

 STRUCTURE

massive

 slightly schistose

 very schistose

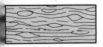 plicated schistose

 augen gneiss

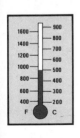 PRESUMED TEMPERATURE OF RECRYSTALLIZATION

 PRESUMED PRESSURE OF RECRYSTALLIZATION

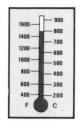

**Type**   Regional metamorphic rock.

**Chemistry**   Siliceous arenaceous.

**Components**   *Essentials*: quartz. *Accessories*: mica (muscovite, biotite, fuchsite), feldspar (orthoclase, microcline, albite plagioclase), "heavy minerals" of detrital origin (apatite, zircon, pyrite, magnetite, ilmenite, etc.). *Accidentals*: garnet, graphite, calcite, sulfides, etc.

**Appearance**   Color very white if composed exclusively of quartz, fairly patchy gray if other minerals are present, to black if there is abundant biotite, graphite or magnetite dispersed as very tiny grains; texture granoblastic, mosaic-type, with mainly minute grain size, sometimes also saccharoidal; structure massive, foliated or schistose depending on the abundance of mica, grading in some cases to "arenaceous schists" or "quartz schists."

**Metamorphic grade**   Present in metamorphic series of every grade, given that quartz is stable throughout the metamorphic range. Other components vary, depending on the conditions of temperature and pressure.

**Genetic environment**   Derived from arenaceous, fairly pure, sedimentary rocks (orthoquartzite, graywacke, arkose), from quartz-rich siltstones, jaspers and flints; can also be derived from aplites and pegmatites.

**Occurrence**   Widespread in metamorphic series derived from sedimentary rocks of every type. Common in the Scottish Highlands. Precambrian Scottish quartzites are present in the Moine Series in northeast Scotland and in the Dalradian Series. A typical quartzite is shown in thin section (see photomicrograph). It contains interlocking and oriented laminae of light-colored mica with vivid interference colors. The itabirites of the USA (North and South Carolina) and Brazil (Minas Gerais) are large masses of quartzite impregnated with iron minerals (hematite, magnetite). In India (Deccan), Guyana and Venezuela, quartzite is associated with manganese minerals (gondites).

**Uses**   Supplies construction material, often used for flooring or facing, both worked and polished; also used in the glass and ceramics industries, and in the manufacture of refractories.

▲ Quartzite (ca. ×1). Barge, Cuneo, Italy.

▲ Quartzite (Sondrio, Italy): photomicrograph of a thin section (ca. ×20; crossed Nicols).

**Type** Regional metamorphic rock.

**Chemistry** Pelitic.

**Components** *Essentials*: quartz, sericite mica, chlorite. *Accessories*: albite, apatite, tourmaline, pyrite, magnetite hematite, ilmenite, graphite. *Accidentals*: manganiferous garnet, chloritoid, stilpnomelane, carbonates (calcite, ankerite), epidote, pyrophyllite, etc.

**Appearance** Color light, silvery gray, lead gray or greenish; texture from granoblastic (quartzose phyllites) to markedly lepidoblastic, grading into porphyroblastic (albitic and garnetiferous phyllites). Grain size very fine, with single micaceous lamellae irresolvable with the naked eye. Schistosity is very accentuated, often banded alternately granoblastic and lepidoblastic, commonly wavy.

**Metamorphic grade** Low, up to greenschist facies. Many phyllites are the result of retrograde metamorphism of rocks of a higher grade; as a result they contain relics of high temperature minerals and medium to high pressure (phyllonites).

**Geotectonic environment** Developed from clayey or clayey-to-sandy sedimentary rocks with a residue of organic material, especially vegetable, that is transformed into graphite. The phyllonites may be derived from many rock types, in particular from mica schist and gneiss, because of the effect of retrograde metamorphism.

**Occurrence** Very common in the Eastern Alps (Bressanone phyllites), less so in the Central Alps (Ambria phyllites), especially in the southern alpine basement. Phyllonites are found in all areas with nappes, above all in the Central Alps (Bormio phyllites) and in the Moine Thrust (northwest Scotland). They are found commonly in New England (USA). Other localities include Devon (England), Rhineland Massif (Germany), Ardennes (Belgium), Landeck (Austria), Northern Ireland and Czechoslovakia. As a rule, phyllites occur in all metamorphosed Paleozoic areas. In thin section (see photomicrograph) wavy bands of muscovite (with vivid interference colors) dotted with porphyroblasts of garnet with a bluish chloritic aureole, alternating with discontinuous layers of microgranular gray quartz, depending on the characteristic texture of the alternating bands which are typical of many southern Alpine quartzose phyllites.

**Uses** Rarely, as slabs for covering roofs in local buildings.

▲ Quartziferous phyllite (ca. ×1). Sondrio, Italy.

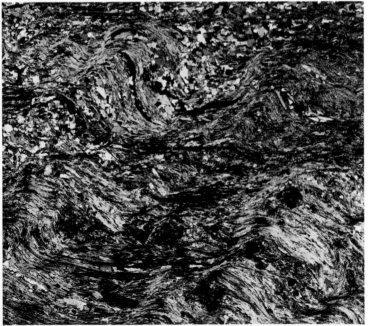

▲ Phyllite (Sondrio, Italy): photomicrograph of a thin section (ca. ×20; crossed Nicols).

## 354 MICA SCHIST

**Type**    Regional metamorphic rock.

**Chemistry**    Pelitic.

**Components**    *Essentials*: quartz, mica (biotite, muscovite, phengite, paragonite). *Accessories*: chlorite, apatite, tourmaline, zircon, pyrite, ilmenite, magnetite, graphite. *Accidentals*: andalusite, cordierite, kyanite, sillimanite, garnet (almandine), staurolite, epidote, calcite, plagioclase, glaucophane, etc.

**Appearance**    Color variable, depending on type of mica; silvery or gray for muscovitic types, brown or black for biotitic types; texture granoblastic and lepidoblastic, often in alternating beds, grading into porphyroblastic (garnet, kyanite, staurolite, andalusite) and poikiloblastic (plagioclase, cordierite, staurolite, etc.); grain size very variable, from very fine to markedly coarse, with micas clearly distinguishable to the naked eye (the criterion used to distinguish them from the phyllites, in which a single micaceous lamella can only be detected microscopically); very evident schistosity; flat or slightly wavy, often with well developed lineation. Distinguished from gneiss by the thickness of the laminae, which are thin in the case of the mica schists and thick in gneisses.

**Metamorphic grade**    From medium to high (greenschist facies, amphibolites, glaucophane schists). In the mica schists from areas that have been regionally metamorphosed one can find a characteristic zonal distribution, due to a regular relationship between temperature and pressure and the presence of certain indicator minerals of a restricted metamorphic grade, associated with others which are not indicative. In the Scottish Highlands we find the sequence chlorite-biotite-garnet-staurolite-kyanite-sillimanite, and a regular change of the rock types from phyllite to mica schist with gradually increasing grain size. In the Abakuma (Japan) region the sequence is conversely biotite-andalusite-cordierite-sillimanite-orthoclase. The Scottish Highlands sequence indicates a uniformly medium- to high-pressure environment as temperature increases; in the Abakuma sequence there is consistently low pressure and increasing temperature.

**Genetic environment**    Mica schists are derived from clayey and clayey-arenaceous and also slightly calcareous rocks, and are often associated with gneisses if the original rocks were an alternation of shales and arkoses. They are very common in "basement" areas of mountain chains with markedly eroded folds.

**Occurrence**    Very common in all areas of regional metamorphism throughout the world. Distinctive examples are the Alps (especially west and central), the Moine Schists and Dalradian Schists of Scotland, the schists of the eastern USA (New England to Georgia), the Otago and Westland Schists of New Zealand and the Sanbagawa Schists of Japan. In the Alps the commonest variety is garnetiferous mica schist characterized by the presence of garnet crystals varying in size from 1 mm to 5 cm (.04–2 in), sometimes with staurolite and kyanite, well known in many areas and especially the upper reaches of Lake Como and Val Malenco (Sondrio, Italy). In thin section (see

Garnetiferous mica schist (ca. ×1). Val Passiria, Bolzano, Italy.

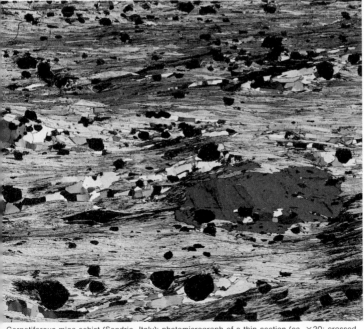

Garnetiferous mica schist (Sondrio, Italy): photomicrograph of a thin section (ca. ×20; crossed cols).

photomicrograph), there are roundish crystals of garnet (dark in the cross polarized light), poikiloblasts of gray albite and lenticles of light quartz. *Sillimanite mica schist,* sometimes cordierite-bearing, outcrops in the Valle della Mera (Sondrio); *amphibole-bearing mica schist,* derived from the metamorphism of dolomitic rocks occurs in St. Gotthard Pass; *tourmaline mica schist* occurs in the eastern Alps (Tyrol). *Paragonite mica schist,* with kyanite and staurolite (Ticino, Switzerland), is also very distinctive; other types of mica schists are limited to compositions rich in a particular component, such as *chloritoid schist,* typical of ferruginous rocks, *ottrelite schist* typical of manganiferous rocks and *graphitic schist,* typical of rocks originally containing vegetable remains transformed into this mineral. Mica schists are among the most common rocks of the crystalline schistose shields of Canada (Grenville Belt extending into the northeastern USA), Norway and Sweden (the Caledonides) and Australia. They are also very common in California (USA), Chile and Japan in association with glaucophane schists. In the latter region only the presence of sporadic minerals, such as glaucophane, chloritoid and jadeite, makes it possible to distinguish the Mesozoic mica schists from the much older ones of the Abakuma region, which contain cordierite.

**Uses**   Rarely used in buildings, especially as slabs covering roofs; sometimes used as gravel.

---

## 355   CHLORITE SCHIST

**Type**   Regional metamorphic rock.
**Chemistry**   Mafic.
**Components**   *Essentials*: chlorite. *Accessories*: pyrite, calcite, magnetite, ilmenite, titanite. *Accidentals*: albite, epidote, actinolite, glaucophane, quartz, muscovite, magnesite, talc, chloritoid.
**Appearance**   Color fairly light green; texture from lepidoblastic to feltlike (chloritite) grading to poikiloblastic or nematoblastic; structure massive or schistose, often with mineralogical and structural relics.
**Metamorphic grade**   Low (greenschist facies and glaucophane schists).
**Genetic environment**   Derived from lava, tuff, mafic rocks and thus with different textures and frequently relict structures. Sometimes derived from marl mixed with tufaceous material.
**Occurrence**   Found in the early Paleozoic schists of New Hampshire and Vermont, the Mother Lode belt of California and in North Carolina (USA). The Dalradian Schists of central Scotland contain much chlorite schist. It is also widespread in the Western Alps in association with prasinite.
**Uses**   Used as unfinished slabs, as roof covering in Alpine regions; "potstone" can be worked on the lathe to make pots and other rustic crockery, and is the raw material for decorative objects made by local craftsmen.

▲ Mica schist with two kinds of mica (ca. ×1). Val dei Ratti, Sondrio, Italy.

▲ Chlorite schist (ca. ×1). Austria.

**Type** Regional metamorphic rock.

**Chemistry** Felsic (from arenaceous to pelitic-arenaceous).

**Components** *Essentials*: feldspar (microcline, perthitic microcline, albite or plagioclase of varied composition), mica (muscovite, phengite, biotite). *Accessories*: epidote, apatite, tourmaline, allanite, magnetite, ilmenite, zircon, monazite, titanite, pyrite, pyrrhotite. *Accidentals*: quartz, chlorite, kyanite, sillimanite, andalusite, cordierite, garnet (almandine), hornblende, augite.

**Appearance** Color is usually light in rock types derived from granites (*orthogneiss*) or from fairly pure feldspathic arenaceous rocks (arkose), grading into fairly dark rocks which contain a considerable pelitic proportion derived from impure sandstones (clayey sandstones, graywacke, etc.). Texture prevalently granoblastic with medium or coarse grain size, grading to porphyroblastic and poikiloblastic (*augen gneiss*), or even nematoblastic (*striped* or *lineated gneiss*); schistosity poorly developed and only in types rich in mica with a fairly fine grain size (*microgranular* and tabular gneiss). What distinguishes a mica schist from a gneiss is not the mineralogical composition, although gneiss is usually richer in feldspar, but the type of fracturing: gneiss breaks into coarser pieces and also breaks with difficulty into small cubes or thick slabs, whereas mica schist fractures easily and into thin units because of its greater schistosity.

**Metamorphic grade** From medium to high (amphibolite facies, up to migmatites); some gneiss is also low grade (albitic gneiss).

**Genetic environment** Most gneiss forms from sedimentary rocks (*paragneiss*) and is formed in a prograde metamorphic environment from schists with dehydrated muscovite and which often transform into feldspar and aluminum silicates. Gneiss thus represents a very advanced stage of the transformation of fine-grained rocks, and represents the stage immediately prior to the start of anatectic remelting which gives rise to migmatites. So these are rocks of high metamorphic grade. Other gneisses are developed from the metamorphism of granitic, granodioritic and tonalitic rocks (*orthogneisses*) and sometimes show relict textures. In this case they may also be of medium or even fairly low metamorphic grade, because they maintain the gneissic composition and texture for primary compositional reasons, while indicating dehydration occurring at high temperatures.

**Occurrence** Orthogneiss of granitoid origin is present throughout the Alps as a product of Alpine metamorphism at the expense of ancient plutonic rocks and migmatites. Augen gneiss includes those of the Antigorio and Monte Leone strata and the Gran Paradiso, Monte Rosa and Mont Blanc massifs as well as those of Tavern, sometimes containing hornblende as well as biotite. In thin section it shows (see photomicrograph) porphyroblasts of perthitic microcline and muscovite (with vivid interference colors), which stand out in a fine-grained ground-

▲ Augen gneiss (ca. ×1). Sondrio, Italy.

▲ As above: photomicrograph of a thin section (ca. ×20; crossed Nicols).

mass consisting of quartz, plagioclase and muscovite. Tabular gneiss includes that of Beura (Val d'Ossola), which is rich in tourmaline. Light-colored augen gneiss is found in the southern Alps. The paragneiss of the Sesia-Lanzo area (Piedmont), the garnetiferous-sillimanitic gneiss of the Lake Series (Piedmont-Lombardy) as well as the conglomeratic gneiss, often with relict pebbles which are still evident, of the Lebendum strata in Val Formazza (Novara) and La Thuile (Val d'Aosta) are all of sedimentary origin. Streaked and lineated gneiss, of doubtful origin, is found in the Monte Rosa strata in Val d'Ossola ("silver stone") and in the Sila ("white schists" of Tiriolo, Catanzaro). Locally migmatitic and amphibole-bearing gneiss is found in Aspromonte and in the Peloritani. The gneiss of Morbegno (Sondrio) is nodular and poikiloblastic, rich in albite and sometimes contains two micas. In other parts of the world gneissic terraces are found in the Precambrian shields of Scandinavia, Canada, Brazil and Australia, where they often grade into migmatites and anatectic granites, without leaving a trace of the distinction between these rock types. In the Vosges (France) and in the Black Forest (Germany) biotite gneisses rich in garnet and cordierite ("kinzigites") constitute the deepest level of a pelitic series which is differentiated on the one hand into migmatites and on the other into typical granulites. In the USA gneiss is well developed in the northwestern Adirondacks of New York and in New England and Georgia. It also occurs in the Rocky Mountains. Other localities include Fiordland in southwest New Zealand, the Central Highlands, Sri Lanka and the "high temperature belt" (Ryoke-Abakuma) of Japan. The "Ollo de sapo" of Spain and Portugal is distinctive. These blastomylonitic gneisses are found occasionally in the central-southern Alps as well.

**Uses**  Some gneiss is used as a building material, both as rough stone and as polished slabs; no longer much used in the form of slabs for road pavements or as blocks for enclosing stairways or flowerbeds.

▲ Fine-grained gneiss (ca. ×1). Val Masino, Sondrio, Italy.

▲ Gneiss (ca. ×0.5). Val d'Ossola, Novara, Italy.

**PRASINITE OR CHLORITE-AMPHIBOLE-EPIDOTE-ALBITE SCHIST**

**Type**   Regional metamorphic rock.

**Chemistry**   Mafic.

**Components**   *Essentials*: chlorite, actinolite, albite, epidote. *Accessories*: calcite, quartz, titanite, rutile, muscovite, phengite, magnetite, ilmenite. *Accidentals*: glaucophane, garnet, hornblende, lawsonite, ankerite.

**Appearance:**   Color light green, grading into yellowish or bluish; texture variable, mainly nematoblastic and lepidoblastic, with albite in small poikiloblastic ocelli ("ocellar albite"): grain size mainly very fine; structure massive or zoned with alternate bands enriched by a particular component, never schistose. Relict gabbroic or diabasic fragments may be found.

**Metamorphic grade**   Rocks present in the mafic series of low to medium grade (greenschist facies and glaucophane schists).

**Genetic environment**   Derived by retrograde metamorphism of greenschist facies, of glaucophane schist rocks, or by prograde metamorphism in a hydrated environment containing basaltic lava, diabase and gabbro. To a small extent may also be derived from tuff or ferruginous marl. Much of the prasinite in the Alps is polymetamorphic rock formed from masses of mafic rocks that have been transformed, first into glaucophane schist, during a period of compression at a low temperature, and later into prasinite, during a later period of pressure release with or without a limited rise in temperature.

**Occurrence**   Prasinite is characteristic of the western and eastern Alps in the areas of calc-schists and greenstones. It is found in Italy in Val di Susa, Val Soana (Turin), Val Varaita (Cuneo), in the Voltri Group (Alessandria) and in Valtellina, where there are prasinitic types containing (see photomicrograph of thin section) albite with ocellar poikiloblasts, prismatic crystals of amphibole and epidote with yellow and blue interference colors and muscovite and chlorite in small sheets. It is well known in eastern Corsica, in the Bagnes Valley (Vallese, Switzerland) and in the Zillertal (Austria). The name "prasinite" is used exclusively for Alpine rocks. Outside Europe they are known as chlorite-amphibole-epidote-albite schists in California, Japan, the Philippines, Venezuela, etc.

**Uses**   Sometimes associated with copper deposits.

▲ Prasinite (ca. ×1). Val Varaita, Cuneo, Italy.

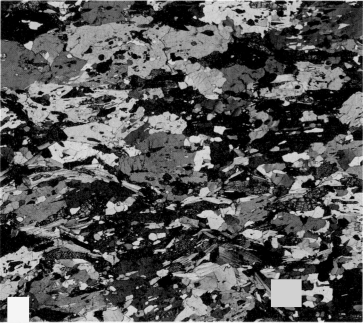

▲ Prasinite (Val Malenco, Sondrio, Italy): photomicrograph of a thin section (ca. ×15; crossed Nicols).

## 358    TALC SCHIST

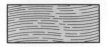

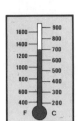

**Type**   Regional metamorphic rock.

**Chemistry**   Ultramafic.

**Components**   *Essentials*: talc. *Accessories*: magnetite, magnesite, calcite, dolomite, chlorite. *Accidentals*: brucite, serpentine (antigorite), tremolite.

**Appearance**   Color grayish-white, sometimes speckled with green; greasy to the touch; texture lepidoblastic with fine-grained granoblastic nodules; schistosity evident, often surrounding relics of olivine, pyroxene and carbonate. A feltlike variety is *steatite*.

**Metamorphic grade**   Regional metamorphism of low grade at the expense of ultramafic rocks; thermal contact metamorphism at low pressure of dolomitic rocks.

**Genetic environment**   Derived by the transformation of olivine to talc in a water-rich environment at low to medium temperatures, or by the silicification of magnesite and dolomite at high temperature, in the presence of water.

**Occurrence**   Large masses are found in the Appalachians, Vermont, Georgia and California (USA). Other occurrences are found in the Central and Eastern Alps, the Pyrenees, Tuscany (Italy), Sweden, USSR, Australia, Korea and India.

**Uses**   The main source of industrial talc; see under "minerals" for uses.

---

## 359    SERPENTINE

**Type**   Regional metamorphic rock.

**Chemistry**   Ultramafic.

**Components**   *Essentials*: serpentine (chrysotile, lizardite). *Accessories*: magnetite, magnesite, talc. *Accidentals*: brucite, tremolite, dolomite, chlorite.

**Appearance**   Color light green or yellowish green, variegated; texture cellular and feltlike with distinctive veining; structure massive with frequent blastomylonitic peridotitic relics.

**Metamorphic grade**   Low, limited to the zeolite facies and to the upper part of the greenschist facies.

**Genetic environment**   Derived by the transformation of olivine and pyroxene in peridotite to serpentine, in a predominantly static environment. Often contains relics of the original mineralogy.

**Occurrence**   Serpentine (see photo) is found in the Tuscan Appennines, Ligure and Emiliano (Italy).

**Uses**   Sometimes associated with small copper deposits; often also used in buildings as polished slabs for facing.

**Note**   In order to distinguish between massive lizardite-bearing Appennine rocks with relict structures, from antigorite-bearing Alpine rocks which are often schistose and totally recrystallized, there is a tendency in Italy to use the name *serpentine* for the former and the name *serpentinite* for the latter. In North America the term serpentine is used for a mineral group, and serpentinite is for the rock group regardless of the type of dominant serpentine mineral.

▲ Talc schist (ca. ×1). Val Malenco, Sondrio, Italy.

▲ "Serpentine" (ca. ×1). Tuscany, Italy.

**Type**   Regional metamorphic rock.

**Chemistry**   Ultramafic.

**Components**   *Essentials*: serpentine (antigorite, lizardite), magnetite. *Accessories*: serpentine in veins (chrysotile) often in the form of asbestos, talc, garnierite, brucite, chlorite. *Accidentals*: magnesite, dolomite, calcite, tremolite, garnet (demantoid, hessonite), relics of various types (olivine, pyroxene).

**Appearance**   Color from dark green to black; texture lamellar or feltlike, with frequent zonations, sometimes cellular because of the presence of relics, especially of pyroxene (bastite); structure massive, with intercalations of veins and rodingitic seams and enrichment in iron, copper and nickel minerals.

**Metamorphic grade**   Low (greenschist facies, glaucophane schists and, in part, zeolite facies); sometimes also only partial recrystallization, resulting in rocks grading into the amphibolite facies.

**Genetic environment**   Derived from the metamorphism of peridotite, pyroxenite and lherzolite, sometimes also of amphibolite and gabbro, in a water-rich environment. During transformation, segregations are formed giving rise to lenses of rodingite, and magnetite crystals are also developed.

**Occurrence**   Widespread in all orogenic belts. It is common in the Alps, where it contains antigorite and lizardite. In Great Britain the best known occurrence is the Lizard Complex in Cornwall. In the USA serpentinite is found in the Stillwater Complex in Montana and in Oregon, California and Maine. A notable occurrence is the Great Serpentinite Belt, which stretches along the eastern coast of New South Wales (Australia). Serpentinite also occurs in Québec (Canada), Cuba, New Caledonia, New Zealand, Norway, Yugoslavia, and in the Himalayas. The thin section (see photomicrograph) shows bluish serpentinite with minute crystals of magnetite and relict or recrystallized olivine and pyroxene having vivid interference colors. Chrysotile serpentinite commonly contains veins of asbestos.

**Uses**   Serpentinite is associated with important deposits of copper, iron, nickel, asbestos, "potstone," talc, etc. When split into thin slabs it is used in buildings for covering roofs and outside facing; when cut into slabs and polished it makes a very effective ornamental material, although it does not stand up for a long time to the weather.

Serpentinite (ca. ×1). Val Malenco, Sondrio, Italy.

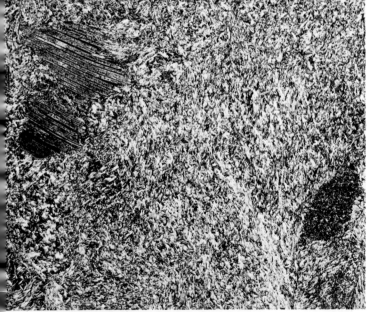

As above: photomicrograph of a thin section (ca. ×20; crossed Nicols).

## 361    GLAUCOPHANE SCHIST

**Type**   Regional metamorphic rock.

**Chemistry**   Mafic.

**Components**   *Essentials*: glaucophane, lawsonite. *Accessories*: calcite, magnetite, chlorite, albite, quartz. *Accidentals*: jadeite, phengite mica, actinolite, ferrocarpholite, ankerite, garnet, chloritoid, kyanite, talc.

**Appearance**   Color green with bluish or violet hues; texture nematoblastic or lepidoblastic grading into ocellar; structure not very schistose, often with structural relicts.

**Metamorphic grade**   Low (glaucophane schist facies) and typical of high pressure.

**Genetic environment**   Derived from basaltic lava, hyaloclastites, tuff mixed to some extent with clays, and also diabase and gabbro in a high pressure metamorphic environment, associated with dynamic metamorphism.

**Occurrence**   Found in the Channel Islands and Northwest Scotland, and in the Coast Ranges of California (USA). Other localities include New Caledonia, Norway (Svalbard), Japan (Sanbagawa and Sangun) and the Western Alps, where it is a product of the initial phase of Alpine metamorphism.

**Uses**   Sometimes associated with mineral deposits of copper and nickel.

---

## 362    EPIDOTE AMPHIBOLITE

**Type**   Regional metamorphic rock.

**Chemistry**   Mafic.

**Components**   *Essentials*: amphibole (actinolite), plagioclase (albite-oligoclase), epidote (clinozoisite, pistacite). *Accessories*: titanite, apatite, magnetite, ilmenite, biotite, quartz. *Accidentals*: calcite, dolomite, chlorite.

**Appearance**   Color light green streaked with greenish yellow; texture nematoblastic grading into diablastic; structure with bands differentiated by color, grain size, orientation and mineralogical composition.

**Metamorphic grade**   Low (greenschist facies).

**Genetic environment**   Derived from lava, tuff, mafic units and sometimes also ferruginous marl recrystallized in a regional metamorphic environment or by retrograde metamorphism from glaucophane schists.

**Occurrence**   Common in the Grenville Series of the Adirondack Complex (USA), the Scottish Highlands and most other parts of the world.

**Uses**   Of no commercial interest.

▲ Glaucophane schist (ca. ×1). Norway.

▲ Epidote amphibolite (ca. ×1). Val Malenco, Sondrio, Italy.

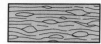

## 363 AMPHIBOLITE

**Type** Regional metamorphic rock.

**Chemistry** Mafic.

**Components** *Essentials*: amphibole (hornblende, anthophyllite, cummingtonite), plagioclase (andesine-bytownite). *Accessories*: ilmenite, magnetite, titanite, epidote, quartz. *Accidentals*: diopside, garnet (almandine-pyrope), biotite, muscovite.

**Appearance** Color fairly dark green, often patchy or speckled with white and yellow; texture diablastic or nematoblastic, grading into porphyroblastic (garnet) or lepidoblastic (biotite), sometimes poikiloblastic (plagioclase); structure massive or alternately banded, sometimes with relics of previous igneous rocks (metagabbro, meta-pillow lava). They grade, with increasing plagioclase, at the expense of amphibole, first to *leucoamphibolites,* then to *amphibole gneiss*.

**Metamorphic grade** Medium (amphibolite facies).

**Genetic environment** Derived from basaltic lava and tuff, or gabbroic and diabasic masses (*orthoamphibolite*) in a regional metamorphic environment. Some may be derived from marly sediment, probably mixed with tufaceous material (*paraamphibolite*). Leucoamphibolite and amphibole gneiss may be derived from the metamorphism of intermediate plutonic rock (tonalite, diorite) and from the metamorphic transformation of tufaceous rock. Numerous areas of amphibole-rich composition are present as inclusions in granitoid rocks.

**Occurrence** In Italy amphibolite is extensively represented in the pre-Alpine basements of the central and eastern Alps, particularly in the high Valtellina (Tirano) and in the Orobian Alps. These rocks have a typically diablastic texture with polysynthetically twinned plagioclase, amphibole in broad prisms with vivid interference colors, and with gray quartz (see photomicrograph). Occasionally it contains garnet or diopside. Abundant amphibolite is also present in the Hercynian and pre-Hercynian basement of Germany, France, England, and in the Baltic shield (Norway), as well as in the Brazilian (São Paulo) and American (Canada, USA) shields, where amphibolite often contains large garnet crystals. Relict structures indicative of an origin by metamorphism of gabbro are found in the Oetztal (Austria) and in older localities in several parts of the world.

**Uses** Often linked to copper deposits. Rarely used as an ornamental stone.

Amphibolite (ca. ×1). Val Malenco, Sondrio, Italy.

Amphibolite (Sondrio, Italy); photomicrograph of a thin section (ca. ×15; crossed Nicols).

## 364　EMBRECHITE (MIGMATITE)

**Type**　Ultrametamorphic rock.

**Chemistry**　Felsic or intermediate.

**Components**　*Essentials*: quartz, potassic feldspar (micro cline, perthite), plagioclase (oligoclase), biotite. *Accessories* zircon, apatite, magnetite. *Accidentals*: muscovite, cordierite sillimanite, garnet.

**Appearance**　Color light, white or grayish; texture gneissic striped or augen, with clear separation of folded paleosomatic beds and neosomatic patches or bands; structure massive with frequent relics.

**Metamorphic grade**　Medium-high (amphibolite facies).

**Genetic environment**　Beginning of the remelting of the felsic components of a pelitic or granitic rock in a water-saturate regional metamorphic environment.

**Occurrence**　Widespread in the large Precambrian shield (Finland; Canada; Massif Central, France; Scottish Highlands In Scotland it is found associated with the Moine Thrust from Loch Eriboll in the north southward to Skye.

**Uses**　Locally used as building stone, more rarely in the form of polished slabs as ornamental stone.

---

## 365　ANATEXITE (MIGMATITE)

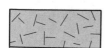

**Type**　Ultrametamorphic rock.

**Chemistry**　Felsic or intermediate.

**Components**　*Essentials*: quartz, potassic feldspar (micro cline, perthite), plagioclase (oligoclase-andesine), biotite *Accessories*: zircon, apatite, magnetite. *Accidentals*: garnet cordierite, sillimanite, amphibole.

**Appearance**　Color dark to light gray; texture granoblast grading into lepidoblastic with dark beds (paleosome), pe meated by light beds (neosome). There are relics and inclu sions present, especially rich in biotite and amphibole.

**Metamorphic grade**　High (amphibolite facies and others).

**Genetic environment**　The most advanced stage of anatex or remelting, in a regional metamorphic environment, of water-saturated pelitic rock.

**Occurrence**　Common in the Precambrian shields and Pre cambrian basements throughout the world.

**Uses**　Locally used as building stone, more rarely used a polished slabs for ornamental purposes.

▲ Embrechite (migmatite) (ca. ×1). Sondrio, Italy.

▲ Anatexite (migmatite) (ca. ×0.5). Val Codera, Sondrio, Italy.

## 366 NEBULITE (MIGMATITE)

**Type** Ultrametamorphic rock.

**Chemistry** Felsic or intermediate.

**Components** *Essentials*: quartz, potassic feldspar (microcline, orthoclase, perthite), plagioclase (albite-andesine), biotite. *Accessories*: apatite, zircon, magnetite. *Accidentals*: sillimanite, cordierite, garnet, amphibole.

**Appearance** Color from light to dark gray; texture granoblastic grading into lepidoblastic; structure rather homogeneous with neosome closely graded and permeating the paleosome, in which the zones can only be detected locally.

**Metamorphic grade** High (amphibolite facies and others).

**Genetic environment** Very advanced stage of the anatexis of a pelitic rock, almost fading into "anatectic granite."

**Occurrence** Common in the Precambrian shields and pre-Paleozoic basements, generally associated with and grading into the anatexites or with rocks of granitic appearance.

**Uses** Locally used as building stone, sometimes as polished slabs for ornamental purposes.

---

## 367 AGMATITE (MIGMATITE)

**Type** Ultrametamorphic rock.

**Chemistry** Variable.

**Components** *Essentials*: quartz, alkaline feldspar (microcline, albite), in the neosome; the paleosome may consist of a gneissic or amphibolitic metamorphic rock, with the appropriate essential, accessory and accidental components.

**Appearance** This is a typical *heterogeneous migmatite*, formed by two clearly distinct parts, almost breccialike in appearance, of which the paleosome constitutes the clasts and the neosome the cement invariably intersecting as veins. The color is mainly dark with very light veins; the structure is massive.

**Metamorphic grade** Medium-high (amphibolite facies).

**Genetic environment** Initial stage of the formation of migmatites, at the edge of plutons or in mafic rocks that are not very fusible and contain localized fluids.

**Occurrence** Common at the contact between intrusive granites of anatectic origin and metamorphic rocks and surrounding migmatites, especially in the Precambrian Shields (Finland) and in the Paleozoic basement (Black Forest, Germany).

**Uses** Used as building stone.

▲ Nebulite (migmatite) (ca. ×0.5). Novate Mezzola, Sondrio, Italy.

▲ An outcropping of agmatite (migmatite). Val Malenco, Sondrio, Italy.

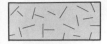

## 368    GRANULITE

**Type**   Regional metamorphic rock.
**Chemistry**   Felsic to mafic.
**Components**   *Essentials*: perthitic orthoclase, antiperthitic plagioclase, quartz, garnet. *Accessories*: rutile, spinel, magnetite, corundum. *Accidentals*: orthopyroxene, clinopyroxene, cordierite, kyanite or sillimanite, brown amphibole, scapolite.

**Appearance**   Color variable from light to dark depending on the mineralogy. Texture granoblastic, frequently with flattened quartz; structure massive with variable grain size.
**Metamorphic grade**   High (granulite facies).
**Genetic environment**   Product of high temperature, variable pressure, anhydrous regional metamorphism, occurring in the lower crust.
**Occurrence**   Very widespread in the Precambrian shields (Scandinavia, Brazil, Canada, USA). Granulite is found in the Moine Series of northwest Scotland and the Grenville Series of the Adirondack Complex in the USA. The numerous granulitic rocks of India are called *charnockites*. Other charnockites are found in Sri Lanka, Zaire, Uganda, Natal, Central Sahara, Malagasy Republic, Siberia and the Ukraine (USSR).

**Uses**   Locally used as construction material; some varieties are also used in the form of polished slabs.

---

## 369    ECLOGITE

**Type**   Regional metamorphic rock.
**Chemistry**   Mafic.
**Components**   *Essentials*: omphacite pyroxene, garnet, quartz. *Accessories*: rutile, pyrite. *Accidentals*: hornblende (pargasite), kyanite, phengite, paragonite, zoisite, glaucophane, dolomite, corundum.

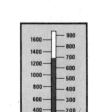

**Appearance**   Color from green to red, often variously speckled; texture granoblastic, sometimes also porphyroblastic; structure massive, grain size from very fine to coarse, often with compositional zoning.
**Metamorphic grade**   Low or medium (glaucophane schist and amphibolite facies), at high pressure.
**Genetic environment**   Derived from lava and basaltic tuff or from gabbroic masses metamorphosed in a continental plate subduction zone at low temperature (*ophiolitic eclogites*) or in anhydrous regional metamorphism (*common eclogites*). To be distinguished from eclogite found in kimberlites and torn from the upper mantle, which are of a different origin.
**Occurrence**   Ophiolitic eclogites are associated with a zone of compression in the Western Alps, California (USA), Japan and the Caribbean. Common eclogites are part of the ancient crystalline basements (Caledonian Chain of Norway; Fichtelgebirge, Germany; France). Eclogite is found in the Precambrian Lewisian rocks around Glenelg, Scotland and in the Jurassic Franciscan Formation of California (USA).
**Uses**   Of no commercial value.

▲ Granulite (ca. ×1). Strona Valley, Novara, Italy.

▲ Eclogite (ca. ×1). Weissenstein, Germany.

**Type**   Regional metamorphic rock.

**Chemistry**   Calcareous-pelitic or calcareous-arenaceous.

**Components**   *Essentials*: calcite, mica (muscovite, phengite, paragonite), chlorite, quartz. *Accessories*: epidote, graphite, ilmenite, albite. *Accidentals*: fuchsite, biotite, garnet (grossularite, andradite), tremolite, chloritoid, glaucophane, actinolite, vesuvianite, zoisite, dolomite, ankerite.

**Appearance**   Color from gray to brownish-gray, to brown, to almost black; texture granoblastic and lepidoblastic, sometimes grading into diablastic; structure schistose, frequently plicated and with alternations of granular and lamellar units; very corroded and cavernous surfaces, altered by the selective reaction of water with the carbonates.

**Metamorphic grade**   Low or medium (glaucophane schist facies, greenschist facies and the beginnings of the amphibolite facies).

**Genetic environment**   Derived by the dynamic metamorphism of calcareous-pelitic or calcareous-arenaceous sedimentary rocks (marl, clay, marly clay, tufaceous marl). Sometimes subjected to high pressure and regional metamorphism, and later retransformed to low to medium temperature regional metamorphic conditions, as was typical of the final phase of Alpine orogenesis.

**Occurrence**   The "formation of calc schists with greenstones" is typical of the western Alps, and less so for the central and eastern Alps. It refers to a sequence of *cipollins* (marbles rich in bands and zones consisting mainly of chlorite, epidote and quartz with accessory minerals), calc schists properly called "quartz schists" (markedly schistose quartzites) and prasinites (including a type particularly rich in chlorite and albite called *ovardite*) deposited partly in a marginal basin and partly in a pelagic environment. This sequence has been successively compressed in a subduction zone of high pressure and low temperature then returned to the surface and reequilibrated at medium pressure and low to medium temperatures (especially in the Central Alps). In a typical calc schist, in thin section we can see (see photomicrograph) twinned calcite, abundant small sheets of mica (with vivid interference colors), opaque graphite and scattered quartz grains. Rocks similar to calc schists are common in the USA: Vermont (particularly in the Castleton area) and New Hampshire; the Stuart Fork Formation in northwestern California; and the Huronian iron formation of northern Michigan. They are also found in the Otago Schists of New Zealand and the Sierra Nevada of Spain.

**Uses**   Of no commercial use.

▲ Calc schist (ca. ×1). Cogoleto, Genoa, Italy.

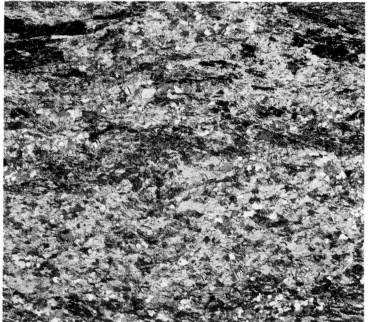

▲ As above: photomicrograph of a thin section (ca. ×15; crossed Nicols).

**Type** Regional and contact metamorphic rock.

**Chemistry** Calcareous.

**Components** *Essentials*: calcite. *Accessories*: sometimes none, sometimes graphite, pyrite, ilmenite. *Accidentals*: dolomite, quartz, mica (muscovite, phlogopite, fuchsite), chlorite, plagioclase, epidote, diopside, fassaite pyroxene, tremolite, wollastonite, vesuvianite, forsterite olivine, talc, brucite, serpentine, periclase.

**Appearance** Color from very white to variously patched or streaked with green, gray, brown and red; texture granoblastic grading into diablastic, nematoblastic, poikiloblastic in the silicate-rich varieties; structure massive or zoned with grain size from fine to very coarse (saccharoidal marble).

**Metamorphic grade** From low to high (from zeolite facies to granulite facies).

**Genetic environment** Derived from fairly pure limestones recrystallized either by dynamic metamorphism, regional metamorphism or contact metamorphism. It is difficult to distinguish which type occurred from this rock (if pure), but this problem can usually be solved by studying coexisting rocks.

**Occurrence** Very heterogeneous distribution in all metamorphic regions, as large masses or thin intercalations. The most famous are the statuary marbles of Carrara (Liguria, Italy), used by Michelangelo and other sculptors ever since. In the USA statuary marble is found in parts of the 60-meter-thick (18 yds) marble beds at Sylacauga (Talladega County, Alabama). Coarse-grained marble occurs in Georgia, and in Vermont there is a great marble belt stretching for 130 km (80 mi) containing marble of many colors. In the Tennessee River Valley there is a gray-pink-red fossiliferous marble containing stylolites (locally called "crowfoot"). In Maryland ornamental verde antique marble is found at Cardiff in Harford County. In Scotland marble is found in limestone of the Dalradian Series. Streaked varieties rich in chlorite or serpentinite are the so-called "cipollini" of Tuscany (Italy) and the calc-schistose formations of the western Alps. Dolomitic marbles are found in Ticino (Italy); brucite marbles (*predazzite*) and periclase marbles (*pencatite*) are found in contact aureoles in the Tyrol.

**Uses** Of great importance in building, both as rough stone and as polished slabs. Commercially, marbles are named either after the place from which they come or on the basis of their color of zonation. Even today the commercial name of marble refers improperly to calcareous rocks that are nonmetamorphic but compact and good for working and polishing. Marbles are also the commonest raw material for sculpture (statuary marbles). Rarely used in the manufacture of lime and the chemical industry.

▲ Pink marble (ca. ×1). Candoglia, Novara, Italy.

▲ Marble (Val Malenco, Sondrio, Italy): photomicrograph of a thin section (ca. ×20; polarizer only).

**Type**   Cataclastic metamorphic rock.

**Chemistry**   From felsic to ultramafic.

**Components**   Variable, depending on the type of original rock, partly made up of clasts (fairly profoundly fractured relicts of the original rock) and partly of blasts (minerals recrystallized during or immediately after the deforming action).

**Appearance**   Variable, depending on the intensity of the deformation. The first stage (*cataclasite*), formed by weak mechanical action, develops by shattering the rock into lenticular forms cemented by microgranular zones (with ''concrete'' texture), along which one can see the differential slips. The second stage (*cataclastic augen gneiss*) leads to extensive microgranular zones intercalated by lenses, eyes, relicts (mainly quartzofeldspathic), with an augenclastic, phacoidal texture. The third stage (*mylonite*) brings about a microgranulation of all the components; the rock appears homogeneous, schistose and blackish from the dispersal of iron oxides and graphite. In thin section, with just the polarizer (see photomicrograph), it appears markedly schistose, with small relict microgranular strips and relict clastic crystals. Finally one may find an actual isotropization of the components (*ultramylonites, pseudotachylites*) with the formation of glass; the rock appears massive, compact, and horny black. The recrystallization of new minerals, mostly idioblastic because they occurred by finite deformation, brings about the granular, light gray, greenish or pink *blastomylonites*.

**Metamorphic grade**   Very low, with strong shear pressures; temperature fairly low or acting for short periods of time.

**Genetic environment**   Along aligned strips with folds and dislocation lines, with the intensity of the cataclastic metamorphism diminishing toward the outside of these strips.

**Occurrence**   Large zones of mylonite are associated with lines of tectonic movement. Examples are along the San Andreas fault, California (USA) and the Insubrica Line in the Central Alps. Blastomylonites are common in the contact zone at the foot of Alpine folds (Dent Blanche and Ortles, Switzerland) or in areas of granulitic metamorphics (Spain, Portugal). Mylonites are also noted in dunite at Milford Sound (New Zealand), in the Laurel Gneiss in Maryland (USA) and in Saskatchewan (Canada). Pseudotachylite is found in the Vredefort dome, South Africa.

**Uses**   Of no commercial importance.

▲ Mylonite (ca. ×0.5). Val Masino, Sondrio, Italy.

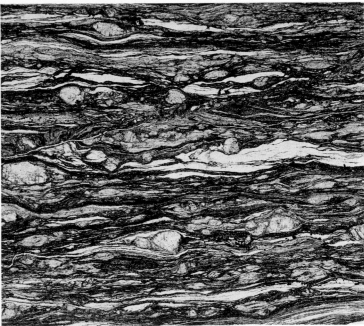

▲ As above: photomicrograph of a thin section (ca. ×20; polarizer only).

## 373 SLATE

**Type** Contact metamorphic rock.

**Chemistry** Pelitic (or pelitic-arenaceous).

**Components** *Essentials*: mica (muscovite, biotite), cordierite, andalusite (chiastolite). *Accessories and accidentals*: the same as for the original rock.

**Appearance** Color dark gray, shiny; texture lepidoblastic-granoblastic, grading into poikiloblastic; structure schistose with local nodules produced by aggregates of mica or elongated poikiloblasts of cordierite and chiastolite.

**Metamorphic grade** Low or medium (upper part of the hornfels facies, hornblende hornfels subfacies).

**Genetic environment** Outer edge of a contact aureole, mainly developed in clays or shales.

**Occurrence** The classic area is around the granite at Barr-Andlau (Vosges, France) and another is in the district of Orijarvi (Finland). It is well developed in the Sierra Nevada of California (USA). It is found in the Aberfoyle Slate in Perthshire (Scotland), and in the Coniston Flags and Skiddaw Slate in the Lake District (England).

**Uses** Used as shingles for roofing, for flooring and for blackboards.

## 374 HORNFELS

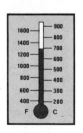

**Type** Contact metamorphic rock.

**Chemistry** Pelitic (and pelitic-arenaceous), sometimes with carbonate.

**Components** *Essentials*: andalusite, cordierite, sillimanite, biotite, potassic feldspar (orthoclase), plagioclase. *Accessories*: corundum, spinel. *Accidentals*: garnet, epidote, hornblende, diopside, vesuvianite, fassaite pyroxene, hypersthene, cummingtonite, anthophyllite, carbonate.

**Appearance** Color light, pink, brown, violet and green; texture granoblastic, poikiloblastic; structure massive or with residual schistosity and mimetic motifs on the structures prior to the contact metamorphism.

**Metamorphic grade** Medium to high (lower part of the hornfels facies, pyroxene hornfels subfacies).

**Genetic environment** Inner edge of a contact aureole in clayey or shistose rocks; inclusions in some plutonic masses.

**Occurrence** Found in the internal part of the Barr-Andlau aureole (Vosges, France) and the zone of contact metamorphism at Oslo (Norway). It is found at Comrie and Lochnager in Scotland and the Sierra Nevada of California (USA). It is also included in the lavas of Laacher See (Germany).

**Uses** Of no commercial importance.

▲ Spotted slate (ca. ×1). Vosges Mountains, France.

▲ Hornfels (ca. ×0.5). Sondalo, Sondrio, Italy.

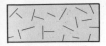

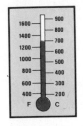

**Type**   Contact metamorphic rock.

**Chemistry**   Calcareous or calcareous-dolomitic.

**Components**   *Essentials*: calcite, wollastonite, garnet (grossularite, andradite), plagioclase (labradorite-anorthite), scapolite, monticellite, diopside, fassaite pyroxene. *Accessories*: titanite, ilmenite, xanthophyllite (calcium mica), humite, graphite. *Accidentals*: anthophyllite, forsterite olivine, phlogopite, periclase, brucite, fluorite, quartz, melilite, celsian, vesuvianite, zoisite.

**Appearance**   Color light, yellow-brown, with pink, light green, white and dark red zonations; texture granoblastic grading into poikiloblastic and idioblastic, sometimes with a mosaic tendency; structure massive or zoned, often conserving the stratification and other sedimentary motifs, but sometimes also with a dense pleating due to the plastic behavior of the calcite subjected to heat and pressure.

**Metamorphic grade**   Medium to high (hornfels facies).

**Genetic environment**   Derived by contact metamorphism from calcareous, calcareous-dolomitic and marly rocks, both in an aureole and in rock slivers (volcanic projectiles) torn from lava rising from the walls of a pipe. Similar rocks (more commonly called ''calc-silicate rocks'') are found in high metamorphic grade, low-pressure series.

**Occurrence**   Noted in richly mineralized blocks within the lavas of Monte Somma, Vesuvius and Laziali (Italy), in the diabase of Scawt Hill (Ireland), in the lavas of Mull (Scotland) and at Jan Mayen (Norway). In some of these occurrences it approaches a state of the maxiumum possible temperature under conditions of low pressure, producing rare minerals such as larnite and tilleyite (sanidinite facies). Large aureoles around granitic rocks are found in the Adamello Complex in the Tyrol (Italy), at Crestmore (California, USA) and at Ben Bullen (Australia). In many instances calcium-bearing minerals poor in silica and of great rarity are formed. It is also common to find pure marbles if derived from calcareous rocks, or periclase and brucite marbles (*pencatites* and *predazzities*) if derived from dolomitic rocks. Calciphyres in regional environments are also common in northern Italy. In Bohemia, where they are known as *erlan*, rocks with calcite, diopside and quartz of regional metamorphic origin are common.

**Uses**   Variegated marbles sometimes highly sought after, although in fairly limited amounts.

▲ Calciphyre (ca. ×0.5). Candoglia, Novara, Italy.

▲ Calciphyre (ca. ×0.5). Albonico, Como, Italy.

## 376 SKARN

**Type**  Metasomatic and contact metamorphic rock.

**Chemistry**  Calcareous-ferriferous or calcareous-manganiferous.

**Components**  *Essentials*: calcite, pyroxene (diopside, hedenbergite, johannsenite, hypersthene), garnet (andradite, almandine), sulfides (pyrite, chalcopyrite, sphalerite), oxides (magnesite). *Accessories and accidentals*: ilvaite, rhodonite, bustamite.

**Appearance**  Color dark brown, black, reddish, violet; texture granoblastic often with very large grain size; structure massive or zoned, with the minerals concentrated in bands, nodules and radiating masses.

**Metamorphic grade**  High (hornfels facies).

**Genetic environment**  Contact metamorphism by granitic-type rocks, rich in volatile elements, of fairly impure limestones, with metasomatic migration of iron, magnesium, manganese and silicon, resulting in recrystallization.

**Occurrence**  Common in Sweden, the USA, Mexico, Peru, Bolivia and Japan. A notable locality is at Crestmore, California (USA). Well developed at Scawt Hill (Ireland); on Skye in the Tertiary Beinn on Dubhaich contact aureole in the Durness Limestone; and the contact aureole of the Dartmoor granite (England). A striking occurrence is at Doubtful Sound in New Zealand.

**Uses**  Frequently the parent rock in mineralized areas producing copper, iron, manganese and molybdenum.

## 377 RODINGITE

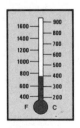

**Type**  Metasomatic metamorphic rock.

**Chemistry**  Calcareous-siliceous.

**Components**  *Essentials*: chlorite, garnet (grossularite, andradite), pyroxene (diopside, fassaite). *Accessories*: titanite, titanclinohumite. *Accidentals*: amphibole, vesuvianite, epidote.

**Appearance**  Color light, pink, brownish, reddish, greenish; texture granoblastic or diablastic, often with large grain size; structure zoned in parallel strips, each enriched in a particular mineral. Geodes are frequent.

**Metamorphic grade**  Low (greenschist facies).

**Genetic environment**  Derived by contact metasomatism of peridotite in the process of serpentinization, by fluids rich in calcium and other elements.

**Occurrence**  Present in New Zealand (at Roding, the type locality), California and Oregon (USA), and especially in the Western Alps (Val d'Orba, Val della Gava, Val d'Ala, Val d'Aosta) where splendid mineralization of vesuvianite, pyroxene and garnet occurs. (The old name for rodingite was "garnetite.")

**Uses**  Of interest to collectors only.

▲ Skarn (ca. ×1). Isle of Elba, Italy.

▲ Rodingite (ca. ×0.5). Valle dell'Orba, Savona, Italy.

# GLOSSARY

**accessory minerals** components usually present in an igneous rock, but in minor abundance; their presence is not essential to the definition of the rock.

**accidental minerals** minerals present, even in large quantity, but not usually found in a particular rock type; may be foreign mineral engulfed by invading magma (xenocryst).

**acicular crystals** crystals with a very elongated habit; needle-shaped.

**aggregate** assemblage of crystals of one or more species.

**allochromatic mineral** mineral which may exhibit various colors; hence color is not diagnostic for that mineral.

**allotriomorphic** texture of an igneous rock, as seen under the microscope, and made up of anhedral crystals.

**alluvial deposit** sedimentary deposit formed from deposition by river waters, often at the base of a mountain range.

**ameboid** a texture formed by minute crystals with lobate outlines.

**amorphous** minerallike material which is noncrystalline; e.g., opal.

**amygdule** secondary filling of minerals in roundish (once gas-filled) cavities in volcanic rocks.

**analyzer** second (upper) polarizer in a polarizing (petrographic) microscope; when inserted, a thin section is viewed under "crossed Nicols."

**anatexis** a process of fusion (melting) on a large scale at depth in the earth's crust; usually on crustal material of granitic composition that has been subducted to considerable depth.

**Ångstrom** a unit of measure (symbol Å) equal to one ten millionth of a millimeter.

**anhedral crystals** crystals which do not display their crystal form; usually presumes they are formed late in the history of igneous rock. Synonyms: allotriomorphic, xenomorphic.

**anion** a negatively charged ion; e.g., oxygen.

**anisotropism** the variation of a physical property with direction in a mineral.

**aphanitic** texture of an igneous rock made up of crystals too small to be seen with the naked eye; e.g., rhyolite.

**apophysis** a small dike- or sill-like body originating from a larger igneous intrusive mass.

**aqua regia** mixture of nitric and hydrochloric acids (in the ratio of 3 to 1), capable of dissolving gold.

**asterism** a starlike (4, 6 or 12 rays) play of light from a polished mineral due to crystallographically arranged minute inclusions; e.g., in sapphire.

**augen texture** a texture typical of the metamorphic

594

rock gneiss, which is rich in crystals or crystalline aggregates of relatively large size and elliptical form.

**authigenic mineral** a mineral formed directly from within a sediment; not a mineral which was carried in.

**autochthonous** something which remains where it was formed; e.g., a portion of a major fault zone which remained in place and was overridden by a moving rock mass.

**basement** the older crystalline rock mass beneath a series of sedimentary rocks; underlies continental masses. Usually consists of metamorphic and plutonic igneous rocks.

**bastite** a poorly defined mixture of serpentinized minerals.

**batholith** a large plutonic intrusive body at least partially igneous, the bottom of which is not clearly demarked, and with margins that are often gradational and that widen with depth.

**bitumen** a mixture of semi-solid hydrocarbons, like asphalt; a major constituent of bituminous coal and a minor constituent in black shales (argillites).

**blast** a prefix or suffix for large minerals grown in the solid state in a metamorphic environment.

**blastomylonitic texture** a texture of recrystallized mylonitic (tectonically sheared) rock which has grown new crystals of larger size in the solid state (a metamorphic process).

**blueschist** a metamorphic facies and rock type that forms at high pressure and low temperature.

**botryoidal aggregate** a globular growth of minerals with a form similar to a bunch of grapes; e.g., hematite.

**boudinage** a tectonic structure formed in a metamorphic environment as a result of the stretching of rigid strata interlayered with plastically deforming strata; the result is the formation of elongated lenses of the rigid strata, sometimes with mineral lineation.

**cabochon** a cut of a gem or mineral in a convex, more or less round surface.

**cataclastic texture** a texture typical of tectonically deformed rocks exhibiting broken and flattened mineral grains; e.g., volcanic breccia, impact breccia.

**cation** a positively charged ion; e.g., iron.

**clast** a fragment of mineral or rock derived by fragmentation; in sedimentary rocks often as a consequence of transportation. Also produced in volcanic and meteoritic impact processes.

**cleavage** the tendency of some minerals to split or break along characteristic planes corresponding to directions of minimum cohesion.

**colloform** rounded or reniform (kidney-shaped) masses produced by colloidal precipitation of matter.

**colloid** a suspension of very fine particles in a liquid phase (usually water).

**columnar crystal** crystals with an elongated prismatic habit.

**comagmatic** two or more igneous rock groups which are supposedly derived from the same magma in the course of their evolution.

**conchoidal fracture** a type of fracture having long curved surfaces, like the inside of a shell; commonly observed in obsidian, or manmade glass.

**concretion** a concentration of authigenic mineral growth around a nucleus (matrix) in sedimentary rocks.

**contact aureole** a rock zone in contact with a hot igneous intrusive that undergoes thermal metamorphism.

**continental plate** the thick terrestrial parts of the crust "floating" on the mantle.

**coprolite** fossilized excrement incorporated in sedimentary rock.

**cumulate rock** igneous rock derived by the accumulation of early formed minerals in a large mafic igneous body, by means of crystal settling.

**decrepitation** the breaking apart of a mineral by expansion caused by the instantaneous liberation of gases, usually producing a popping noise; produced in some minerals by interaction with atmospheric humidity.

**dendrite** skeletal crystals, with arborescent form, usually found on the fracture surface of a rock.

**density** the mass per unit volume of a body ($g/cm^3$).

**diablastic texture** a metamorphic texture consisting of acicular or fibrous minerals oriented in the same direction and intimately interpenetrated.

**diagenesis** the final process of forming a sedimentary rock from a sediment.

**diamagnetic mineral** a mineral which is very weakly repelled by a magnet; typical of most minerals.

**diaplectic glass** glass derived from a mineral by impact metamorphism.

**dichroism** pleochroism in two directions in a mineral.

**differentiation** the processes of chemical and mechanical evolution of a magma in the course of its crystallization such that different rock types are formed from the same original magma.

**dike** igneous body injected into a fracture in the "country" rocks.

**dislocation, line of** the intersection of a fault, or a large fracture, with the Earth's surface.

**dome** the upper structure of a cupola; a portion of a minor igneous intrusive.

**druse** a crust of fine crystals filling a cavity or coating a matrix.

**epiplutonic rock** igneous body crystallized at intermediate depth (0.5–4 km; 0.8–2.5 mi).

**epitaxy** the crystallographically oriented association between crystals of two different minerals with one growing on or around the other.

**essential mineral** a mineral whose presence is essential to the definition of an igneous rock type.

**euhedral mineral** a mineral with well-formed crystal faces, used when referring to the microscopic texture of a rock (synonym: idiomorphic).

**evaporite**   sedimentary rock or mineral of chemical origin formed by the evaporation of a solution, usually sea water; e.g., halite (rock salt).

**exsolution**   separation, in the solid state, of two or more mineral phases from a single previously homogeneous crystalline compound or mineral.

**extrusive rock**   fine-grained igneous rock formed by the crystallization of magma on the Earth's surface; e.g., basalt.

**facies, metamorphic**   a set of temperature and pressure conditions under which a group of metamorphic rocks is formed from preexisting rock of any type.

**fault**   a major fracture in the Earth's crust with relative movement of the rock masses on both sides; e.g., the San Andreas fault (California).

**felsic**   pertaining to rocks composed largely of quartz and feldspar ($SiO_2 > 65$ percent); also used as an adjective for these minerals; e.g., felsic minerals.

**felsitic texture**   a microcrystalline to cryptocrystalline texture in fine-grained extrusive igneous rocks.

**ferromagnetic mineral**   mineral strongly attracted to a magnet; e.g., magnetite.

**fissility**   the property of a rock or mineral of breaking up into long, thin, flat subparallel sheets.

**fissure volcano**   a volcano which erupted with other volcanos along a fissure.

**fluidal texture**   a texture of lavas (fine-grained extrusive igneous rocks) characterized by the orientation of elongated crystals as the result of flow motion.

**fluorescence**   the phenomenon of photoluminescence under ultraviolet light as displayed by some minerals; e.g., fluorite.

**fluvioglacial**   deposits derived from the action of rivers originating from the melting of glaciers (ice).

**fold**   a plastic deformation of rock strata such that it has the appearance of folding, as a result of compression or shear; e.g., fold sedimentary rocks of the Appalachian Mountains (Pennsylvania).

**foliated**   typical of mineral aggregates that flake easily.

**form**   the overall crystalline form of a crystal.

**formula**   the chemical composition of a substance (mineral) expressed in chemical symbols; e.g., NaCl, sodium chloride.

**fractionation**   the segregation of minerals or chemical components during magmatic crystallization producing different rock types from the same magma (synonym: differentiation).

**fracture**   the manner in which a rock or mineral breaks.

**fumarole**   a vent from which volcanic gases and, in some instances, solid particles issue.

**gangue**   the noneconomic mineral components of a mineral deposit; e.g., quartz is often a gangue mineral.

**gel**   a semisolid colloidal solution; e.g., opal is a solidified silica gel.

**geode** roundish cavity in a sedimentary rock lined with crystallized minerals, often in layers of different color.

**geological thermometer** a mineral or group of minerals giving an indirect measure of the temperature of formation of the enclosing rock.

**geosuture** the contact between two rock masses extending in depth as far as the mantle; contact between continental plates that have either split apart or collided.

**geyser** thermal spring with intermittent steam emissions produced by the heating of water by volcanic action.

**graded bedding** structure involving the superposition of grains of decreasing diameter developed in sediments deposited from a fluid (water).

**grade, metamorphic** the level of metamorphism as determined by the conditions of pressure and temperature characteristic of a metamorphic rock.

**granoblastic texture** a texture formed by minerals of nearly the same dimensions in a metamorphic rock.

**granoclastic texture** a texture defined by mineral grains of nearly the same dimensions, and formed by fracture during a tectonic process.

**granophyric texture** a texture of crystals of quartz and feldspar having variable and irregular interpenetration; formed in the late gas-rich stages of igneous crystallization.

**graphic intergrowth** a texture due to the simultaneous crystallization of quartz and potassic feldspar in the form of small strips folded at an angle; resembles script, hence the term "graphic."

**greisen** igneous rock which has been altered by the action of fluids rich in volatile elements.

**habit** the characteristic appearance of a crystal as determined by its predominant form.

**hardness** the resistance of a mineral to scratching and abrasion; the Mohs scale (1–10) is a relative measure of this property.

**heavy minerals** minerals with a specific gravity greater than 2.9, present as residual grains in clastic sedimentary rocks.

**hemihedral class** crystallographic class of reduced symmetry, as opposed to holohedral class.

**holocrystalline texture** a texture formed exclusively of crystals, free from glass; typical of plutonic rocks.

**holohedral class** crystallographic class of maximum symmetry.

**hyaline** a textural term for a glassy substance.

**hyaloclastic texture** a texture characteristic of volcanic rocks consisting of broken glassy fragments and formed by rapid cooling in water.

**hyalopilitic texture** a texture in volcanic rocks with abundant glass accompanied by small crystals of more or less oriented feldspar.

**hybrid rock** a rock derived from the mixing of one magma with another.

**hydrothermal vein**   a vein formed by the crystallization of minerals from predominantly hot water solutions of igneous origin.

**hypidiomorphic texture**   texture formed by euhedral minerals associated with anhedral minerals occupying the remaining space.

**hyposilicic rock**   igneous rock with lower content of silica ($SiO_2 < 52$ percent); relatively mafic.

**idioblastic texture**   a texture typical of metamorphic rocks in which nearly all the minerals have euhedral crystalline form.

**idiochromatic mineral**   a mineral which has diagnostic color; e.g., sulfur is always yellow.

**igneous rock**   a rock formed by the crystallization of a magma or lava.

**inclusion**   either a solid (inorganic or organic, crystalline or amorphous), a liquid, or a gas included in a mineral or rock.

**interference colors**   color seen in a nonisotropic mineral when viewed through crossed polarizers or crossed nicols.

**intersertal texture**   texture consisting of interlocking microcrystals of feldspar in a fine-grained groundmass.

**intumescence**   a swelling or bubbling of a mineral specimen when heated or fused.

**ion**   an atom in an ionized (charged) state.

**iridescence**   the light phenomenon associated with the play of colors from a mineral surface.

**isomorphism**   a property of minerals having the same or similar mineral crystal form; sometimes used as solid solution series.

**isotropic body**   a body in which the physical properties do not change with direction.

**isotypes**   minerals having the same crystal structure; i.e., corresponding atoms in different minerals are in the same positions.

**kidney**   a nodular concretionary form found in some minerals; e.g., hematite (synonym: reniform).

**kilobar**   unit of measure of pressure equal to 1000 bars, about 1000 atmospheres.

**Kobell scale**   the fusibility scale of minerals.

**labradorescence**   typical iridescence of labradorite feldspar.

**laccolith**   lenslike igneous pluton, more or less convex, intruded into discontinuous planes of surrounding rocks.

**lamellar crystals**   crystals prevalently developed in two dimensions (flattened).

**laterite**   a highly leached soil found in tropical climates, usually enriched in iron and aluminum.

**leaching zone**   zone in the lower part of a mineral deposit where chemical components are leached and then redeposited in a secondary zone of enrichment.

**lepidoblastic texture**   texture typical of metamorphic rocks containing lamellar minerals, as in a schistose texture.

**leucocratic**   containing rock components of light color, i.e., felsic minerals such as quartz and feldspar.

**lineation**   orientation of mineral components or of tectonic structures along parallel lines.

**lithoclase**   a fracture in a rock, usually filled with crystals; e.g., Alpine lithoclases.

**luminescence**   an emission of light due to a stimulus such as directed pressure, heating, rubbing, irradiation with x-rays, ultraviolet light, etc.

**luster**   a reflective property of mineral surfaces.

**luxullianite**   a variety of granite containing tourmaline formed by metasomatic action.

**mafic**   pertaining to rocks composed predominantly of ferromagnesian minerals—olivine, pyroxene, amphibole (44–52 percent $SiO_2$); also used as an adjective for these minerals, e.g., mafic minerals.

**magma**   liquid or molten rock material; called lava when it reaches the Earth's surface.

**mammillary**   aggregate of minerals of rounded structure; breastlike.

**mantle**   intermediate thick layer of the Earth's major zonation between the thin crust and the iron-nickel core.

**matrix**   fine-grained portion of a rock; also used for the rock and mineral material on which a mineral grows.

**melanocratic rock**   rock composed of 69–90 percent mafic minerals.

**mesosilicic rock**   rock containing an intermediate quantity of silica (52–65 percent $SiO_2$) (synonym: intermediate).

**metamict state**   alteration of the crystal structure of a mineral toward an amorphous state as the result of damage due to the radioactive decay process of some chemical elements such as uranium and thorium.

**metamorphic rock**   rock produced by metamorphic processes.

**metasomatism**   process of changing the composition of minerals and adding or removing minerals by the addition and removal of chemical elements or components in an exchange process involving fluids, but not magma.

**metastable form**   a mineral outside its normal field of stability.

**meteorite**   a solid body coming from outer space and falling to the Earth's surface.

**miarolitic cavity**   small angular cavities of various forms in igneous rocks, sometimes lined with mineral crystals; most common in felsic plutonic rocks.

**microcrystalline**   association of crystals on a microscopic scale.

**mimetic twins**   pseudosymmetry caused by twinning; e.g., aragonite.

**mineral deposit**   concentration of one or more economically valuable minerals.

**Mohs scale**   a relative scale of the hardness of minerals, arbitrarily reading from 1 to 10.

**mylonite**   fine-grained rock caused by crushing of rock

in a fault zone.

**nematoblastic texture** texture due to orientation of prismatic to fibrous minerals; used in reference to fibrous schistosity.

**neosome** the part of a migmatite that is of granitic composition and formed during the process of anatexis.

**ocellar texture** a texture of metamorphic rocks due to the presence of radiating groups of platy minerals with indistinct outlines.

**ochre** a mass of earthy appearance consisting of minute crystals, often of a particular color (yellow, brown, red) and used as a pigment.

**oölite** spherical grains, less than 2 mm (.08 in) in diameter, composed of a mineral or organic nucleus covered by successive concentric layers of various minerals; term also used for a rock composed of oölites.

**ophiolite** mafic or ultramafic rocks derived from the emergence of the mantle or of the oceanic crust; usually formed in a marine environment.

**ophitic texture** a texture of some mafic igneous rocks which has euhedral crystals of plagioclase cemented by anhedral crystals of pyroxene or amphibole.

**orbicular structure** a rock structure consisting of large rounded units containing alternating layers of mineral of different kind and color; found in igneous and metamorphic rocks.

**orogenesis** complex of phenomena which, by means of the deformation of the Earth's crust, leads to the formation of mountain ranges.

**outcrop** surface exposure of the underlying rock body (Earth's crust) through the soil and detritus.

**oxidation** chemical phenomenon consisting of the addition of oxygen to a compound or process; the opposite of reduction.

**paleosome** rock components preexisting before the process of anatexis brings in new melt (neosome) of granitic composition, in a migmate.

**paragenesis** order of formation of minerals in a rock; also used as a term for the association of minerals in a rock or mineral deposit.

**paramorphism** when a substance passes from one structural state to another without modification of the crystal morphology.

**pelagic sediment** sediment formed at depth in mid-ocean, hence with only slight clastic content.

**pelitic rock** clastic sedimentary rock with grains less than 1/16 mm (.0025 in) in diameter, a mudstone or shale.

**periglacial** processes acting at the margin of a glacier.

**perlitic texture** texture found in some felsic volcanic (igneous) rocks consisting of small, subspherical, glassy masses due to concentric fissuring caused by rapid cooling.

**persilicic rock** rock with a high content of silica ($SiO_2$ > 65 percent) (synonym: felsic rock).

**perthite** fine-scale intergrowth of sodic and potassic feldspar (e.g., albite, microcline) along sinuous planes caused by exsolution.

**phenocryst** mineral of large dimension compared to the average grain size of an igneous rock; implies that it probably crystallized earlier under slower cooling conditions.

**phosphorescence** the phenomenon of photoluminescence which lasts even when excitation (usually by ultraviolet light) ceases.

**piezoelectric** a property of polar minerals that produces electrically charged extremities when strained.

**pillow lava** a structure with spherical forms (like pillows) taken on by some mafic lavas when suddenly cooled under water.

**pilotaxitic texture** in glass-free volcanic rocks, defined by feltlike masses of minute crystals.

**pipe** a volcanic structure, usually tubular, through which magma (or lava) reaches the surface; as in volcanic pipe or kimberlite pipe.

**pisolite** spherical or subspherical grains with a diameter > 2 mm (.08 in) and with a mineral or organic nucleus and concentric layers of various minerals.

**placer** an alluvial or glacial deposit where heavy minerals are recovered from the lowest portions of the sediment.

**pluton** deep-seated coarse-grained igneous body.

**poikilitic texture** a texture found in some igneous rocks due to the presence of crystals enclosing inclusions of numerous smaller crystals of a different mineral.

**poikiloblastic texture** a texture in metamorphic rocks in which larger crystals enclose inclusions of smaller crystals of a different mineral; analogous to poikilitic texture in igneous rocks.

**polysynthetic twinning** multiple twinning of a crystal, usually on a fine scale, and causing pseudosymmetry.

**prograde metamorphism** progressively increasing metamorphism hence increasing temperature and/or pressure.

**protrusion** pushing up of a solid or semisolid magmatic mass by volcanic pressure.

**pseudomorphism** the ability of one mineral to replace another mineral (of similar but different composition) while maintaining the same crystal habit.

**pycnometer** an instrument for measuring the specific gravity of a mineral or rock.

**pyroelectric** a property of polar minerals that produce electrically charged extremities when heated.

**radioactivity** spontaneous disintegration of the atomic nucleus of some chemical elements with the emission of energetic and charged particles ($\alpha$, $\beta$, $\gamma$ rays).

**reduction** chemical phenomenon consisting of stripping of electrons from a cation, equivalent to the removal of oxygen from a compound; the opposite of oxidation.

**relicts** unfused or remnant parts of a rock or any original rock or mineral which persists after some alteration process.

**residual mineral**   a mineral that has resisted the alteration processes that disintegrate rocks.

**retrograde metamorphism**   a metamorphic process whereby a lower grade of metamorphism is superimposed on a rock that has undergone a higher grade of metamorphism.

**rift**   sinking of a narrow strip of the crust between two parallel faults, caused by a spreading process (synonym: graben).

**saccharoidal**   association of equidimensional mineral grains barely visible to the naked eye and resembling sugar.

**saturated rock**   igneous rock containing silica (quartz or tridymite); saturated with respect to $SiO_2$ so that free silica can form.

**scalar properties**   the physical properties of minerals independent of direction.

**schistosity**   characteristic of metamorphic rocks due to the disposition of mineral components (usually micas and other phyllosilicates) in parallel or subparallel layers.

**secondary enrichment zone**   a portion of a mineralized zone immediately beneath the gossan (uppermost surficial zone) of a mineral deposit; a zone of redeposition and enrichment in ore minerals, economically very valuable.

**secondary minerals**   minerals formed by alteration at the expense of preexisting minerals.

**sectility**   property of some minerals and rocks able to be cut with a knife blade.

**sedimentary rock**   rock formed by sedimentary processes.

**shield**   large Precambrian part of the stable continental basement (synonym: craton).

**sill**   tabular igneous body injected between two surfaces of stratification or schistosity.

**simple crystal form**   the set of physically equivalent faces of a crystal, related to one another by the crystal's symmetry elements.

**solid solution**   the ability of two isostructural minerals to form a complete series of minerals of any intermediate composition.

**space lattice**   a three-dimensional lattice of atoms making up a crystal.

**specific gravity**   a measure of density; ratio of the mass of a mineral to the mass of an equal volume of water.

**spherulitic texture**   a texture consisting of spherical grains with fibrous-radiating structure of various minerals.

**stock**   discordant (cuts aross "country" rocks) pluton with steep contacts; smaller than a batholith.

**structure**   the overall character of a rock observable on a hand specimen or outcrop scale.

**subduction zone**    plane along which the oceanic crust moves into the mantle as a result of plate motion.

**sublimate**    passing directly from the solid state to a gaseous state.

**subsidence**    slow sinking of either a continental or submarine region.

**syntexis**    process of assimilation of preexisting rocks by hot magma, thereby changing the magmatic composition and resultant mineralogy and rocks; e.g., assimilation of limestone by basaltic magma.

**tabular crystals**    crystals in which the faces of a pinacoid are dominant, i.e., flat crystals.

**tenacity**    resistance of a mineral to breakage.

**texture**    the combination of form, dimensions and disposition of mineral grains in a rock.

**thermal areas**    areas with a high rate of heat flow in the Earth's crust.

**transgression**    invasion of the land by the sea.

**trap**    general term for dark lava flow, usually basalt.

**triboluminescence**    luminescence caused by rubbing a mineral.

**trichroism**    pleochroism in three directions.

**turbidity current**    submarine landslides, along the continental slope, of large masses of incoherent or only slightly coherent sediment.

**twin**    a nonsymmetrically oriented proper association of two or more individuals of the same mineral; e.g., swallow-tail twinning of gypsum.

**ultramafic rocks**    rocks which are very high in mafic minerals, usually containing < 45 percent silica; e.g., peridotite, dunite.

**ultraviolet light**    radiation with a wavelength slightly less than that of visible light.

**undersaturated rock**    igneous rock deficient in silica so that a silica mineral such as quartz could not be formed and a mineral like olivine (with a low $SiO_2$ content) might form.

**unit cell**    the smallest portion of a crystal lattice that represents all of its properties when repeated in any direction.

**vacuolar texture**    texture typical of igneous rocks rich in (air) bubbles and roundish cavities.

**vectorial properties**    physical properties of a mineral which are variable with direction.

**vein**    an irregular tabular igneous intrusion.

**vicarious elements**    elements which take each other's place in trace amounts in a mineral's crystal structure.

**xenolith**    "country rock" enclosed in a magma.

**x-rays**    electromagnetic radiation with a wavelength between ultraviolet light and $\gamma$ rays (0.1–150 Å).

# INDEX OF ENTRIES

# Illustration credits

The photographs which appear in this volume are by Rodolfo Crespi—with the following exceptions:

Pages 2: E. Genovesi; 6–7: Photo Archive B/L. Ricciarini; 412–413: Tomaš Miček; 426: Attilio Montrasio; 431: Foto Pictor; entries 6, 112, 218: L. Ricciarini (C. Bevilacqua); entry 11 (color photograph in small square): A. Cozzi; entries 63, 246: Julius Weber; entries 131, 132, 152, 216, 217, 273: Jeffery Ravodowitz; entries 16, 33, 70: Stelvio Andreis; entries 41, 68, 78, 101, 108, 120, 137, 145, 146, 219, 267: Giovanni Pinna; the beginning of the section on sedimentary rocks: Van Phillips.

Drawings are by Vincenzo Saletti and Raffaelo Segattini.

The samples of rocks and minerals illustrated in the book come in part from the collections of the Institute of Mineralogy at the University of Milan, and the Civic Museum of Natural History in Verona and Milan, Italy.